Theo Fett

Stress Intensity Factors – T-Stresses – Weight Functions

Schriftenreihe des Instituts für Keramik im Maschinenbau
IKM 50

Institut für Keramik im Maschinenbau
Universität Karlsruhe (TH)

Stress Intensity Factors
T-Stresses
Weight Functions

by
Theo Fett

Institute of Ceramics in Mechanical Engineering (IKM),
University of Karlsruhe (TH)

universitätsverlag karlsruhe

Impressum

Universitätsverlag Karlsruhe
c/o Universitätsbibliothek
Straße am Forum 2
D-76131 Karlsruhe
www.uvka.de

Universitätsverlag Karlsruhe 2008
Print on Demand

ISSN: 1436-3488
ISBN: 978-3-86644-235-1

Dedicated to

Dr. Sheldon M. Wiederhorn

on the occasion of his

75th Birthday

by the staff of the

Institute of Ceramics in Mechanical Engineering (IKM)

Head Michael J. Hoffmann

University of Karlsruhe

Preface

Failure of cracked components is governed by the stresses in the vicinity of the crack tip. The singular stress contribution is characterised by the stress intensity factor K, the first regular stress term is represented by the so-called T-stress. Sufficient information about the stress state is available, if these two parameters are known.

While stress intensity factor solutions are reported in handbooks for many crack geometries and loading cases, T-stress solutions are available only for a small number of test specimens and simple loading cases as for instance pure tension and bending.

One of the most frequently applied methods for the computation of stress intensity factors and T-stress even under highly complicated loading is the Green's function or weight function method. These procedures are explained in detail and are extensively applied.

The computations quoted in this booklet were performed since 1997 when the book on "Weight functions and stress intensity factors" by T. Fett and D. Munz appeared. The results are compiled in form of figures, tables, and approximate relations.

The author has to thank his colleagues Gabriele Rizzi (Forschungszentrum Karlsruhe, IMF) for supplementary Finite Element computations and Michael Politzky (Forschungszentrum Karlsruhe, IKET) for his support in the field of computer application.

Universität Karlsruhe

Karlsruhe, April 2008   Theo Fett

CONTENTS

Part A

THE STRESS FIELD NEAR A CRACK

Part B

EFFECTS OF THE T-STRESS IN BRITTLE MATERIALS

Part C

COMPENDIUM OF SOLUTIONS   87

Part D

2-DIMENSIONAL CRACKS   331

Nomenclature:

a	crack length
a_0	depth of a notch
c	crack length in case of indentation cracks
d	distance
e	eccentricity
f	angular function
h_I	mode-I weight function
h_II	mode-II weight function
ℓ	length of a small crack ahead of a notch
p	crack face pressure
r	radial coordinate
t	thickness, Green's function for T-stresses
u_n, u	normal (radial) displacement
v, v	circumferential displacement, crack face displacement in crack direction
x, y	Cartesian coordinates
z	complex coordinate ($x+iy$)

$A_\mathrm{n}, \widetilde{A}_n$	coefficients of the symmetric Williams stress function (integer exponents)
$\hat{A}_n$	coefficients of the antisymmetric Williams stress function (integer exponents)
B	thickness
$B_\mathrm{n}, \widetilde{B}_n$	coefficients of the symmetric Williams stress function
$\hat{B}_n$	coefficients of the antisymmetric Williams stress function
$C_\mathrm{n},$	coefficients for T-stress Green's function
D_n	coefficients for weight function expansions
D	disk diameter
E	Young's modulus
F, F_I	geometric function related to mode-I stress intensity factors
F_II	geometric function related to mode-II stress intensity factors
G	shear modulus
H	plate length, half height of a bar
K_I	mode-I stress intensity factor
K_II	mode-II stress intensity factor
L	length of a bar
L_n	notation of boundaries
M	bending moment
P	concentrated normal force
Q	concentrated shear force
R	disk radius

T	T-stress term
W	specimen width
Y	geometric function $Y=F\sqrt{a}$
Z	Westergaard stress function

α	relative crack size
α_n	coefficients of the symmetric Williams stress function (integer exponents)
α_T	thermal expansion coefficient
β	biaxiality ratio
β_n	coefficients of the symmetric Williams stress function
γ	angle (region under external load)
δ	crack opening displacement, Dirac delta function
λ	normalised crack opening displacement; distance of load application points
ν	Poisson's ratio
ρ	abbreviation for x/a
σ	stress
σ_0, σ^*	characteristic stresses
σ_∞	remote stress
τ	shear stress
ω	angle

Ω	angle between crack and force direction
Φ	Airy stress function, Westergaard stress function
Φ_a	antisymmetric part of the Airy stress function
Φ_s	symmetric part of the Airy stress function
Θ	temperature, Heavyside step function, angle between crack and force in a Brazilian disk test

PART A

THE STRESS FIELD NEAR A CRACK

Fracture behaviour of cracked structures is dominated mainly by the near-tip stress field. In linear-elastic fracture mechanics interest is focussed mostly on stress intensity factors which describe the singular stress field ahead of a crack tip and govern fracture of a specimen when a critical stress intensity factor is reached. The usefulness of crack tip parameters representing the singular stress field was shown very early by numerous investigations. Nevertheless, there is experimental evidence that also the stress contributions acting over a longer distance from the crack tip may affect fracture mechanics properties. The constant stress contribution (first "higher-order" term of the Williams stress expansion, denoted as the T-stress term) is the next important parameter. Sufficient information about the stress state is available, if the stress intensity factor and the T-stress are known. In special cases, it may be advantageous to know also higher coefficients of the stress series expansion. This is desirable e.g. for the computation of stresses over a somewhat wider distance from a crack tip. For this purpose, additional higher-order terms are necessary.

While stress intensity factor solutions are reported in handbooks for many crack types and loading cases, T-stress terms and higher-order stress solutions are available only for a small number of fracture mechanics test specimens and simple loading cases as for instance pure tension and bending.

In real applications the stresses in a component can be highly non-linear. A method which allows stress intensity factors and T-stresses for such complicated loading cases to be determined is the Green's function or weight function method. In part A the stress intensity factors, and T-stresses are defined and the weight function methods are explained in detail.

A1

Stresses in a cracked body

A1.1   Airy stress function

The complete stress state in a cracked body is known, if a related stress function is known. In most cases, the Airy stress function Φ is an appropriate tool, which results as the solution of

$$\Delta\Delta\Phi = 0 \ , \quad \Phi = \Phi_s + \Phi_a \tag{A1.1.1}$$

For a cracked body a series representation for Φ was given by Williams [A1.1]. The solutions of (A1.1.1) are of the type

$$\Phi = \{r^{\lambda+2}\cos(\lambda+2)\varphi, \ r^{\lambda+2}\cos\lambda\varphi, \ r^{\lambda+2}\sin(\lambda+2)\varphi, \ r^{\lambda+2}\sin\lambda\varphi\} \tag{A1.1.2}$$

written in polar coordinates r, φ with the crack tip as the origin. The symmetric part of the Airy stress function, Φ_s, reads for a crack with surfaces free of tractions (with values λ multiple of ½)

$$\begin{aligned}
\Phi_s &= \sum_{n=0}^{\infty} r^{n+3/2}\alpha_n\left[\cos(n+\tfrac{3}{2})\varphi - \frac{2n+3}{2n-1}\cos(n-\tfrac{1}{2})\varphi\right] \\
&+ \sum_{n=0}^{\infty} r^{n+2}[\beta_n\cos(n+2)\varphi + \gamma_n\cos n\varphi]
\end{aligned} \tag{A1.1.3}$$

In this representation, the coefficients α_n, β_n, and γ_n are proportional to the applied loading and contain a length in their dimension. For traction free crack faces it holds $\gamma_n=-\beta_n$. In order to obtain dimensionless coefficients it is sometimes of advantage to normalise the crack-tip distance r on either the component width W, $r{\rightarrow}r/W$ resulting in

$$\Phi_s = \sigma^* W^2 \sum_{n=0}^{\infty} (r/W)^{n+3/2} A_n\left[\cos(n+\tfrac{3}{2})\varphi - \frac{n+\tfrac{3}{2}}{n-\tfrac{1}{2}}\cos(n-\tfrac{1}{2})\varphi\right]$$

$$+ \sigma^* W^2 \sum_{n=0}^{\infty} (r/W)^{n+2} B_n[\cos(n+2)\varphi - \cos n\varphi] \tag{A1.1.4}$$

or on the crack length a, $r{\rightarrow}r/a$ with

$$\Phi_s = \sigma^* a^2 \sum_{n=0}^{\infty} (r/a)^{n+3/2} \widetilde{A}_n\left[\cos(n+\tfrac{3}{2})\varphi - \frac{n+\tfrac{3}{2}}{n-\tfrac{1}{2}}\cos(n-\tfrac{1}{2})\varphi\right]$$

$$+ \sigma^* a^2 \sum_{n=0}^{\infty} (r/a)^{n+2} \widetilde{B}_n[\cos(n+2)\varphi - \cos n\varphi] \tag{A1.1.5}$$

where σ* is a characteristic stress. The geometric data are explained in Fig. A1.1. In the following the formulation according to (A1.1.4) will be used predominantly.

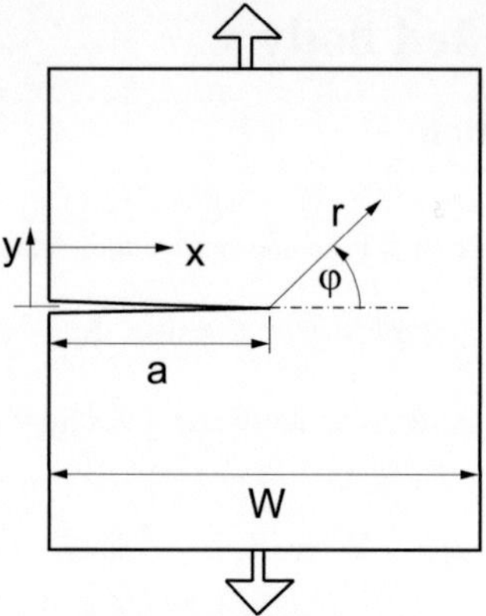

Fig. A1.1 Geometrical data of a crack in a component.

The tangential, radial and shear stresses result from the stress function by

$$\sigma_\varphi = \frac{\partial^2\Phi}{\partial r^2}, \; \sigma_r = \frac{1}{r}\frac{\partial\Phi}{\partial r} + \frac{1}{r^2}\frac{\partial^2\Phi}{\partial\varphi^2}, \; \tau_{r\varphi} = \frac{1}{r^2}\frac{\partial\Phi}{\partial\varphi} - \frac{1}{r}\frac{\partial^2\Phi}{\partial r\partial\varphi} \tag{A1.1.6}$$

In the symmetric case, the stress components are given by

$$\frac{\sigma_\varphi}{\sigma^*} = \sum_{n=0}^{\infty} A_n (r/W)^{n-1/2}(n+3/2)(n+1/2)\left[\cos(n+3/2)\varphi - \frac{n+3/2}{n-1/2}\cos(n-1/2)\varphi\right]$$

$$+ \sum_{n=0}^{\infty} B_n (r/W)^n (n+2)(n+1)[\cos(n+2)\varphi - \cos n\varphi] \tag{A1.1.7}$$

$$\frac{\sigma_r}{\sigma^*} = \sum_{n=0}^{\infty} A_n (r/W)^{n-1/2}(n+3/2)\left[\frac{n^2-2n-5/4}{n-1/2}\cos(n-1/2)\varphi - (n+1/2)\cos(n+3/2)\varphi\right]$$

$$+ \sum_{n=0}^{\infty} B_n (r/W)^n [(n^2-n-2)\cos n\varphi - (n+2)(n+1)\cos(n+2)\varphi] \tag{A1.1.8}$$

$$\frac{\tau_{r\varphi}}{\sigma^*} = \sum_{n=0}^{\infty} A_n (r/W)^{n-1/2}(n+3/2)(n+1/2)[\sin(n+3/2)\varphi - \sin(n-1/2)\varphi]$$

$$+ \sum_{n=0}^{\infty} B_n (r/W)^n (n+1)[(n+2)\sin(n+2)\varphi - n\sin n\varphi] \tag{A1.1.9}$$

The coefficients α_n, A_n, $\tilde{A}_n$, β_n, B_n, and $\tilde{B}_n$ are simply related by

$$\alpha_n = \sigma * \frac{A_n}{W^{n-1/2}} = \sigma * \frac{\tilde{A}_n}{a^{n-1/2}} \tag{A1.1.10}$$

$$\beta_n = \sigma * \frac{B_n}{W^n} = \sigma * \frac{\tilde{B}_n}{a^n} \tag{A1.1.11}$$

In all mode-I considerations the symmetric part has to be used exclusively. For pure mode-II loadings the anti-symmetric part must be applied. The anti-symmetric part Φ_a reads, e.g.

$$\Phi_a = \sigma * W^2 \sum_{n=0}^{\infty} (r/W)^{n+3/2}\, \hat{A}_n [\sin(n+\tfrac{3}{2})\varphi - \sin(n-\tfrac{1}{2})\varphi]$$

$$+ \sigma * W^2 \sum_{n=0}^{\infty} (r/W)^{n+2}\, \hat{B}_n \left[\sin(n+3)\varphi - \frac{n+3}{n+1}\sin(n+1)\varphi \right] \tag{A1.1.12}$$

A1.2   Stress intensity factor

The stress intensity factor K is a measure of the singular stress term occurring near the tip of a crack and defined by

$$\sigma_{ij} = \frac{K_{\mathrm{I,II}}}{\sqrt{2\pi r}}\, f_{ij}(\varphi) \tag{A1.2.1}$$

with r and φ according to Fig. A1.1. The angular functions are for mode I:

$$f_{xx} = \cos\left(\frac{\varphi}{2}\right)\left[1 - \sin\left(\frac{\varphi}{2}\right)\sin\left(\frac{3\varphi}{2}\right)\right] \tag{A1.2.2a}$$

$$f_{yy} = \cos\left(\frac{\varphi}{2}\right)\left[1 + \sin\left(\frac{\varphi}{2}\right)\sin\left(\frac{3\varphi}{2}\right)\right] \tag{A1.2.2b}$$

$$f_{xy} = \cos\left(\frac{\varphi}{2}\right)\sin\left(\frac{\varphi}{2}\right)\cos\left(\frac{3\varphi}{2}\right) \tag{A1.2.2c}$$

and for mode II:

$$f_{xx} = \sin\left(\frac{\varphi}{2}\right)\left[2 + \sin\left(\frac{\varphi}{2}\right)\sin\left(\frac{3\varphi}{2}\right)\right] \tag{A1.2.3a}$$

$$f_{yy} = \sin\left(\frac{\varphi}{2}\right)\cos\left(\frac{\varphi}{2}\right)\cos\left(\frac{3\varphi}{2}\right) \tag{A1.2.3b}$$

$$f_{xy} = \cos\left(\frac{\varphi}{2}\right)\left[1 - \sin\left(\frac{\varphi}{2}\right)\cos\left(\frac{3\varphi}{2}\right)\right] \tag{A1.2.3c}$$

The stress intensity factors K_I and K_{II} are expressed as

$$K_I = \sigma^* \sqrt{\pi a}\, F_I(a/W) \qquad\qquad (A1.2.4a)$$

$$K_{II} = \tau^* \sqrt{\pi a}\, F_{II}(a/W) \qquad\qquad (A1.2.4b)$$

where a is the crack length, W is the width of the component, and σ^*, τ^* are characteristic stresses in the component, e.g. the outer fibre stress in a bending bar. F_I and F_{II} are functions of the ratio of crack length to specimen width as well as of the type of load applied.

In terms of the coefficients α_n, A_n, $\tilde{A}_n$, the stress intensity factor K_I reads

$$K_I = \alpha_0 \sqrt{18\pi} = \sigma^* A_0 \sqrt{18\pi W} = \sigma^* \tilde{A}_0 \sqrt{18\pi a} \qquad\qquad (A1.2.5)$$

and the geometric function F_I

$$F_I = \frac{\alpha_0}{\sigma^*} \sqrt{18/a} = A_0 \sqrt{18/\alpha} = \tilde{A}_0 \sqrt{18} \qquad\qquad (A1.2.6)$$

with the relative crack depth $\alpha = a/W$.

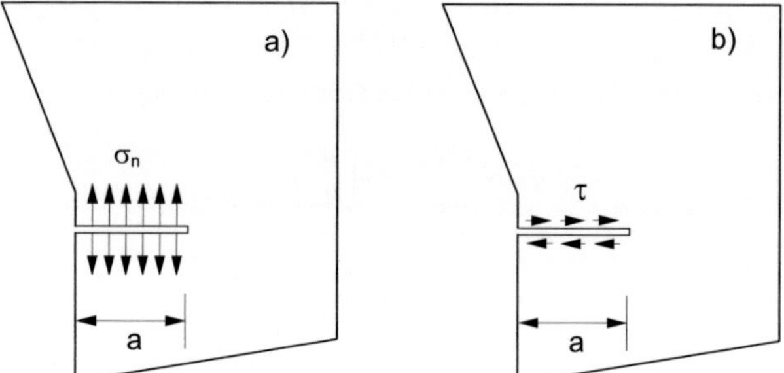

Fig. A1.2 Edge crack in a non-symmetric component under normal and shear stresses.

Mode-I stress intensity factors are not necessarily caused by normal stresses. Also shear stresses can be responsible for mode-I stress intensity factors. For demonstration purposes a crack in a non-symmetric plate is given in Fig. A1.2. Under a pure crack-face pressure σ_n, a stress intensity factor of

$$K_I^{(1)} = \sigma_n F_I^{(1)} \sqrt{\pi a} \qquad\qquad (A1.2.9)$$

and a mode-II contribution of

$$K_{II}^{(1)} = \sigma_n F_{II}^{(1)} \sqrt{\pi a} \qquad\qquad (A1.2.10)$$

are generated. Under constant shear stress τ acting in the crack face direction, two stress intensity factors occur

$$K_I^{(2)} = \tau F_I^{(2)} \sqrt{\pi a} \qquad (A1.2.11)$$

$$K_{II}^{(2)} = \tau F_{II}^{(2)} \sqrt{\pi a} \qquad (A1.2.12)$$

Only in the case of a crack normal to the surface and located in a symmetric component do the stress intensity factors $K_{II}^{(1)}$ and $K_I^{(2)}$ disappear.

A1.3   T-stress term for traction free crack faces

The first higher-order stress term ($n=0$) is given by the dependency r^0 (i.e. the stress component is independent on the distance from the crack tip) with the coefficients β_0, B_0, and $\widetilde{B}_0$. From relations (A1.1.7-A1.1.9) it results for the case of loading by remote stresses

$$\sigma_{\varphi,0} = -4\,\beta_0 \sin^2 \varphi \qquad (A1.3.1)$$

$$\sigma_{r,0} = -4\,\beta_0 \cos^2 \varphi \qquad (A1.3.2)$$

$$\tau_{r\varphi,0} = 2\,\beta_0 \sin(2\varphi) \qquad (A1.3.3)$$

By the general stress transformation from polar to Cartesian coordinates (rotated by an arbitrary angle γ) according to

$$\sigma_{xx} = \sigma_r \cos^2 \gamma + \sigma_\varphi \sin^2 \gamma - 2\tau_{r\varphi} \sin \gamma \cos \gamma \qquad (A1.3.4)$$

$$\sigma_{yy} = \sigma_r \sin^2 \gamma + \sigma_\varphi \cos^2 \gamma + 2\tau_{r\varphi} \sin \gamma \cos \gamma \qquad (A1.3.5)$$

$$\tau_{xy} = (\sigma_r - \sigma_\varphi)\sin \gamma \cos \gamma + \tau_{r\varphi}(\cos^2 \gamma - \sin^2 \gamma) \qquad (A1.3.6)$$

and identifying γ with $\pi+\varphi$, it yields

$$\sigma_{xx,0} = -4\,\beta_0 = -4\sigma * B_0 \qquad (A1.3.7)$$

$$\sigma_{yy,0} = 0 \qquad (A1.3.8)$$

$$\tau_{xy,0} = 0 \qquad (A1.3.9)$$

Since for remote stresses the constant stress components $\tau_{xy,0}=\tau_{yx,0}$ and $\sigma_{yy,0}$ disappear (traction free crack faces), the stress tensor reads

$$\sigma_{ij,0} = \begin{pmatrix} \sigma_{xx,0} & 0 \\ 0 & 0 \end{pmatrix} = \begin{pmatrix} -4\sigma * B_0 & 0 \\ 0 & 0 \end{pmatrix} \underset{def}{=} \begin{pmatrix} T & 0 \\ 0 & 0 \end{pmatrix} \qquad (A1.3.10)$$

where T is the so-called "T-stress".

The T-stress in the different stress function representations is

$$T = -4\beta_0 = -4\sigma * B_0 = -4\sigma * \widetilde{B}_0 \ . \qquad\qquad (A1.3.11)$$

Leevers and Radon [A1.2] proposed a dimensionless representation of T by the stress biaxiality ratio β, defined as

$$\beta = \frac{T\sqrt{\pi a}}{K_I} = \frac{T}{\sigma * F} \qquad\qquad (A1.3.12)$$

Different methods were applied in the past to compute the T-stress term for fracture mechanics standard test specimens. Regarding one-dimensional cracks, Leevers and Radon [A1.2] carried out a numerical analysis based on a variational method. Kfouri [A1.3] applied the Eshelby technique. Sham [A1.4, A1.5] developed a second-order weight function based on a work-conjugate integral and evaluated it for the straight edge notched (SEN) specimen using the FE method. In [A1.6, A1.7] a Green's function for T-stresses was determined on the basis of boundary collocation results. Wang and Parks [A1.8] extended the T-stress evaluation to two-dimensional surface cracks using the line-spring method. A compilation of results from literature was given by Sherry et al. [A1.9].

Most of the T-stress solutions derived by the author were obtained with the boundary collocation method (BCM) and Green's function technique. Therefore, these methods shall be described in more detail in Section A2.2. The boundary collocation method can provide a large number of coefficients of a Williams expansion of the stress function. Therefore, additional coefficients are reported in some cases.

A1.4 T-stress in case of crack faces loaded by tractions

From eq.(A1.3.10) it can be concluded that

- the T-stress is identical with the constant x-stress term, $T=\sigma_{xx,0}$, or

- the T-stress is identical with the first coefficient of the regular part of the Williams expansion, namely, $T=-4\beta_0$ (or $-4\sigma*B_0$, $-4\sigma*\widetilde{B}_0$).

It has to be emphasized that these two definitions of the T-stress are equivalent only in the case of crack faces free of tractions (e.g. Fig. A1.3a), the case for which the Williams expansion was derived [A1.1]. However, in practical problems also traction loaded crack faces are of interest. Examples are for instance:

- Cracks in walls of tubes under internal gas or liquid pressure,

- crack bridging stresses due to crack-face interactions in coarse grained ceramics (Fig. A1.3b).

In this context the question has to be answered what the appropriate definition of T-stress in such cases is. In the following considerations the modifications in the Airy stress function are addressed.

A1.4.1 The Airy stress function

The stress state of a component containing a crack can be described in terms of the Airy stress function Φ that results by solving the bi-harmonic equation (A1.1.1) with the solutions (A1.1.2).

The symmetric Airy stress function for a crack with surfaces free of shear tractions reads

$$\Phi = \sum_{n=0}^{\infty} r^{n+3/2} \alpha_n \left[\cos(n+\tfrac{3}{2})\varphi - \frac{2n+3}{2n-1} \cos(n-\tfrac{1}{2})\varphi \right]$$
$$+ \sum_{n=0}^{\infty} r^{n+2} [\beta_n \cos(n+2)\varphi + \gamma_n \cos n\varphi] \qquad (A1.4.1)$$

The tangential stresses σ_φ result from (A1.1.6) yielding

$$\sigma_\varphi = \sum_{n=0}^{\infty} r^{n-\frac{1}{2}} (n+\tfrac{3}{2})(n+\tfrac{1}{2}) \alpha_n [\cos(n+\tfrac{3}{2})\varphi - \tfrac{2n+3}{2n-1} \cos(n-\tfrac{1}{2})\varphi]$$
$$+ \sum_{n=0}^{\infty} r^n (n+2)(n+1)[\beta_n \cos(n+2)\varphi + \gamma_n \cos n\varphi] \qquad (A1.4.2)$$

A1.4.2 Crack surfaces free of tractions

The case of a crack with traction free surfaces requires that $\sigma_\varphi=0$ for $\varphi=\pi$. This condition is automatically fulfilled for the first sum of (A1.4.2). From the second sum

$$\gamma_n = -\beta_n \qquad (A1.4.3)$$

is obtained. This yields the well-known Williams stress function [A1.1]

$$\Phi = \sum_{n=0}^{\infty} r^{n+3/2} \alpha_n \left[\cos(n+\tfrac{3}{2})\varphi - \frac{2n+3}{2n-1} \cos(n-\tfrac{1}{2})\varphi \right]$$
$$+ \sum_{n=0}^{\infty} r^{n+2} \beta_n [\cos(n+2)\varphi - \cos n\varphi] \qquad (A.1.4.4)$$

from which the radial stresses result as

$$\sigma_r = \sum_{n=0}^{\infty} \alpha_n r^{n-\frac{1}{2}} (n+\tfrac{3}{2}) \left[\frac{n^2 - 2n - \tfrac{5}{4}}{n-\tfrac{1}{2}} \cos(n-\tfrac{1}{2})\varphi - (n+\tfrac{1}{2}) \cos(n+\tfrac{3}{2})\varphi \right]$$
$$+ \sum_{n=0}^{\infty} \beta_n r^n [(n^2 - n - 2)\cos n\varphi - (n+2)(n+1)\cos(n+2)\varphi] \qquad (A1.4.5)$$

The constant stress term related to $n=0$ is

$$\sigma_{r,0} = -2\beta_0 (1 + \cos 2\varphi) \qquad (A1.4.6)$$

and the constant x-stress results by setting $\varphi=\pi$ as

$$\sigma_{xx,0} = -4\beta_0 \tag{A1.4.7}$$

This relation is one possibility to define the T-stress, namely by

$$T = -4\beta_0 \tag{A1.4.8}$$

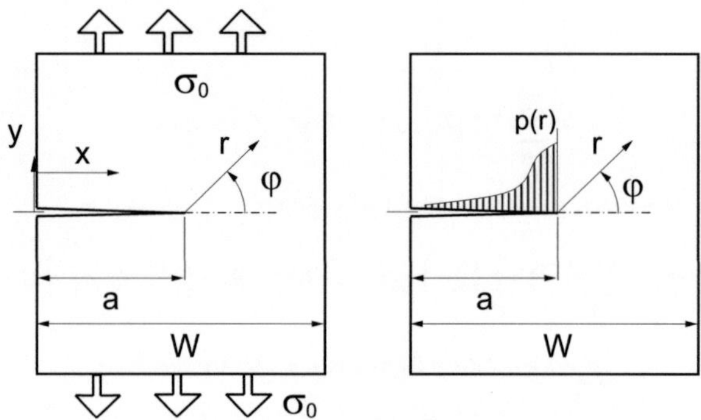

Fig. A1.3 a) A crack with traction free surfaces loaded by externally applied tractions σ_0, b) crack, loaded by crack-face tractions $p(r)$.

A1.4.3 Cracks loaded by crack-face tractions

Next, the case of cracks with loaded surfaces is considered. A crack-face pressure distribution along the crack faces $p(r)$ is assumed to be described by the power series expansion

$$p(r) = p_0 + p_1 r + p_2 r^2 + \dots + p_n r^n + \dots = -\sigma_y(r) \tag{A1.4.9}$$

with positive p for pressure on the crack. In order to satisfy these traction boundary conditions it is sufficient to consider the part of the Airy stress function containing integer exponents exclusively. By adjusting the terms with same exponent of r^n in eqs.(A1.4.2) and (A1.4.9), we obtain

$$(n+2)(n+1)[\beta_n \cos(n+2)\pi + \gamma_n \cos n\pi] = -p_n \tag{A1.4.10}$$

This allows eliminating the coefficient γ_n

$$\gamma_n = (-1)^{n+1}\frac{p_n}{(n+1)(n+2)} - \beta_n \tag{A1.4.11}$$

resulting in

$$\Phi = \sum_{n=0}^{\infty} r^{n+3/2} \alpha_n \left[\cos(n+\tfrac{3}{2})\varphi - \frac{2n+3}{2n-1}\cos(n-\tfrac{1}{2})\varphi \right]$$
$$+ \sum_{n=0}^{\infty} r^{n+2} \left\{ \beta_n [\cos(n+2)\varphi - \cos n\varphi] + (-1)^{n+1}\frac{p_n}{(n+1)(n+2)}\cos n\varphi \right\} \tag{A1.4.12}$$

From (A1.4.12), the tangential stresses

$$\sigma_\varphi = \sum_{n=0}^{\infty} r^{n-\frac{1}{2}}(n+\tfrac{3}{2})(n+\tfrac{1}{2})\alpha_n [\cos(n+\tfrac{3}{2})\varphi - \tfrac{2n+3}{2n-1}\cos(n-\tfrac{1}{2})\varphi]$$
$$+ \sum_{n=0}^{\infty} r^{n} \{ (n+2)(n+1)\beta_n[\cos(n+2)\varphi - \cos n\varphi] + (-1)^{n+1}p_n\cos n\varphi \} \tag{A1.4.13}$$

the radial stresses

$$\sigma_r = \sum_{n=0}^{\infty} r^{n-\frac{1}{2}}(n+\tfrac{3}{2})\alpha_n \left[\frac{n^2-2n-\tfrac{5}{4}}{n-\tfrac{1}{2}}\cos(n-\tfrac{1}{2})\varphi - (n+\tfrac{1}{2})\cos(n+\tfrac{3}{2})\varphi \right]$$
$$+ \sum_{n=0}^{\infty} r^{n} \{ (n+1)\beta_n[(n-2)\cos n\varphi - (n+2)\cos(n+2)\varphi] - (-1)^{n+1}p_n\frac{n-2}{n+2}\cos n\varphi \} \tag{A1.4.14}$$

and the shear stresses

$$\tau_{r\varphi} = \sum_{n=0}^{\infty} r^{n-\frac{1}{2}}(n+\tfrac{3}{2})(n+\tfrac{1}{2})\alpha_n \left[\sin(n+\tfrac{3}{2}\varphi) - \sin(n-\tfrac{1}{2})\varphi \right]$$
$$+ \sum_{n=0}^{\infty} r^{n} \{ (n+1)\beta_n[(n+2)\sin(n+2)\varphi - n\sin n\varphi] + (-1)^{n+1}p_n\frac{n}{n+2}\sin n\varphi \} \tag{A1.4.15}$$

result. Equations (A1.4.12) to (A1.4.15) are appropriate for the application of the boundary collocation procedure to problems with crack-face loading.

The constant stress terms related to $n=0$ for crack-face loading are given by

$$\sigma_{ij,0} = \begin{pmatrix} \sigma_{xx,0} & \sigma_{xy,0} \\ \sigma_{yx,0} & \sigma_{yy,0} \end{pmatrix} = \begin{pmatrix} -4\beta_0 - p_0 & 0 \\ 0 & -p_0 \end{pmatrix} \tag{A1.4.16}$$

For traction free crack faces the T-stress could be defined equivalently by the total constant x-stress $\sigma_{xx,0}$, eq. (A1.4.7), or by the coefficient β_0 of the Williams expansion, eq.(A1.4.8). In the case of crack-face loading, these two possibilities of defining the T-stress result in different values of T. In order to distinguish the two attempts, the T-stresses are indicated as $T^{(1)}$ and $T^{(2)}$. Similar to eq. (A1.4.8) we still can identify again T with the Williams coefficient β_0 resulting in

$$T^{(1)} = -4\beta_0 \tag{A1.4.8a}$$

On the other hand one can define T via the total constant x-stress, i.e.

$$T^{(2)} = \sigma_{xx,0} = -4\beta_0 - p_0 \qquad (A1.4.17)$$

as for instance done in [A1.10]. The different definitions of T-stress are not a physical problem because the x-stress as the physically relevant quantity is the same in both cases.

A1.5 T-stress under crack loading by residual stresses

The T-stress term is the first regular stress term in the Williams series expansion [A1.1]. As outlined in Section A1.3, the Cartesian components of the first regular term of traction-free crack faces are

$$\sigma_{xx} = \frac{K_{\mathrm{I}}}{\sqrt{2\pi r}} f_{xx}(\varphi) + T + O(r^{1/2}) \qquad (A1.5.1)$$

$$\sigma_{yy} = \frac{K_{\mathrm{I}}}{\sqrt{2\pi r}} f_{yy}(\varphi) + O(r^{1/2}) \qquad (A1.5.2)$$

with the angular functions f_{xx} and f_{yy} that are identical for $\varphi=0$, namely, $f_{xx}=f_{yy}=1$. Due to the r-independence of the T-term in eq.(A1.5.1), it is often used to represent the stress at the crack faces over a wide crack-tip distance. This is correct for e.g. an internal through-the-thickness crack in an infinite body under tension for which $\sigma_x=T$ holds along the whole crack. Unfortunately, this is not a general result. Especially in the case of residual stresses this assumption is misleading as will be shown in the following considerations.

Let us now consider a more complicated loading case. In a limited zone of a component a residual stress field may occur (Fig. A1.4a). Such a stress field may be created by a local martensitic phase transformation in zirconia ceramics due to the high stresses near a crack tip. Another possibility is the development of an ion exchange layer at the crack surfaces of soda-lime glass which is in contact with water or humid air. The volumetric expansion in these zones might be $\varepsilon_{\mathrm{vol}}$. The two relevant stress components generated in this zone are denoted by $\sigma_{\mathrm{res},x}$ and $\sigma_{\mathrm{res},y}$.

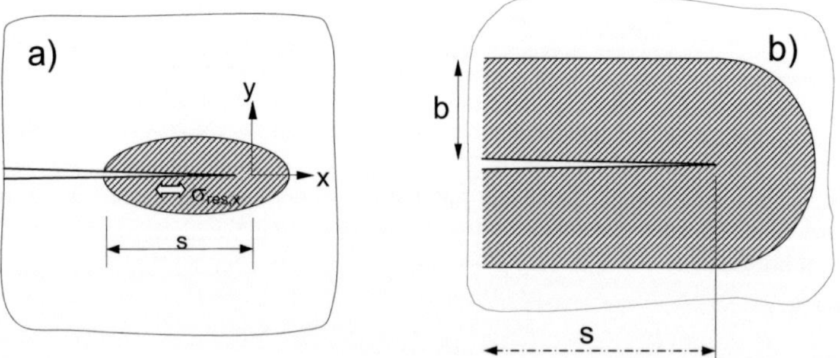

Fig. A1.4 Residual stresses caused in crack-tip zones by a volumetric expansion strain $\varepsilon_{\mathrm{vol}}$, a) zone of finite length s, b) limit case of an infinitely long zone.

The x-stresses in such a volumetric expansion zone at $y=0$ are superimposed by the residual stresses $\sigma_{res,x}$, $\sigma_{res,y}$ and the contributions by the stress intensity factor and the T-stress.

Let us consider a phase transformation zone at the crack surfaces as shown in Fig. A1.4a. The T-stress of this configuration was obtained from a finite element study. For this purpose, the transformation strains were replaced by thermal strains. The related volume strain ε_{vol} is then given for a temperature increase $\Delta\Theta$ in the zone

$$\varepsilon_{vol} \cong 3\alpha\Delta\Theta \tag{A1.5.3}$$

with the thermal expansion coefficient α.
The x-stresses in an infinitely long layer is under plane strain or generalized plane strain conditions

$$\sigma_x = -\tfrac{1}{3}\frac{\varepsilon_{vol}E}{1-\nu} \tag{A1.5.4}$$

Computations were carried out with ABAQUS Version 6.2, providing the individual stress intensity factors K_I and K_{II} as well as the T-stress. For this purpose, ABAQUS employs an interaction integral according to Shih and Asaro [A1.11]. This quantity was used in [A1.12]. For an infinitely long zone of $s\to\infty$, Fig. A1.4b, the T-stress evaluation in [A1.12] yields

$$T \cong -\tfrac{1}{6}\frac{\varepsilon_{vol}E}{1-\nu} \tag{A1.5.5}$$

where ε_{vol} is the volume strain, E is Young's modulus, and ν Poisson's ratio. For the negative sign in (A1.5.5) see the remark in Section C22.4.

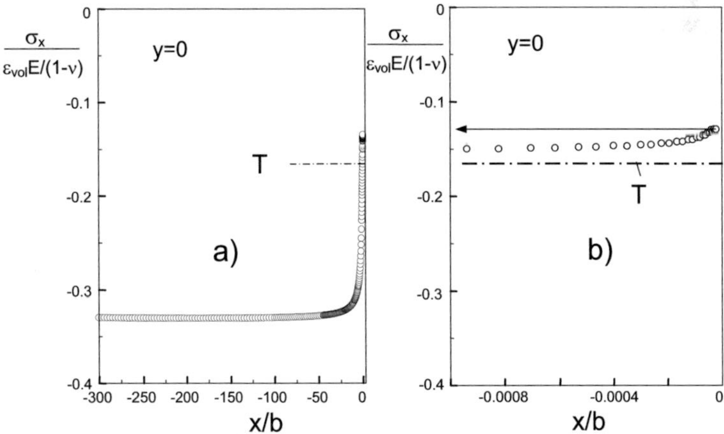

Fig. A1.5 a) Finite element results of x-stresses for an expansive zone along the free surface in the crack wake according to Fig. A1.4b, b) detail near $x/b=0$.

13

Figure A1.5 shows the x-stress component for $y=0$ from finite element computations. At $x=0$, the total x-stress is roughly

$$\sigma_x\big|_{x=0} \approx -0.136\frac{\varepsilon_{vol}E}{1-\nu}$$

(A1.5.6)

as indicated in Fig. A1.5b, and reaches a value of -1/3 $\varepsilon_{vol}E/(1-\nu)$ for $x/b\ll1$ as becomes obvious from Fig. A1.5a and would be expected from eq.(A1.5.4). It is obvious that the stress value T is hardly representative for the x-stresses in the crack wake.

References A1

[A1.1] Williams, M.L., On the stress distribution at the base of a stationary crack, J. Appl. Mech. **24**(1957), 109-114.

[A1.2] Leevers, P.S., Radon, J.C., Inherent stress biaxiality in various fracture specimen geometries, Int. J. Fract. **19**(1982), 311-325.

[A1.3] Kfouri, A.P., Some evaluations of the elastic T-term using Eshelby's method, Int. J. Fract. **30**(1986), 301-315.

[A1.4] Sham, T.L., The theory of higher order weight functions for linear elastic plane problems, Int. J. Solids and Struct. **25**(1989), 357-380.

[A1.5] Sham, T.L., The determination of the elastic T-term using higher order weight functions, Int. J. Fract. **48**(1991), 81-102.

[A1.6] Fett, T., A Green's function for T-stresses in an edge-cracked rectangular plate, Eng. Fract. Mech. **57**(1997), 365-373.

[A1.7] Fett, T., T-stresses in rectangular plates and circular disks, Engng. Fract. Mech. **60**(1998), 631-652.

[A1.8] Wang, Y.Y., Parks, D.M., Evaluation of the elastic T-stress in surface-cracked plates using the line-spring method, Int. J. Fract **56**(1992), 25-40.

[A1.9] Sherry, A.H., France, C.C., Goldthorpe, M.R., Compendium of T-stress solutions for two and three- dimensional cracked geometries, Engng. Fract. Mech. **18**(1995), 141-155.

[A1.10] Pham, V.-B., Bahr, H.-A., Bahr, U., Fett, T., Balke, H., Crack paths and the problem of global directional stability, Int. Journal of Fracture **141**(2006), 513-534.

[A1.11] Shih, C.F., Asaro, R.J., Elastic-plastic analysis of cracks on biomaterial interfaces: Part I – Small scale yielding, J. Appl. Mech. (1988), 299-316.

[A1.12] T. Fett, G. Rizzi, Fracture mechanics parameters of crack surface zones under volumetric strains, Int. J. Fract. **127**(2004), L117-L124.

<u>A2</u>

Methods for the determination of *K* and *T*

Numerous methods are described in literature for the determination of stress intensity factors and T-stress. Analytical and semi-analytical approaches are applicable mostly in the case of special geometries and simple loading cases as for instance internally cracked infinite and edge-cracked semi-infinite bodies under remote tractions. As an example, the analytical procedure reported by Wigglesworth [A2.1] may be briefly addressed.

For cracks in finite components, numerical procedures have to be used predominantly. For this purpose a few of the available methods may be mentioned, namely

- Finite element methods with evaluation of *K* and *T* from the stresses or by the evaluation of energy integrals,

- Boundary element procedure,

- Boundary collocation method

In the K-T-compendium (Section C) some results are given which were obtained with ABAQUS [A2.2] Versions 6.2 and 6.3 which provide stress intensity factors K_I, K_{II}, and *T* directly. For the analysis of *T*, ABAQUS employs an interaction integral according to Shih and Asaro [A2.3].

Most of the numerical data were obtained by application of the Boundary collocation procedure. Since this method also provides the next higher-order stress terms it is described in Section A2.2.

A2.1 Analytical method by Wigglesworth

Very early, Wigglesworth [A2.1] derived an analytical solution for the edge-cracked half-space under a constant crack-face pressure *p* (Fig. A2.1a) or, equivalently, by a remote tension σ_∞. The procedure enables to determine the coefficients of a series expansion of the stress distribution (here denoted as $\overline{A}_n$ and $\overline{B}_n$). Unfortunately, the Wigglesworth analysis is somewhat fallen in oblivion. One reason might be the iterative procedure necessary for the evaluation.

As the main result Wigglesworth [A2.1] showed that the coefficients $\overline{A}_n$ and $\overline{B}_n$ could be obtained from the asymptotic expansion of a function $q(m)$ for large integer numbers m, where $q(m)$ is defined by

$$q(m) = -2h(1)\frac{\Gamma(\frac{3}{2})\Gamma(\frac{m}{2})}{\Gamma(\frac{m}{2}+\frac{3}{2})}h(m) \qquad (A2.1.1)$$

with

$$\log h(m) = \frac{m}{\pi} \int\limits_0^\infty \frac{\psi(v)}{m^2 + v^2} dv \qquad (A2.1.2)$$

and the function $\psi(v)$

$$\psi(v) = \log[1 / (1 - v^2 \operatorname{cosech}^2 \tfrac{1}{2} \pi v)] \qquad (A2.1.3)$$

The asymptotic expansion of $\log h(m)$ reads

$$\log h(m) \rightarrow \alpha_0 / m - \alpha_1 / m^3 + \ldots = \sum_{n=0}^\infty (-1)^n \frac{\alpha_n}{m^{2n+1}} \qquad (A2.1.4)$$

where

$$\alpha_n = \frac{1}{\pi} \int\limits_0^\infty v^{2n} \psi(v) dv \qquad (A2.1.5)$$

The function $h(m)$ can now be expressed by using (A2.1.4) as

$$h(m) = \exp(\alpha_0 / m - \alpha_1 / m^3 + \ldots) \qquad (A2.1.6)$$

with the asymptotic expansion

$$h(m) \rightarrow 1 + \beta_1 / m + \beta_2 / m^2 + \beta_3 / m^3 + \ldots \qquad (A2.1.7)$$

The coefficients are interrelated by

$$\beta_1 = \alpha_0, \quad \beta_2 = \alpha_0^2 / 2!, \quad \beta_3 = \alpha_0^3 / 3! - \alpha_1, \text{ etc.} \qquad (A2.1.8)$$

with the α_n known from (A2.1.5).

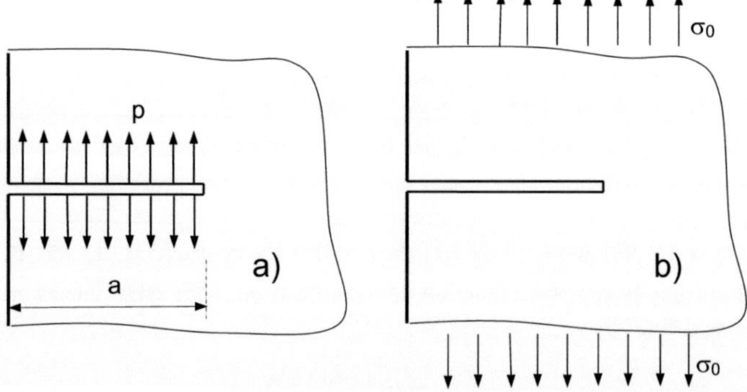

Fig. A2.1 Edge-cracked semi-infinite body, a) crack loaded by an internal pressure p, b) crack loaded by remote y-tractions σ_0.

The first coefficients α_n and β_n, obtained by application of Mathematica 3.0 [A2.4], are compiled in Table A2.1.

Table A2.1 Coefficients α_n and β_n according to eqs.(A2.1.5) and (A2.1.8)

n	α_n	β_n
0	0.1594228254	1
1	0.1157911013	0.1594228254
2	0.3171665716	0.0127078186
3	1.7462768020	-0.1151157958
4	15.774232536	-0.0184328297
5	210.33303560	0.31569597745
6	3874.8986328	0.05718920898
7	94197.687783	-1.7411806856
8	2.9202579843×10^6	-0.3148221728
9	1.1243271×10^8	15.745940668

The coefficients $\overline{A}_n$ are then iteratively given by

$$\overline{A}_0 = \sqrt{8}h(1),$$

$$\overline{A}_n = \sqrt{8}h(1)(d_n^0 + \beta_1 d_{n-1}^1 + ... + \beta_n d_0^n)$$

(A2.1.9)

with $h(1)\cong1.12152$, identical with the geometric function F for the stress intensity factor of the edge-cracked semi-infinite body, and the coefficients

$$d_n^0 = (-\tfrac{1}{2})^n \binom{\tfrac{1}{2}}{n},$$

$$d_n^{r+1} = \frac{d_0^r + d_1^r + ... + d_n^r}{n+r+\tfrac{3}{2}}$$

(A2.1.10)

A similar procedure is described in [A2.1] for the coefficients $\overline{B}_n$. Compared with the stress coefficients in (A1.1.5), which are slightly different from those defined by Wigglesworth, it holds (after dropping the tilde symbol)

$$A_n = \sqrt{8}\,h(1)\overline{A}_n = \sqrt{8}\,\overline{A}_n\,F$$

(A2.1.11)

$$B_n = \sqrt{8}\,h(1)(n+1)\overline{B}_{n+1} = \sqrt{8}\,(n+1)\overline{B}_{n+1}\,F$$

(A2.1.2)

The Wigglesworth procedure was carried out for 150 coefficients by using Mathematica Version 3.0. In Table A2.2, the first coefficients for A_n and B_n are compiled. It results for the geometric function of the stress intensity factor of an edge crack in a semi-infinite body under remote tension

$$F = 1.12152225523... \qquad (A2.1.13)$$

The coefficient B_0, representing the T-stress according to eq.(A1.3.11) is

$$B_0 = 0.131491901... \qquad (A2.1.14)$$

and, consequently

$$T = -4\sigma_\infty B_0 = -0.5259676026\,\sigma_\infty \qquad (A2.1.15)$$

It should be mentioned, that a very similar value of -0.526 was already reported in literature [A2.5, A2.6]. In [A2.7], this value was given with an increased accuracy by $-4B_0 = -0.52596 \pm 0.00003$.

Table A2.2 Coefficients A_n and B_n according to eqs.(A2.1.6) and (A2.1.9).

n	$(-1)^{n+1}\dfrac{\sqrt{2}}{F^{\frac{n+3/2}{n-1/2}}}A_n$	$4(-1)^n(n+1)B_n$
0	1	0.5259676026
1	-0.143718116	-0.384982976
2	0.0199655992	-0.214309639
3	0.0196651671	-0.086876065
4	0.0118558588	-0.0142437609
5	0.0062538226	0.0193802356
6	0.0029935128	0.0305459991
7	0.0012562099	0.0306628565
8	0.0003899590	0.0263270316
9	-9.7144×10^{-6}	0.0208939094
10	-0.0001717008	0.0158426721
11	-0.0002189425	0.0116784377
12	-0.0002148644	0.0084544707
13	-0.0001908770	0.0060469457
14	-0.0001617433	0.0042866037
15	-0.0001338496	0.0030144163
16	-0.0001095285	0.0021000215
17	-0.0000892583	0.0014436503
18	-0.0000727470	0.0009718567
19	-0.0000594446	0.0006317771
20	-0.0000487713	0.0003858288

The total constant x-stress is for loading by remote tractions σ_∞

$$\sigma_{xx,0} = T \qquad (A2.1.16)$$

and in the case of the constant crack-face pressure p

$$\sigma_{xx,0} = T + p \tag{A2.1.17}$$

The reason for these different values is explained in Section A1.4.

A2.2 Boundary collocation procedure

A2.2.1 Boundary conditions

A simple possibility to determine the coefficients A_n and B_n is the application of the boundary collocation method [A2.8-A2.10]. For practical application of eq.(A1.1.3), which is used to determine A_n and B_n, the infinite series for the Airy stress function Φ must be truncated after the Nth term, for which an adequate value must be chosen. The still unknown coefficients are determined by fitting the stresses and displacements to the specified boundary conditions. The stresses result from the relations A1.1.6.

Stresses firstly derived by Williams [A2.11] are given in eqs.(A1.1.7-A1.1.9). The displacements in radial and tangential direction, u and v, respectively, read

$$\frac{u}{\sigma^* W} = \frac{1+v}{E} \sum_{n=0}^{\infty} A_n \left(\frac{r}{W}\right)^{n+1/2} \frac{2n+3}{2n-1} [(n+4v-\tfrac{5}{2})\cos(n-\tfrac{1}{2})\varphi - (n-\tfrac{1}{2})\cos(n+\tfrac{3}{2})\varphi] +$$

$$+ \frac{1+v}{E} \sum_{n=0}^{\infty} B_n \left(\frac{r}{W}\right)^{n+1} [(n+4v-2)\cos n\varphi - (n+2)\cos(n+2)\varphi] \tag{A2.2.1}$$

$$\frac{v}{\sigma^* W} = \frac{1+v}{E} \sum_{n=0}^{\infty} A_n \left(\frac{r}{W}\right)^{n+1/2} \frac{2n+3}{2n-1} [(n-\tfrac{1}{2})\sin(n+\tfrac{3}{2})\varphi - (n-4v+\tfrac{7}{2})\sin(n-\tfrac{1}{2})\varphi] +$$

$$+ \frac{1+v}{E} \sum_{n=0}^{\infty} B_n \left(\frac{r}{W}\right)^{n+1} [(n+2)\sin(n+2)\varphi - (n-4v+4)\sin n\varphi] \tag{A2.2.2}$$

(v=Poisson's ratio), from which the Cartesian components u_x and u_y result as

$$u_x = u\cos\varphi - v\sin\varphi \tag{A2.2.3a}$$

$$u_y = u\sin\varphi + v\cos\varphi \tag{A2.2.3b}$$

In the special case of an internally cracked circular disk of radius R (Fig. A2.2), the stresses at the boundaries are

$$\sigma_n = \tau_{r\varphi} = 0 \tag{A2.2.4}$$

along the quarter circle. Along the perpendicular symmetry line, the boundary conditions are:

$$u_x = \text{const.} \;\rightarrow\; \frac{\partial u_x}{\partial y} = 0 \qquad\qquad (A2.2.5)$$

$$\tau_{xy} = 0 \qquad\qquad (A2.2.6)$$

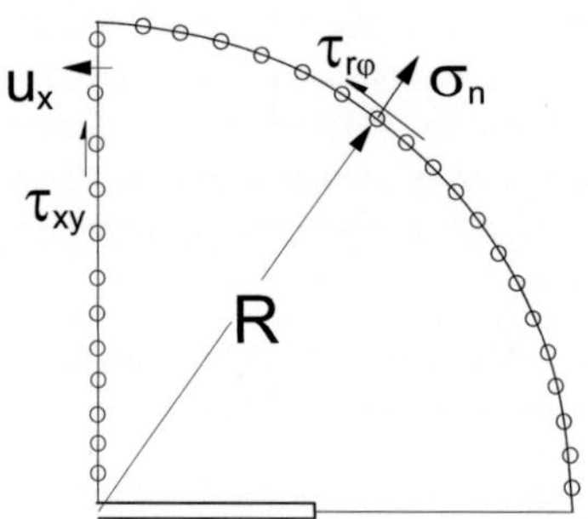

Fig. A2.2 Node selection and boundary conditions for an internally cracked disk.

About 100 coefficients for eq.(A1.1.3) were determined from 600-800 stress and displacement equations at 400 nodes along the outer contour (symbolised by the circles in Fig. A2.2). For a selected number of (N+1) collocation points, the related stress components (or displacements) are computed, and a system of $2(N+1)$ equations allows to determine up to $2(N+1)$ coefficients. The computation expenditure can be reduced by the selection of a clearly larger number of edge points and by subsequently solving the then over determined system of equations using a least-squares routine.

In the case of the edge-cracked rectangular plate of width W and height $2H$ (Fig. A2.3), the stresses at the border are

$$\sigma_x = 0, \; \tau_{xy} = 0 \qquad \text{for } x = 0 \qquad\qquad (A2.2.7a)$$

$$\sigma_y = \sigma^*, \; \tau_{xy} = 0 \qquad \text{for } y = H \qquad\qquad (A2.2.7b)$$

$$\sigma_x = 0, \; \tau_{xy} = 0 \qquad \text{for } x = W \qquad\qquad (A2.2.7c)$$

and in the case of the double-edge-cracked plate (Fig. A2.4), it holds

$$\sigma_x = 0, \; \tau_{xy} = 0 \qquad \text{for } x = 0 \qquad\qquad (A2.2.8a)$$

$$\sigma_y = \sigma^*, \; \tau_{xy} = 0 \qquad \text{for } y = H \qquad\qquad (A2.2.8b)$$

$$\frac{\partial u_x}{\partial y} = 0, \quad \tau_{xy} = 0 \qquad \text{for } x = W \tag{A2.2.8c}$$

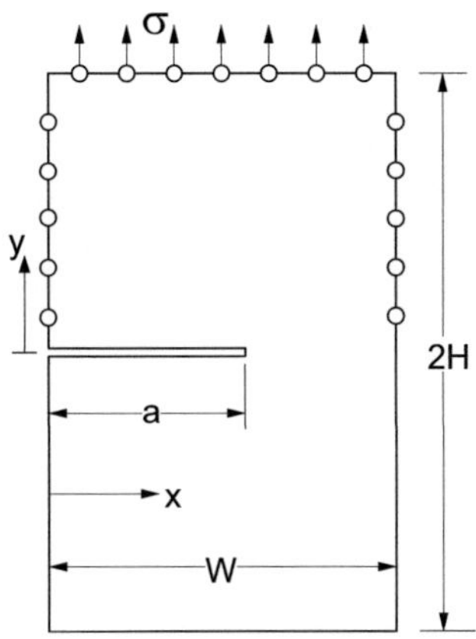

Fig. A2.3 Collocation points for the edge-cracked rectangular plate.

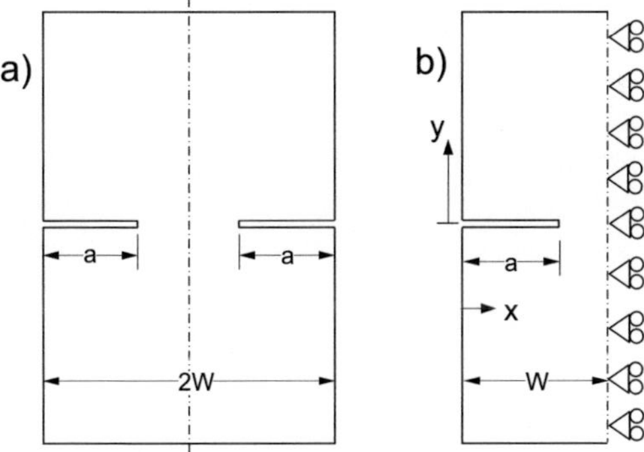

Fig. A2.4 Double-edge-cracked plate a) geometric data, b) half-specimen with symmetry boundary conditions.

A2.2.2 Boundary collocation procedure for point forces

The treatment of point forces at the crack face in case of a finite body shall be illustrated in the following sections for a circular disk with an internal crack loaded by a couple of forces at $x = y = 0$. In order to describe the crack-face loading by concentrated forces, we superimpose two loading cases. First, the singular crack-face loading is modelled by the centrally loaded crack in an infinite body described by the Westergaard stress function

$$Z = \frac{Pa}{\pi} \frac{1}{z\sqrt{z^2 - a^2}} \qquad (A2.2.9)$$

The stresses resulting from this stress function disappear at infinite distances from the crack only. In the finite body the stress-free boundary condition consequently is not fulfilled. To nullify the traction at the outer boundaries, stresses resulting from the Airy stress function, eq.(A1.1.3), are added, which do not superimpose additional stresses at the crack faces. The basic principle used for such calculations, the principle of superposition, shall be illustrated in more detail in Section A2.3.

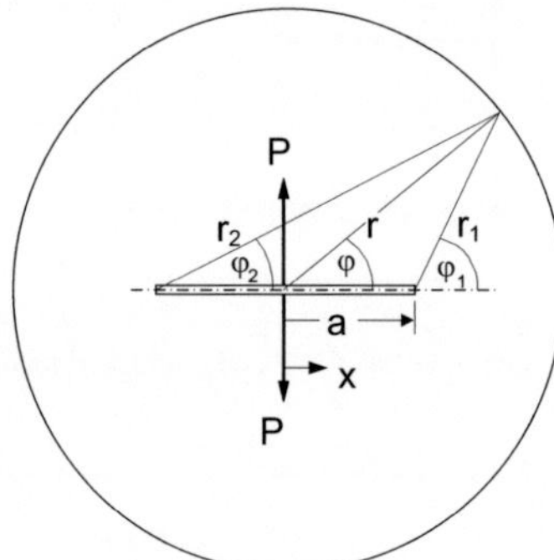

Fig. A2.5 Coordinate system for the application of the Westergaard stress function to a finite component.

The stresses caused by Z are

$$\sigma_x = \operatorname{Re} Z - y \operatorname{Im} Z' \qquad (A2.2.10a)$$

$$\sigma_y = \operatorname{Re} Z + y \operatorname{Im} Z' \qquad (A2.2.10b)$$

$$\tau_{xy} = -y \operatorname{Re} Z' \qquad (A2.2.10c)$$

with

$$Z' = \frac{dZ}{dz} = -\frac{Pa}{\pi} \frac{2z^2 - a^2}{z^2(z^2 - a^2)^{3/2}} \qquad (A2.2.11)$$

For practical use it is of advantage to introduce the coordinates shown in Fig. A2.5. The following geometric relations hold

$$z = r\exp(i\varphi), \quad z - a = r_1\exp(i\varphi_1), \quad z + a = r_2\exp(i\varphi_2) \qquad (A2.2.12)$$

$$r = \sqrt{x^2 + y^2} \ , \quad \tan\varphi = y/x \tag{A2.2.13}$$

$$r_1 = \sqrt{(x-a)^2 + y^2} \ , \quad \tan\varphi_1 = y/(x-a) \tag{A2.2.14}$$

$$r_2 = \sqrt{(x+a)^2 + y^2} \ , \quad \tan\varphi_2 = y/(x+a) \tag{A2.2.15}$$

$$\mathrm{Re}\,Z = \frac{Pa}{\pi r \sqrt{r_1 r_2}} \cos(\varphi + \tfrac{1}{2}\varphi_1 + \tfrac{1}{2}\varphi_2) \tag{A2.2.16}$$

$$\mathrm{Im}\,Z = -\frac{Pa}{\pi r \sqrt{r_1 r_2}} \sin(\varphi + \tfrac{1}{2}\varphi_1 + \tfrac{1}{2}\varphi_2) \tag{A2.2.17}$$

$$\mathrm{Re}\,Z' = -\frac{Pa}{\pi}\left[\frac{2}{(r_1 r_2)^{3/2}} \cos\tfrac{3}{2}(\varphi_1 + \varphi_2) - \frac{a^2}{r^2 (r_1 r_2)^{3/2}} \cos(2\varphi + \tfrac{3}{2}\varphi_1 + \tfrac{3}{2}\varphi_2) \right] \tag{A2.2.18}$$

$$\mathrm{Im}\,Z' = \frac{Pa}{\pi}\left[\frac{2}{(r_1 r_2)^{3/2}} \sin\tfrac{3}{2}(\varphi_1 + \varphi_2) - \frac{a^2}{r^2 (r_1 r_2)^{3/2}} \sin(2\varphi + \tfrac{3}{2}\varphi_1 + \tfrac{3}{2}\varphi_2) \right] \tag{A2.2.19}$$

The stress function Z does not provide any T-stress term. Nevertheless, the equilibrium traction at the circumference acts as a normal external load and may produce a T-stress. Radial and tangential stress components along the contour of the disk are plotted in Fig. A2.6 for a crack with $a/R = 0.4$.

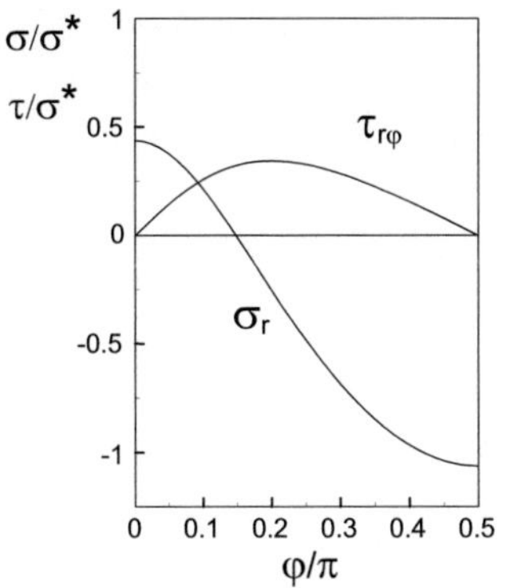

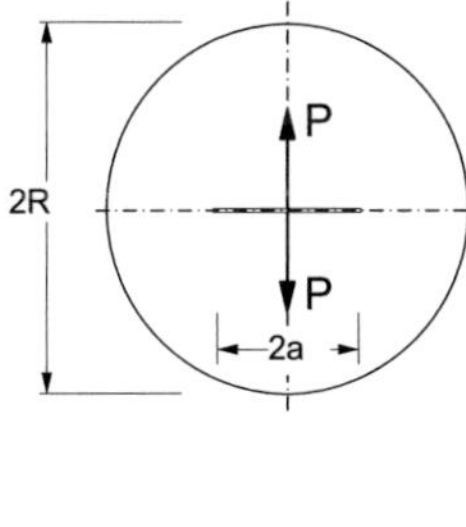

Fig. A2.6 Normal and shear tractions created by the stress function eq.(A2.2.9) along the fictitious disk contour (for φ see Fig. A2.5), $\sigma^* = P/(\pi R t)$, $t =$ thickness.

A2.3 Principle of superposition

The procedure necessary for the computations addressed in Section A2.2 is illustrated below. Disk geometry may be chosen. Figure A2.7 explains the principle of superposition for the case of T-stresses. Part a) shows a crack in an infinite body, loaded by a couple of forces P. The T-stress in this case is denoted as T_0. First, we compute the normal and shear stresses along a contour (dashed circle) which corresponds to the finite disk. We cut out the disk along this contour and apply normal and shear traction at the free boundary, which are identical to the stresses computed before (Fig. A2.7b).

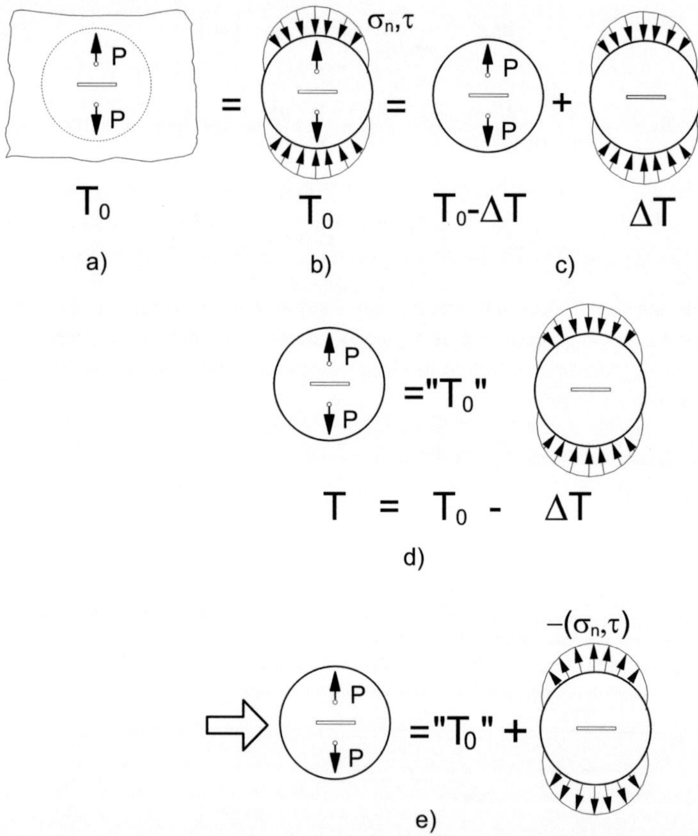

Fig. A2.7 Illustration of the principle of superposition to compute T-stresses for single forces.

The disk loaded by the combination of single forces and boundary traction exhibits the same T-term T_0. Next, we consider the situation b) to be the superposition of the two loading cases shown in part c), namely, the cracked disk loaded by the couple of forces (with T-stress $T-\Delta T$) and a cracked disk loaded by the boundary traction, having the T-term ΔT. As represented by part d), the T-term of the cracked disk is the difference $T=T_0-\Delta T$. If the sign of the boundary traction is changed, the equivalent relation is given by part e).

References A2

[A2.1] Wigglesworth, L.A., Stress distribution in a notched plate, Mathematica **4**(1957), 76-96.

[A2.2] ABAQUS Manual, Version 6.3, Hibitt, Karlsson & Sorensen, Inc.; 2002.

[A2.3] Shih CF, Asaro RJ. Elastic–plastic analysis of cracks on biomaterial interfaces: Part I—Small scale yielding. J Appl Mech **55**(1988), 299–316.

[A2.4] Mathematica 3.0, Wolfram Research Inc., USA.

[A2.5] Fett, T., T-stresses in rectangular plates and circular disks, Engng. Fract. Mech. **60** (1998) 631-652.

[A2.6] Broberg, K.B., quoted in: S. Melin, The influence of the T-stress on the directional stability of cracks, Int. J. Fracture **114** (2002) 259-265

[A2.7] Fett, T., T-stress and stress intensity factor solutions for 2-dimensional cracks, VDI-Verlag, 2002, Düsseldorf.

[A2.8] Gross, B., Srawley, J.E., Brown, W.F., Stress intensity factors for a single-edge-notched tension specimen by Boundary Collocation of a stress function, NASA, Technical Note, D-2395, 1965.

[A2.9] Gross, B., Srawley, W.F., Stress intensity factors for a single-edge-notched tension specimen by Boundary Collocation of a stress function, NASA, Technical Note, D-2395, 1965.

[A2.10] Newman, J.C., An improved method of collocation for the stress analysis of cracked plates with various shaped boundaries, NASA TN D-6376, 1971.

[A2.11] Williams, M.L., On the stress distribution at the base of a stationary crack, J. Appl. Mech. **24**(1957), 109-114.

<u>A3</u>

Weight function technique

A3.1 Weight function

Most numerical methods require a separate calculation of the stress intensity factor for each given stress distribution and each crack length. The weight function procedure developed by Bückner [A3.1] simplifies the determination of stress intensity factors. If the weight function is known for a crack in a component, the stress intensity factor can be obtained by multiplying this function by the stress distribution and integrating it along the crack length. The weight function does not depend on the special stress distribution, but only on the geometry of the component.

The method is considered below for the case of an edge crack. If $\sigma_n(x)$ is the normal stress distribution in the uncracked component along the prospective crack line of an edge crack, the stress intensity factors are given by [A3.2]

$$K_{\mathrm{I}}^{(1)} = \int_0^a h_{11}(x,a)\sigma_n(x)dx \;, \quad K_{\mathrm{II}}^{(1)} = \int_0^a h_{21}(x,a)\sigma_n(x)dx \tag{A3.1.1}$$

They define the weight functions h_{11} and h_{21}. For shear stresses τ_{xy} acting at the crack faces it results

$$K_{\mathrm{I}}^{(2)} = \int_0^a h_{12}(x,a)\tau_{xy}(x)dx \;, \quad K_{\mathrm{II}}^{(2)} = \int_0^a h_{22}(x,a)\tau_{xy}(x)dx \tag{A3.1.2}$$

defining the weight functions h_{12} and h_{22}. Under a combined crack-face loading, the stress intensity factors can be superimposed, which results in

$$K_{\mathrm{I}} = \int_0^a [h_{11}(x,a)\sigma_n(x) + h_{12}(x,a)\tau_{xy}(x)]dx \tag{A3.1.3}$$

$$K_{\mathrm{II}} = \int_0^a [h_{21}(x,a)\sigma_n(x) + h_{22}(x,a)\tau_{xy}(x)]dx \tag{A3.1.4}$$

or in matrix form

$$\mathbf{K} = \int_0^a \underline{H}\,\mathbf{S}\,dx \;, \quad \mathbf{K} = (K_I, K_{II})^T, \quad \mathbf{S} = (\sigma_n, \tau)^T \tag{A3.1.5}$$

with

$$\underline{H} = (H_{ij}) = \begin{pmatrix} h_{11} & h_{12} \\ h_{21} & h_{22} \end{pmatrix} \tag{A3.1.6}$$

The weight function $h(x,a)$ can be interpreted as the Green's function for a stress intensity factor problem. This means that the weight function is identical to the stress intensity factor caused by a pairs of forces P and Q acting at the points $x = x'$ and x''. We express the single forces by stress distributions

$$\sigma(x) = \frac{P}{B}\delta(x - x'); \quad \tau(x) = \frac{Q}{B}\delta(x - x'') \tag{A3.1.7}$$

δ=Dirac delta function, B=thickness. Inserting the stress distributions into eq.(A3.1.3) yields

$$K_{\mathrm{I}} = \frac{P}{B}\int_0^a h_{11}(x,a)\delta(x - x')dx + \frac{Q}{B}\int_0^a h_{12}(x,a)\delta(x - x'')dx$$

$$K_{\mathrm{I}} = \frac{P}{B}h_{11}(x',a) + \frac{Q}{B}h_{12}(x'',a) \tag{A3.1.8}$$

and in the same way from eq.(A3.1.4)

$$K_{\mathrm{II}} = \frac{P}{B}h_{21}(x',a) + \frac{Q}{B}h_{22}(x'',a) \tag{A3.1.9}$$

In most practical applications the weight functions h_{12} and h_{21} disappear. In such cases we drop the superscripts and write simply

$$K_{\mathrm{I}} = \int_0^a h_{\mathrm{I}}(x,a)\sigma_n(x)dx \tag{A3.1.10a}$$

$$K_{\mathrm{II}} = \int_0^a h_{\mathrm{II}}(x,a)\tau(x)dx \tag{A3.1.10b}$$

A3.2   Determination of weight functions

The general procedures for the determination of weight functions are described below for the weight function component $h_{\mathrm{I}}(=h_{11})$. The relation of Rice [A3.3] allows to determine the weight function from the crack opening displacement $v_r(x,a)$ under any arbitrarily chosen loading and the corresponding stress intensity factor $K_{\mathrm{Ir}}(a)$ according to

$$h_{\mathrm{I}}(x,a) = \frac{E'}{K_{\mathrm{Ir}}(a)}\frac{\partial v_r(x,a)}{\partial a} \tag{A3.2.1}$$

($E' = E$ for plane stress and $E' = E/(1-v^2)$ for plane strain conditions), where the subscript r stands for the reference loading case. It is convenient to use $\sigma_r(x) = \sigma_0 = $ constant for the reference stress distribution.

One possibility to derive the weight function with eq.(A3.2.1) is the evaluation of numerically determined crack opening profiles which may be obtained by BCM computations. By applying the BCM procedure to a couple of cracks with slightly different lengths a and $a+da$, a large number of coefficients A_n and B_n are obtained. Then, eq.(A2.2.3b) provides the related couple of crack opening displacements $v(a)$ and $v(a+da)$ from which the derivative in eq.(A3.2.1) can be obtained. In order to minimise the numerical effort, approximate methods are often used in literature.

A3.2.1 Petroski-Achenbach procedure

As a consequence of the Williams stress function [A3.4], the crack opening displacement can be expressed as [A3.2]

$$v_r(x,a) = \sum_{n=0}^{\infty} A_n (1 - x/a)^{n+\frac{1}{2}} = A_0 \sum_{n=0}^{\infty} C_n (1 - x/a)^{n+\frac{1}{2}} \tag{A3.2.2}$$

with
$$A_0 = \sqrt{\frac{8}{\pi}} \frac{\sigma_0}{E'} a Y_r(a/W) , \quad C_n = A_n/A_0 , \quad C_0 = 1 \tag{A3.2.3}$$

By introducing eq.(A3.2.2) into eq.(A3.2.1), we obtain

$$h_1(x,a) = \sqrt{\frac{2}{\pi a}} \sum_{n=0}^{\infty} \left[(3 - 2n)C_{n-1} + (2n+1)C_n + 2\frac{\alpha}{Y}\frac{dY}{d\alpha}C_{n-1} + 2\alpha \frac{dC_{n-1}}{d\alpha} \right] (1 - x/a)^{n-\frac{1}{2}} \tag{A3.2.4}$$

with $C_{-1}=dC_{-1}/d\alpha=0$ and $\alpha=a/W$. According to a proposal of Petroski and Achenbach [A3.5], the series expansion (A3.2.4) may be truncated after the term with $n = 2$. The unknown coefficient C_2 can be determined from the energy balance equation

$$\int_0^a K_{Ir}^2(a')da' = E' \int_0^a \sigma_r v_r(x,a)dx \tag{A3.2.5}$$

The basic idea of Petroski and Achenbach was modified and additional conditions were introduced, for instance, the disappearing second and third derivatives of the crack opening displacement at the crack mouth [A3.6]

$$\frac{\partial^2 v_r}{\partial x^2} = \frac{\partial^3 v_r}{\partial x^3} = 0 \quad \text{for} \quad x = 0 \tag{A3.2.6}$$

which provided additional coefficients. For details see references [A3.7] and [A3.2].

A3.2.2 Adjustment of weight functions to reference stress intensity factors

A minor disadvantage of the Petroski-Achenbach procedure (Section A3.2.1) is that numerical integrations are necessary to evaluate the integral over K^2 in eq.(A3.2.5). A direct determination of the weight function coefficients is possible by adjusting the weight function to reference stress intensity factors. As can be seen from eq.(A3.2.4), a representation of the weight function by the power series expansion [A3.8, A3.9]

$$h(x,a) = \sqrt{\frac{2}{\pi a}}\left[\frac{1}{\sqrt{1-x/a}} + \sum_{n=0}^{\infty} D_n(1-x/a)^{n+\frac{1}{2}}\right] \qquad (A3.2.7)$$

is possible. A number of conditions allow determining the coefficients D_n. If a number of m reference loading cases with the stress distributions $\sigma_{r,i}(x)$ are known, the weight function must satisfy a number of m conditions

$$\int_0^a h(x,a)\,\sigma_{r,i}(x)\,dx = \sigma^*_{r,i}\sqrt{a}\,Y_{r,i}\,, \qquad i=1...m \qquad (A3.2.8)$$

In eq.(A3.2.8) $\sigma^*_{r,i}$ denotes a characteristic value of the stress distribution $\sigma_{r,i}(x)$. Additional coefficients can be derived from the conditions (A3.2.6). By use of (A3.2.1), they result in

$$\frac{\partial^2 \mathbf{h}}{\partial x^2} = \frac{E'}{K_r}\frac{\partial}{\partial a}\frac{\partial^2 \mathbf{u}}{\partial x^2} \qquad (A3.2.9)$$

$$\frac{\partial^3 \mathbf{h}}{\partial x^3} = \frac{E'}{K_r}\frac{\partial}{\partial a}\frac{\partial^3 \mathbf{u}}{\partial x^3} \qquad (A3.2.10)$$

where

$$\mathbf{h} = \begin{pmatrix} h_I \\ h_{II} \end{pmatrix}, \quad \mathbf{u} = \begin{pmatrix} v_r \\ u_r \end{pmatrix} \qquad (A3.2.11)$$

Finally, the conditions (A3.2.6) read

$$\left.\frac{\partial^2 \mathbf{h}}{\partial x^2}\right|_{x=0} = 0\,, \quad \left.\frac{\partial^3 \mathbf{h}}{\partial x^3}\right|_{x=0} = 0 \qquad (A3.2.12)$$

A3.2.3　Most general case

The procedure for the determination of weight function coefficients by adjusting to reference stress intensity factors may also be applied to the most general case of mixed-mode problems as addressed in Section A3.1. As outlined in [A3.2], the series expansions for the four weight functions read

$$h_{11} = \sqrt{\frac{2}{\pi a}}\sum_{n=0}^{\infty} D_n^{(11)}(1-x/a)^{(n-1)/2} \qquad (A3.2.13a)$$

$$h_{12} = \sqrt{\frac{2}{\pi a}}\sum_{n=0}^{\infty} D_n^{(12)}(1-x/a)^{(n-1)/2} \qquad (A3.2.13b)$$

$$h_{21} = \sqrt{\frac{2}{\pi a}}\sum_{n=0}^{\infty} D_n^{(21)}(1-x/a)^{(n-1)/2} \qquad (A3.2.13c)$$

$$h_{22} = \sqrt{\frac{2}{\pi a}} \sum_{n=0}^{\infty} D_n^{(22)} (1 - x/a)^{(n-1)/2} \qquad \text{(A3.2.13d)}$$

with

$$D_0^{(11)} = D_0^{(22)} = 1 , \quad D_0^{(12)} = D_0^{(21)} = 0 \qquad \text{(A3.2.13e)}$$

From the stress intensity factors for the two reference loading cases shown in Fig. A3.1 and e.g. from the disappearing second and third derivatives for h_{11} and h_{22} at $x = 0$, a sufficient number of coefficients were determined in [A3.10, A3.11] for eqs.(A3.2.13a-d).

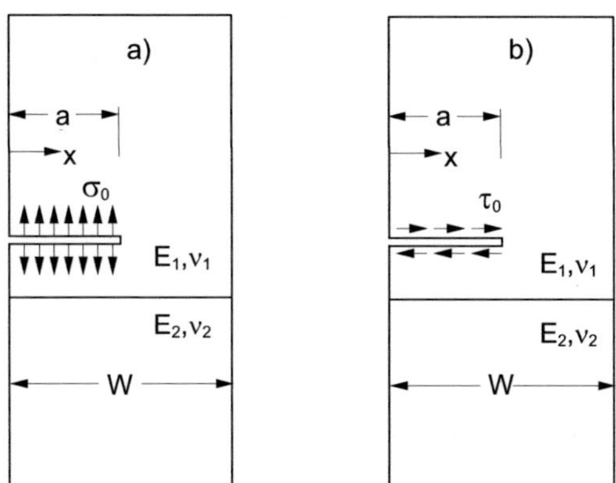

Fig. A3.1 Reference loading cases (constant normal traction and constant shear traction along the crack faces) for an edge-cracked bimaterial joint.

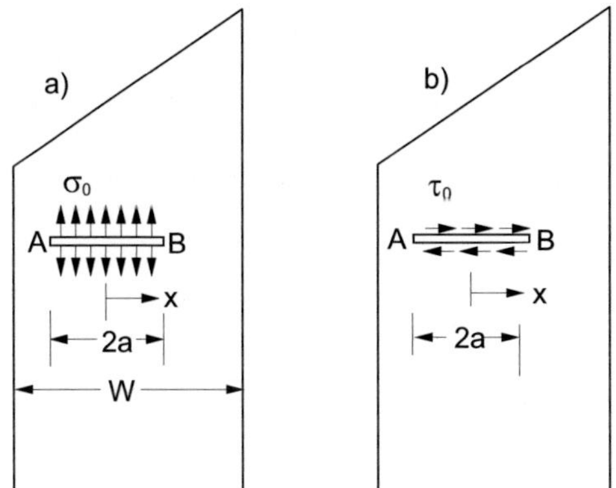

Fig. A3.2 Reference loading cases (constant normal traction and constant shear traction) for an internal crack.

In the case of an internal crack (see Fig. A3.2), the weight function contributions for symmetric loading, $\sigma(x) = \sigma(-x)$, $\tau(x) = \tau(-x)$, can be expressed for point A by

$$h_{11,A} = \frac{2}{\sqrt{\pi a}} \left[\frac{1}{\sqrt{1-\rho^2}} + \sum_{n=0}^{\infty} A_n^{(11)} (1-\rho^2)^{(n+1)/2} \right] \tag{A3.2.14a}$$

$$h_{12,A} = \frac{2}{\sqrt{\pi a}} \sum_{n=0}^{\infty} A_n^{(12)} (1-\rho^2)^{(n+1)/2}, \quad \rho = x/a \tag{A3.2.14b}$$

$$h_{21,A} = \frac{2}{\sqrt{\pi a}} \sum_{n=0}^{\infty} A_n^{(21)} (1-\rho^2)^{(n+1)/2} \tag{A3.2.14c}$$

$$h_{22,A} = \frac{2}{\sqrt{\pi a}} \left[\frac{1}{\sqrt{1-\rho^2}} + \sum_{n=0}^{\infty} A_n^{(22)} (1-\rho^2)^{(n+1)/2} \right] \tag{A3.2.14d}$$

and for point B

$$h_{11,B} = \frac{2}{\sqrt{\pi a}} \left[\frac{1}{\sqrt{1-\rho^2}} + \sum_{n=0}^{\infty} B_n^{(11)} (1-\rho^2)^{(n+1)/2} \right] \tag{A3.2.15a}$$

$$h_{12,B} = \frac{2}{\sqrt{\pi a}} \sum_{n=0}^{\infty} B_n^{(12)} (1-\rho^2)^{(n+1)/2}, \quad \rho = x/a \tag{A3.2.15b}$$

$$h_{21,B} = \frac{2}{\sqrt{\pi a}} \sum_{n=0}^{\infty} B_n^{(21)} (1-\rho^2)^{(n+1)/2} \tag{A3.2.15c}$$

$$h_{22,B} = \frac{2}{\sqrt{\pi a}} \left[\frac{1}{\sqrt{1-\rho^2}} + \sum_{n=0}^{\infty} B_n^{(22)} (1-\rho^2)^{(n+1)/2} \right] \tag{A3.2.15d}$$

where for the coefficients $B \neq A$ in general.

A3.2.4 Numerical evaluation of weight function integrals

The integration of the weight function may lead to numerical problems, since the value of h tends to infinity for $x \to a$. Therefore, it is recommended to split the integral and to apply the mean value theorem to the right hand term of

$$K_I = \int_0^a h(x,a)\sigma(x)\,dx = \int_0^{a-\varepsilon} h(x,a)\sigma(x)\,dx + \sigma_{x=a} \int_{a-\varepsilon}^a h(x,a)\,dx \tag{A3.2.27}$$

with $\varepsilon \ll a$, where the first integral up to $a-\varepsilon$ can be evaluated numerically (e.g. by use of Simpson's rule) and the second one by analytical integration. Having in mind that for $\varepsilon \to 0$ all terms of the weight function can be neglected compared with the singular term, we obtain e.g. for the weight function of the type of eq.(A3.2.27)

$$K_{\mathrm{I}} = \int\limits_{0}^{a-\varepsilon} h(x,a)\sigma(x)\,dx + \sigma_{x=a}\sqrt{8\varepsilon/\pi} \qquad (A3.2.28)$$

References A3

[A3.1] Bückner, H., A novel principle for the computation of stress intensity factors, ZAMM **50** (1970), 529-546.

[A3.2] Fett, T., Munz, D., Stress intensity factors and weight functions, Computational Mechanics Publications, Southampton, 1997.

[A3.3] Rice, J.R., Some remarks on elastic crack-tip stress fields, Int. J. Solids and Structures 8(1972), 751-758.

[A3.4] Williams, M.L., On the stress distribution at the base of a stationary crack, J. Appl. Mech. **24**(1957), 109-114.

[A3.5] Petroski, H.J., Achenbach, J.D., Computation of the weight function from a stress intensity factor, Engng. Fract. Mech. **10**(1978), 257.

[A3.6] Fett, T., Mattheck, C., Munz, D., On the calculation of crack opening displacements from the stress intensity factor, Eng. Fract. Mech. **27**(1987),697-715.

[A3.7] Wu, X.R., Carlsson, A.J., Weight functions and stress intensity factor solutions, Pergamon Press, Oxford 1991.

[A3.8] Fett, T., Munz, D., The weight function method for calculation of stress intensity factors, (in German), Proceedings of "DVM-Arbeitskreis BRUCHVORGÄNGE", Berlin 26./27.03.1991.

[A3.9] Fett, T., Direct determination of weight functions from reference loading cases and geometrical conditions, Engng. Fract. Mech. **42**(1992)435-444.

[A3.10] Fett, T., Munz, D., Tilscher, M., Weight functions for sub-interface cracks, Int. J. Sol. and Struct. **34**(1997), 393-400.

[A3.11] Fett, T., Tilscher, M., Munz, D., Weight functions for cracks near the interface of a bimaterial joint and application to thermal stresses, Eng. Fract. Mech. **56**(1997), 87-100.

A4

Green's function for T-stress

A4.1 Green's function for symmetric crack problems

As a consequence of the principle of superposition, stress fields for different loadings can be added up in the case of single loadings acting simultaneously. This leads to an integration representation of the loading parameters. The method was applied very early to the singular stress field and for the computation of the related stress intensity factor by Bückner. Similarly to stress intensity factors [A4.1], the T-stress contribution can be expressed by an integral [4.2-4.8]

$$T = -\sigma_y\big|_{x=a} + \int_0^a t(x,a)\,\sigma(x)\,dx \qquad (A4.1.1)$$

where the integration has to be performed using the stress field σ_y in the uncracked body (Fig. A4.1a). The stress contributions are weighted by a weight function t as a function of the location x where the stress σ_y acts.

If in the uncracked body a σ_x stress component already exists at the location of the tip of the prospective crack, the total T-value is obtained by adding this stress contribution, i.e.

$$T = \sigma_x\big|_{x=a} - \sigma_y\big|_{x=a} + \int_0^a t(x,a)\,\sigma(x)\,dx \qquad (A4.1.2)$$

For loading by a crack-face pressure distribution $p(x)$, the T-stress results from

$$T_{crack\ face} = \int_0^a t(x,a)p(x)dx \qquad (A4.1.3)$$

(see e.g. [4.7]) since there are no stresses in the uncracked body, i.e. $\sigma_x|_{x=a}=\sigma_y|_{x=a}=0$.

The Green's function t can be interpreted as the T-term for a pair of single forces P acting at the crack face at the location $x=x_0$ (Fig. A4.1b). This can be shown easily for the single forces represented by a singular pressure distribution

$$p(x) = \frac{P}{B}\delta(x - x_0) \qquad (A4.1.4)$$

where δ is the Dirac delta function and B the thickness of the plate (often chosen to be $B=1$). By introducing this stress distribution into eq.(4.1.3), we obtain

$$T_P = \frac{P}{B}\int_0^a \delta(x - x_0)t(x,a)\,dx = \frac{P}{B}t(x_0,a) \qquad (A4.1.5)$$

i.e. the weight function term $t(x_0,a)$ is the Green's function for the T-stress term.

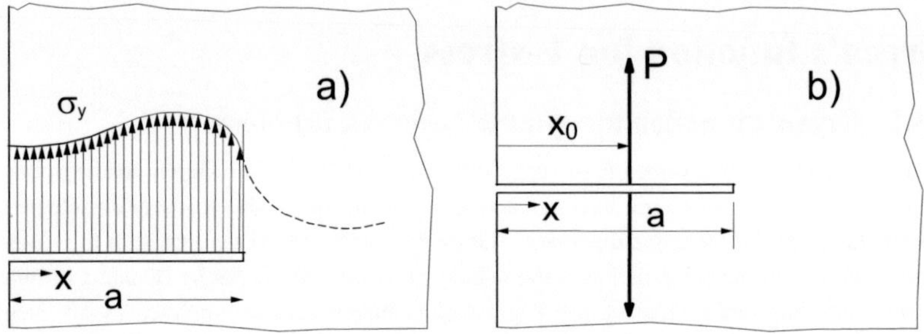

Fig. A4.1 Crack loaded by a) continuously distributed normal traction σ_y (present in the uncracked body), b) by a pair of concentrated forces.

A4.2   Set-up of Green's function

A4.2.1   Type of the Green's function

Following the analysis of Sham (eq.(32) in [4.3]), the T-stress contribution given by the integrals in (A4.1.1-A4.1.3) can be expressed as

$$\int\limits_0^a t(x,a)\,\sigma(x)\,dx \propto \frac{1}{\pi}\int\limits_{Crack\ area}[X(s)u_r + Y(s)v_r]ds \tag{A4.2.1}$$

for a crack face loading (extending over the upper and lower crack surfaces) with tractions X and Y acting in x- and y- directions, respectively.

The relevant displacements in the surroundings of a crack read (see Section A2.2.2)

$$u \propto \sum_{n=0}^{\infty} a_n r^{n+1/2}\frac{2n+3}{2n-1}[(n+4v-\tfrac{5}{2})\cos(n-\tfrac{1}{2})\varphi - (n-\tfrac{1}{2})\cos(n+\tfrac{3}{2})\varphi]+$$

$$+\sum_{n=0}^{\infty} b_n\,r^{n+1}[(n+4v-2)\cos n\varphi - (n+2)\cos(n+2)\varphi \tag{A4.2.2}$$

$$v \propto \sum_{n=0}^{\infty} a_n\,r^{n+1/2}\frac{2n+3}{2n-1}[(n-\tfrac{1}{2})\sin(n+\tfrac{3}{2})\varphi - (n-4v+\tfrac{7}{2})\sin(n-\tfrac{1}{2})\varphi]+$$

$$+\sum_{n=0}^{\infty} b_n\,r^{n+1}[(n+2)\sin(n+2)\varphi - (n-4v+4)\sin n\varphi] \tag{A4.2.3}$$

In the following considerations, we are exclusively interested in tractions $Y(s)$ normally to the crack surface. Therefore, we set $X(s)=0$.

The crack face displacements ($\varphi=\pm\pi$) result as

$$v_r \propto \sum_{n=0}^{\infty} a_n\, r^{n+1/2}\, \frac{2n+3}{2n-1}\, 4(v-1)\,, \quad r = a - x \tag{A4.2.4}$$

From (A4.2.1), we can conclude easily that

$$\int_0^a t(x,a)\,\sigma_y(x)\,dx \propto \int_{Crack}[X(s)u_r + Y(s)v_r]\,ds \propto \int_{Crack} Y(s)v_r\,ds \propto \int_0^a \sigma_y(x)v_r\,dx \tag{A4.2.5}$$

The Green's or weight function is the T-stress caused by a point force P at location x for the 2-dimensional specimen or a line load in thickness direction P/B (in the 3-dimensional case) with

$$\sigma_y(x) = P\delta(x - x') \tag{A4.2.6}$$

(δ=Dirac delta function). It results from inserting (A4.2.6) into (A4.2.5) using the well-known properties of δ

$$t(x',a) \propto v_r(x',a) \tag{A4.2.7}$$

and, with (A4.2.4), a power series expansion for the Green's function

$$t(x',a) = \tfrac{1}{a}\sum_{n=0}^{\infty} C_n(1 - x'/a)^{n+1/2} \tag{A4.2.8}$$

A4.2.2 Determination of the coefficient C_0

In principle, the total Green's function can be determined for instance from finite element computations by application of pairs of concentrated forces P on the crack faces. The results can be fitted to $t(x,a)$ by use of a sufficient number of terms $(1-x/a)^{n+1/2}$ in eq.(A4.2.8) [A4.6-A4.10]. Whereas a polynomial with a restricted number of terms is an appropriate Green's function representation for cracks loaded by tractions over the whole crack length, the use of a more accurate near-tip solution may be of advantage for theoretical considerations. Unfortunately, the accuracy of T-stresses from concentrated forces decreases for a very small crack-tip distance. For the evaluation of t at $(a\text{-}x)\rightarrow0$, it is of advantage to determine the T-stress under constant crack-face tractions distributed over the near-tip region.

The first coefficients C_0 were determined in [A4.10] for several crack geometries. Crack elements very close to the tip were loaded with a constant crack face pressure p_0. The results were determined for a large number of contours and are plotted in Fig. A4.2a versus the contour number. The plateau values are plotted in Fig. A4.2b as a function of the size d of the loaded crack surface. From this plot and Fig. A4.2c, one obtains for the DCB specimen

$$T/p_0 = 0.747(d/H)^{3/2} \tag{A4.2.9}$$

and for the edge crack in a large plate (Fig. A4.2d)

$$T / p_0 = 0.20 \, (d/a)^{3/2} \qquad (A4.2.10)$$

All these results allow the first coefficient of the power series representation to be determined. It results

$$T / p_0 = \frac{1}{a} \int\limits_{a-d}^{a} C_0 \sqrt{1 - x/a} \; dx = \frac{2}{3} C_0 \left(\frac{d}{a} \right)^{3/2} \qquad (A4.2.11)$$

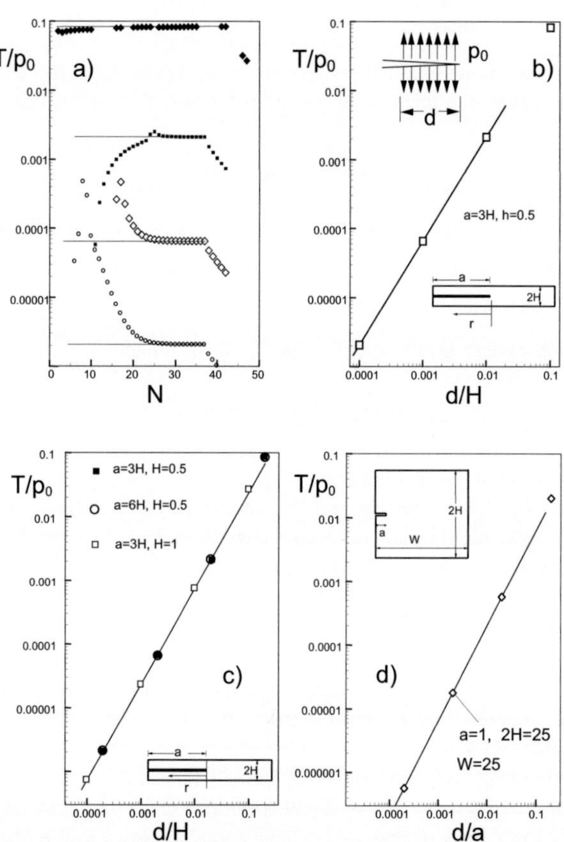

Fig. A4.2 Constant stress term for near-tip loading by a constant pressure p_0, a) results for a DCB specimen versus number of contours, b) plateau values of Fig. A4.2a versus size of the loading region, c) results for DCB specimens with different heights H, d) results for a small crack of a=1 in a large body of H=12.5 and W=25.

When comparing the coefficients of the $d^{3/2}$ dependency of (A4.2.9) and (A4.2.10) with (A4.2.11), we obtain for the DCB specimen

$$C_0 \cong \frac{9}{8}\left(\frac{a}{H}\right)^{3/2} \qquad\qquad\qquad (A4.2.12a)$$

and for the edge crack in the half-space

$$C_0 = 0.3 \qquad\qquad\qquad (A4.2.12b)$$

In the case of the finite CT specimen, the coefficient C_0 depends on the relative crack length a/W as shown in Fig. A4.3a. Figure A4.3b shows the results for the edge-cracked bar often used for bending tests.

The coefficient C_0 of the standard CT specimen can be approximated for $0.3 \leq a/W \leq 0.7$ as

$$C_0 \cong (0.9626 + 0.9981\alpha - 8.532\alpha^2 + 11.45\alpha^3)\alpha^{3/2} \, , \; \alpha = a/W \qquad (A4.2.12c)$$

and of the edge-cracked bar for $0.2 \leq a/W \leq 0.6$ as

$$C_0 \cong 0.308 - 0.0951\alpha - 2.896\alpha^2 + 6.286\alpha^3 \qquad\qquad (A4.2.12d)$$

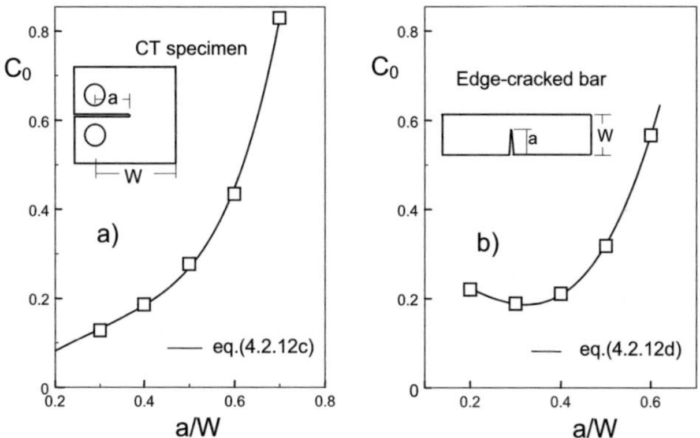

Fig. A4.3 Coefficient C_0. a) Results of the standard CT specimen, b) results of edge-cracked bars.

A4.3 Adjustment to reference T-stress solutions

A4.3.1 General procedure

From the considerations made in the preceding Section, it was found that the Green's function for the T-stress can be expressed in the form

$$t = \frac{1}{a}\sum_{v=0}^{\infty} C_v (1 - x/a)^{v+1/2} \cong \frac{1}{a}\sum_{v=0}^{N} C_v (1 - x/a)^{v+1/2} \qquad (A4.3.1)$$

The determination of the first N coefficients is simply possible if T-stress solutions for special loading cases, so-called reference loading cases, are available. Similar to the direct adjustment method (DAM), applied in [A4.6] for the determination of weight functions for stress intensity factors (see also Section A3.2.2), the coefficients C_v for Green's functions of T-stresses may be determined.

A number of μ reference loading cases is assumed with the stress distributions $\sigma_i(x)$ acting in the uncracked body normally the prospective crack plane. The Green's functions must fulfil a number of μ conditions

$$\frac{1}{a}\int_0^a C_v (1-x/a)^{v+1/2}\, \sigma_{y,i}(x)\,dx = T_i - \sigma_{x,i}\Big|_{x=a} + \sigma_{y,i}\Big|_{x=a} \quad , i=1...\mu \qquad (A4.3.2)$$

Equation (A4.3.2) leads to a system of μ linear equations. Its solution provides a number of μ coefficients.

A4.3.2 Single-term approximation

This procedure may be demonstrated for the case of an edge crack in a rectangular plate of width W and height $2H$. For a long plate of $H/W=6$, the tensile and bending solutions, T_t and T_b, obtained by Sham [A4.3], are given for some relative crack lengths, $\alpha=a/W$, by the data of Table A4.1. The T-stress under tension is scaled with the remote tensile stress σ_t and the bending solution with the outer fibre bending stress σ_b.

Table A4.1 T-stress solutions for an edge-cracked plate under tension and bending loading ($H/W=6$) [A4.3].

α	T_t/σ_t	T_b/σ_b
0	-0.526	-0.526
0.2	-0.5919	-0.2407
0.3	-0.6143	-0.0824
0.4	-0.5853	0.1159
0.5	-0.4314	0.3911
0.6	0.0278	0.8275
0.7	1.3332	1.6609
0.8	5.9755	3.9115

First, it is assumed, that only the tensile reference solution T_t might be available. Introducing this solution into (A4.3.2) gives with $\sigma_x=0$

$$C_0 = \frac{3}{2}\left(\frac{T_t}{\sigma_t}+1\right) \qquad (A4.3.3)$$

The approximation for the Green's function

$$t = \frac{1}{a}C_0\sqrt{1-x/a} \qquad (A4.3.4)$$

is plotted as the dashed curves in Fig. A4.4a for deep cracks and in Fig. A4.4b for the limit case α=0 which represents the edge-cracked half-space. Finite element results are entered as the circles. For the deep cracks, the weight function data were taken from [A4.3]. The results for the edge-cracked half-space are given in [A4.11].

Comparison of the numerical data with the results of eqs.(A4.3.3) and (A4.3.4) shows that the agreement is very poor.

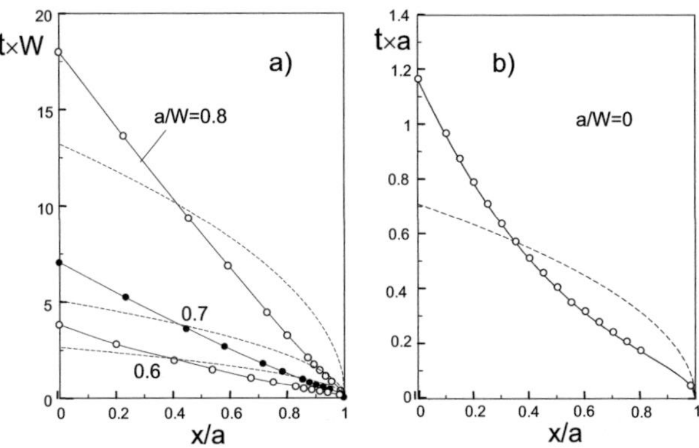

Fig. A4.4 Approximate Green's functions (dashed curves) obtained by direct adjustment to only one reference loading case (tensile load); dashed curve: single-term representation by eq.(A4.3.4), symbols in a) results from [A4.3], in b) finite element results from [A4.11].

A4.3.3 Approximation with two terms

As a second loading case, bending is considered. In this case it holds σ_x=0 and

$$\sigma_y = \sigma_b(1 - 2x/W) \qquad (A4.3.5)$$

This second loading case allows for deriving an improved Green's function

$$t = \tfrac{1}{a}[C_0\sqrt{1 - x/a} + C_1(1 - x/a)^{3/2}] \qquad (A4.3.6)$$

The coefficients follow as

$$C_0 = \frac{15}{16\alpha}\left((7 - 4\alpha)\frac{T_t}{\sigma_t} - 7\frac{T_b}{\sigma_b} + 10\alpha\right) \qquad (A4.3.7a)$$

$$C_1 = -\frac{35}{16\alpha}\left((5 - 4\alpha)\frac{T_t}{\sigma_t} - 5\frac{T_b}{\sigma_b} + 6\alpha\right) \qquad (A4.3.7b)$$

that results in

41

$$C_0 = \frac{15(-0.3889 + 1.8706\,\alpha - 2.0012\,\alpha^2 - 1.0544\,\alpha^3 + 2.283\,\alpha^4 - 0.3932\,\alpha^5)}{8(1-\alpha)^2} \qquad \text{(A4.3.7c)}$$

$$C_1 = \frac{35(0.5487 - 2.1127\,\alpha + 2.1180\,\alpha^2 + 1.1845\,\alpha^3 - 2.0864\,\alpha^4 + 0.3932\,\alpha^5)}{8(1-\alpha)^2} \qquad \text{(A4.3.7d)}$$

The Green's functions according to eq.(A4.3.6) are shown in Fig. A4.5 as the solid curves. For the edge-cracked half-space the bending loading case is not applicable since it is identical with the tensile case. Therefore, increasing deviations have to be expected for decreasing relative crack depths.

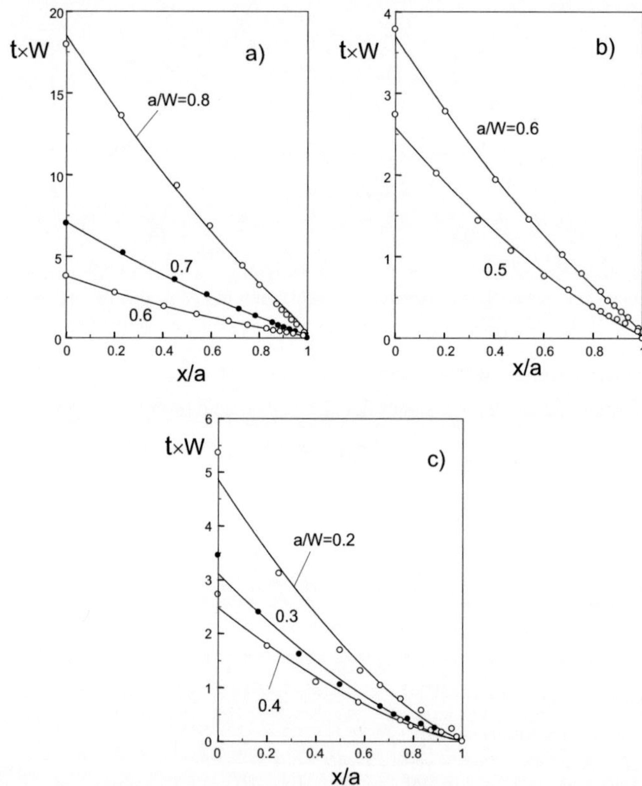

Fig. A4.5 Comparison of the approximate two-term Green's functions (A4.3.6) with results from [A4.3].

If, for instance, the T-stresses for constant stress, T_t, and for a pair of point forces P acting at the crack mouth, T_P, are available with

$$\sigma_1 = \sigma_t \, , \; \sigma_2 = \frac{P}{B}\,\delta(x) \qquad\qquad (A4.3.8)$$

(B=thickness) the coefficients result as

$$C_0 = \frac{15}{4}\left(\frac{T_t}{\sigma_t}+1\right) - \frac{3}{2}\frac{aB}{P}T_p \qquad\qquad (A4.3.9a)$$

$$C_1 = -\frac{15}{4}\left(\frac{T_t}{\sigma_t}+1\right) - \frac{5}{2}\frac{aB}{P}T_p \qquad\qquad (A4.3.9b)$$

A4.3.4 Approximate Green's functions with integer exponents

If only a single reference loading case is available, the determination of the Green's function makes some problems. A set-up according to eq.(A4.3.4) cannot be recommended for this case. From the numerical Green's functions for deep edge cracks, Fig. A4.5a, it can be seen that the square-root-shaped part near $x/a=1$ is very small and the curves are nearly linear. Therefore, a simpler linear set-up for approximate Green's function was proposed in [A4.6] by

$$t \cong \tfrac{1}{a}E_0(1-x/a) \qquad\qquad (A4.3.10)$$

or in a more general representation [A4.12]

$$t = \tfrac{1}{a}\sum_{v=0}^{N} E_v(1-x/a)^{v+1} \qquad\qquad (A4.3.11)$$

It should be mentioned that such a set-up has no deeper theoretical basis; nevertheless the good numerical results seem to justify its application. If we restrict the expansion to the first term, we obtain the unknown coefficient E_0

$$T = -\sigma_y\,|_{x=a} + E_0\,\tfrac{1}{a}\int_0^a (1-x/a)\sigma_y\,dx \qquad\qquad (A4.3.12)$$

Let us assume the T-term T_t of an edge-cracked plate under pure tension σ_t to be known. Introducing this constant stress into eq.(A4.3.10) yields

$$T_t = -\sigma_t + \sigma_t E_0\,\tfrac{1}{a}\int_0^a (1-x/a)dx = \sigma_t\left(-1+\tfrac{1}{2}E_0\right) \qquad\qquad (A4.3.13)$$

and the coefficient E_0 results trivially as

$$E_0 = 2\left(1 + \frac{T_t}{\sigma_t}\right) \tag{A4.3.14}$$

Knowledge of additional reference solutions for T allows further coefficients to be determined.

Figure A4.6 gives a comparison of the FE-results of Sham [A4.3] with the straight-line approximation according to eq.(A4.3.9). The agreement is good especially for deep cracks of $a/W>0.6$ as becomes obvious from Fig. A4.6a.

In order to improve this type of Green's function, the next regular term may be added. Consequently, the Green's function expansion reads for edge cracks

$$t(x) = \tfrac{1}{a}E_0(1 - x/a) + \tfrac{1}{a}E_1(1 - x/a)^2 \tag{A4.3.15}$$

The determination of the two coefficients E_0 and E_1 is possible, if T-stress solutions for two different reference loading cases are available. If, for instance, the T-stresses for constant stress, T_t, and for a pair of point forces P acting at the crack mouth, T_P, are available with reference stresses according to (A4.3.8), the coefficients result as

$$E_0 = 6\left(\frac{T_t}{\sigma_t} + 1\right) - \frac{2aB}{P}T_p \tag{A4.3.16a}$$

$$E_1 = -6\left(\frac{T_t}{\sigma_t} + 1\right) + \frac{3aB}{P}T_p \tag{A4.3.16b}$$

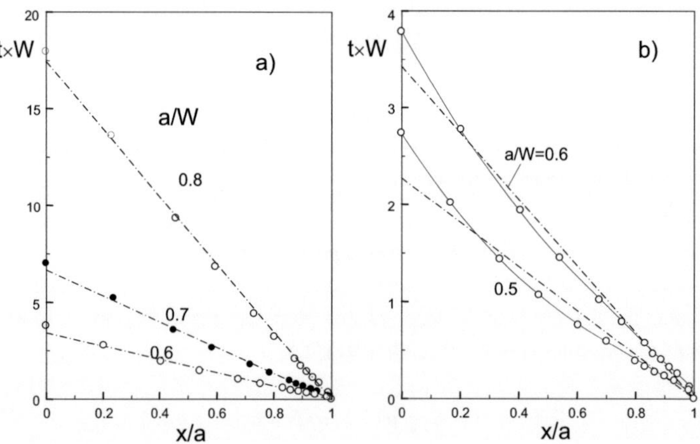

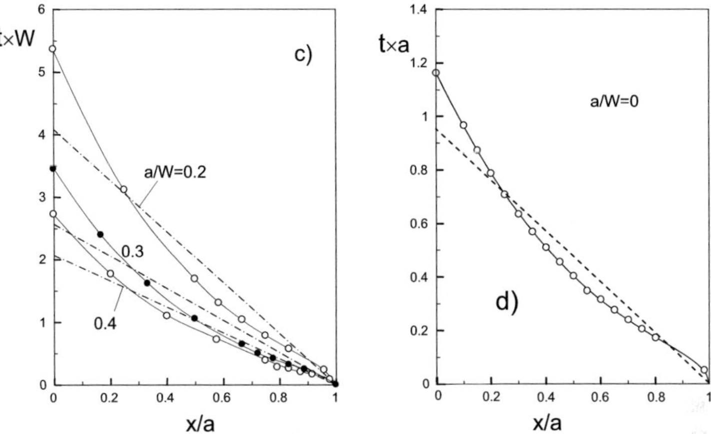

Fig. A4.6 Comparison of the approximate linear single-term Green's functions, eq.(A4.3.10) with FE results from Sham [A4.3].

A4.3.5 Symmetrically loaded internal crack

The derivation of an approximate Green's function for internal cracks is similar to those of edge cracks. Due to the symmetry at $x = 0$, the general set-up must be modified. Improved descriptions, symmetric with respect to $x = 0$, are

$$t = \frac{1}{a} \sum_{v=0}^{\infty} C_v (1 - x^2/a^2)^{v+1/2} \qquad (A4.3.17a)$$

$$t = \frac{1}{a} \sum_{v=0}^{\infty} E_v (1 - x^2/a^2)^{v} \qquad (A4.3.17b)$$

with the first approximations

$$t \cong \frac{1}{a} C_0 (1 - x^2/a^2)^{1/2} \qquad (A4.3.18a)$$

$$t \cong \frac{1}{a} E_0 (1 - x^2/a^2) \qquad (A4.3.18b)$$

In this case, the coefficients C_0 and E_0 result from the pure tension case as

$$C_0 = \frac{4}{\pi}\left(1 + \frac{T_t}{\sigma_t}\right) \qquad (A4.3.19a)$$

$$E_0 = \frac{3}{2}\left(1 + \frac{T_t}{\sigma_t}\right) \qquad (A4.3.19b)$$

45

The two-term representations

$$t = \tfrac{1}{a}C_0(1 - x^2/a^2)^{1/2} + \tfrac{1}{a}C_1(1 - x^2/a^2)^{3/2} \qquad\qquad \text{(A4.3.20a)}$$

$$t = \tfrac{1}{a}E_0(1 - x^2/a^2) + \tfrac{1}{a}E_1(1 - x^2/a^2)^2 \qquad\qquad \text{(A4.3.20b)}$$

yield for a concentrated force in the crack centre (x=0) as the second loading case with

$$\sigma_1 = \sigma_t \;,\; \sigma_2 = \frac{P}{2B}\,\delta(x) \qquad\qquad \text{(A4.3.21)}$$

(Note: the force is different from that of eq.(A4.3.8) by a factor of 1/2 since the total symmetric load P belongs only half to the crack part along the positive x-axis) the coefficients

$$C_0 = \frac{16}{\pi}\left(\frac{T_t}{\sigma_t}+1\right) - \frac{3aB}{P}\,T_p \qquad\qquad \text{(A4.3.22a)}$$

$$C_1 = -\frac{16}{\pi}\left(\frac{T_t}{\sigma_t}+1\right) + \frac{4aB}{P}\,T_p \qquad\qquad \text{(A4.3.22b)}$$

and

$$E_0 = \frac{15}{2}\left(\frac{T_t}{\sigma_t}+1\right) - \frac{8aB}{P}\,T_p \qquad\qquad \text{(A4.3.23a)}$$

$$E_1 = -\frac{15}{2}\left(\frac{T_t}{\sigma_t}+1\right) + \frac{10aB}{P}\,T_p \qquad\qquad \text{(A4.3.23b)}$$

A4.4 Modified Green's functions for non-symmetric crack problems

A4.4.1 Extended set-up

Relation (A4.1.2) used so far is restricted to symmetric crack problems, for instance, an edge crack normal to the free surface of a rectangular plate. In the case of more complicated crack and component geometries (e.g. oblique edge cracks, Fig. A4.7a, and kink cracks, Fig. A4.7b) and loading cases including shear stresses, the T-stress can be computed from [A4.8]

$$T = \int_0^a t^{(1)}(x,a)\,\sigma_y\,dx + \int_0^a t^{(2)}(x,a)\,\tau_{xy}\,dx - \sigma_y\big|_{x=a} + \sigma_x\big|_{x=a} \qquad\qquad \text{(A4.4.1)}$$

similar to stress intensity factors [A4.6].

The weight functions $t^{(1)}$ and $t^{(2)}$ are the T-terms for a pair of single forces P and Q acting normal and parallel to the crack face at the location $x=x'$. This can be shown easily for the case of single forces represented by delta-shaped stress distributions

$$\sigma = \frac{P}{B}\delta(x-x') \; , \; \tau_{xy} = \frac{Q}{B}\delta(x-x') \qquad (A4.4.2)$$

where δ is the Dirac delta function and B the thickness of the plate (often chosen to be $B=1$). By inserting these stress distributions into eq.(A4.4.1), it is obtained that

$$\frac{P}{B}\int_0^a \delta(x-x_0)t^{(1)}(x,a)dx + \frac{Q}{B}\int_0^a \delta(x-x_0)t^{(2)}(x,a)dx = \frac{P}{B}t^{(1)}(x',a) + \frac{Q}{B}t^{(2)}(x',a) \qquad (A4.4.3)$$

i.e. the weight function terms $t^{(1)}$ and $t^{(2)}$ are the Green's functions for the T-stress term. The Green's functions $t^{(1)}$ and $t^{(2)}$ can be expressed by power series expansions

$$t^{(1)} = \frac{1}{a}\sum_{n=1}^{\infty} D_n^{(1)}(1-x/a)^{(2n-1)/2} \; , \; t^{(2)} = \frac{1}{a}\sum_{n=1}^{\infty} D_n^{(2)}(1-x/a)^{(2n-1)/2} \qquad (A4.4.4)$$

with unknown coefficients which have to be determined e.g. by fitting numerically obtained results. For practical applications, the infinite series have to be truncated after the Nth term.

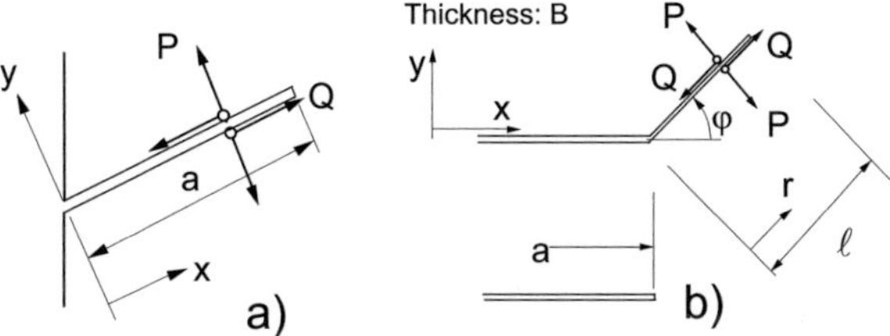

Fig. A4.7 Non-symmetric crack problems, a) oblique edge crack, b) semi-infinite kinked crack in an infinite body, reference crack of length a.

A4.4.2 Numerical results

Finite element results of the T-stress for the semi-infinite crack with a kink, Fig. A4.7b, are shown in Fig. A4.8. The data, normalized on the concentrated normal (P) and shear forces (Q), are identical with the Green's function. The T-stress data were fitted in the Green's function representation according to

$$t^{(1)} = \frac{1}{\ell}\sum_{n=1}^{N} D_n^{(1)}(1-r/\ell)^{(2n-1)/2} \; , \; t^{(2)} = \frac{1}{\ell}\sum_{n=1}^{N} D_n^{(2)}(1-r/\ell)^{(2n-1)/2} \qquad (A4.4.5)$$

for N terms with the coefficients compiled in Tables A4.1 and A4.2.

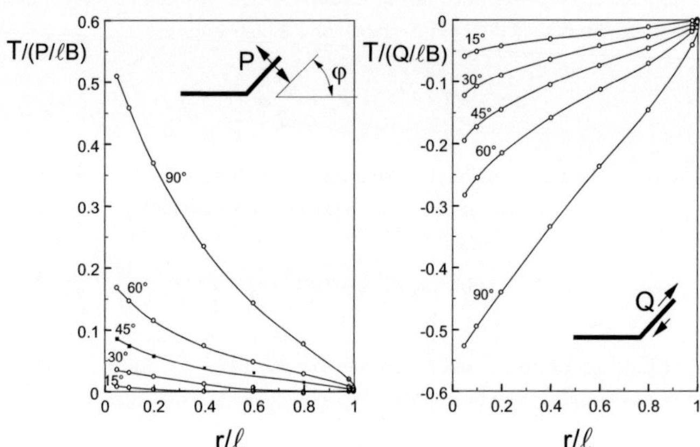

Fig. A4.8 Green's functions for the T-stress of kinked cracks.

Table A4.2 Coefficients for eq.(A4.4.5) as obtained under normal forces P.

φ	$D_1^{(1)}$	$D_2^{(1)}$	$D_3^{(1)}$	$D_4^{(1)}$
15°	-0.002	0.0191	-0.0459	0.0386
30°	0.	0.0327	-0.0368	0.0447
45°	0.017	0.1526	-0.2973	0.2259
60°	0.0437	0.1408	-0.2466	0.2534
90°	0.1326	0.1823	0.0436	0.2063

Table A4.3 Coefficients for eq.(A4.4.5) as obtained under shear forces Q.

φ	$D_1^{(2)}$	$D_2^{(2)}$	$D_3^{(2)}$	$D_4^{(2)}$
15°	0.	-0.2126	0.3384	-0.2026
30°	-0.0516	-0.0776	0.1414	-0.1507
45°	-0.0732	-0.2411	0.4552	-0.3645
60°	-0.1294	-0.1842	0.2507	-0.2464
90°	-0.2873	-0.1696	-0.1486	0.0501

References A4

[A4.1] Bückner, H., A novel principle for the computation of stress intensity factors, ZAMM **50** (1970), 529-546.

[A4.2] Sham, T.L., The theory of higher order weight functions for linear elastic plane problems, Int. J. Solids and Struct. **25**(1989), 357-380.

[A4.3] Sham, T.L., The determination of the elastic T-term using higher order weight functions, Int. J. Fract. **48**(1991), 81-102.

[A4.4] Fett, T., A Green's function for T-stresses in an edge-cracked rectangular plate, Eng. Fract. Mech. **57**(1997), 365-373.

[A4.5] Fett, T., T-stresses in rectangular plates and circular disks, Engng. Fract. Mech. **60**(1998), 631-652.

[A4.6] Fett, T., Munz, D., Stress intensity factors and weight functions, Computational Mechanics Publications, Southampton, 1997.

[A4.7] Wang, X., Elastic *T*-stress solutions for semi-elliptical surface cracks in finite thickness plates. *Engng. Fract. Mech.* **70** (2003), pp. 731–756.

[A4.8] Fett, T., Rizzi, G., Bahr, H.A., Green's functions for the T-stress of small kink and fork cracks, Engng. Fract. Mech. 73(2006), 1426-1435.

[A4.9] Fett, T., Rizzi, G., Weight functions for stress intensity factors and T-stress for oblique cracks in a half-space, Int. J. Fract. **132**(2005), L9-L16.

[A4.10] Fett, T., Rizzi, G., T-stress of cracks loaded by near-tip tractions, Engng. Fract. Mech. 73(2006), 1940-1946.

[A4.11] Fett, T., Rizzi, G., Report FZKA6937, Forschungszentrum Karlsruhe 2004, Karlsruhe.

[A4.12] Fett, T., T-stress and stress intensity factor solutions for 2-dimensional cracks, VDI-Verlag, 2002, Düsseldorf.

A5

Perturbation method

Many stress intensity factor solutions are available in literature for the cases of straight cracks. Computation of stress intensity factors for deviating crack shapes are seldomly reported. For the treatment of such a fracture mechanics problem we can use the so-called perturbation theory, well-known in physics, especially in astronomy and atomic physics. Perturbation theory comprises mathematical methods to find an approximate solution for a problem which cannot be solved exactly with a sufficient effort. The iterative procedure starts with the exact solution of the unperturbed problem and a disturbance that has to be small.

In the fracture mechanics problem of a slightly curved or kinked crack, the disturbance is the small deviation between the crack of interest and an undisturbed crack, for which the exact solution of the stress field is known. In this sense, "small disturbance" means that for the disturbed crack the deviations normal to the straight crack and also the slopes within the disturbance have to be small, but not necessarily the length of the disturbance.

A5.1 Cracks in infinite bodies

An analysis of straight cracks with small perturbations was presented by Goldstein and Salganik [A5.1] and Cotterell and Rice [A5.2]. They computed the mixed-mode stress intensity factors and crack paths under restrictions of a first-order analysis

$$dy/dx \ll 1, \quad y \ll a \tag{A5.1.1}$$

(for x, y, and a, see Fig. A5.1). These procedures were originally derived for internal cracks in an *infinite* body. Recently, a simpler approach was presented by means of the weight function method [A5.3, A5.4]. It was shown in [A5.4] that this procedure can be extended to cover crack configurations in *finite* bodies.

Cotterell and Rice [A5.2] computed the mixed-mode stress intensity factors for slightly curved or kinked semi-infinite cracks in an infinite body. A crack for which the stress intensity factor solution is known is considered as the unperturbed crack. This may be a straight crack of length a_0 (Fig. A5.1a), loaded by remote y-tractions. For this crack, the stress field in the vicinity of the crack is the sum of the singular stresses and the constant stress term, the so-called T-stress (stress field symbolised by the hatched zone in Fig. A5.1b). The perturbed crack is assumed to be again a straight crack of length a_0, but now having a kink of length ℓ at the end, Fig. A5.1c. This perturbed crack is embedded in the stress field caused by the unperturbed crack (Fig. A5.1d). The mixed-mode stress intensity factors K_I and K_{II} for the perturbed crack can be computed by use of the weight function method. It should be mentioned that the unperturbed crack has not necessarily to be a straight crack. The only advantages of the straight crack are the availability of exact stress intensity factor and weight function solutions, necessary for the numerical evaluation.

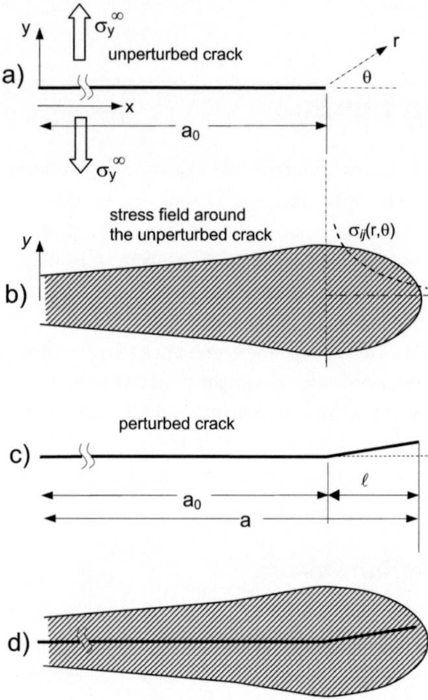

Fig. A5.1 Procedure proposed by Cotterell and Rice [A5.2] (schematic): a) unperturbed crack loaded by remote tractions, b) stress field around the unperturbed crack (perturbed crack removed), c) crack with a kink of length ℓ at the tip, d) perturbed crack embedded in the stress field caused by the unperturbed crack.

A5.2 Perturbation procedure for finite cracks in semi-infinite bodies

A5.2.1 Basic relations

For demonstrating the procedure, a straight edge crack of length a in a half-space is considered that is loaded by a remote y-stress (Fig. A5.2a). Figure A5.2b represents the same crack with a small perturbation behind the crack tip $x_2 < a$. For the computation of stress intensity factors, the small disturbance of the crack is considered to be loaded by the stress field existing in the vicinity of the unperturbed crack of Fig. A5.2a.

The stress intensity factor solutions K_{I0} and K_{II0} of the straight undisturbed crack in Fig. A5.2a are

$$K_{I0} = \sigma_y^\infty F \sqrt{\pi a} \tag{A5.2.1}$$

with the geometric function F, and, trivially

$$K_{\text{II}0} = 0 \qquad\qquad (A5.2.2)$$

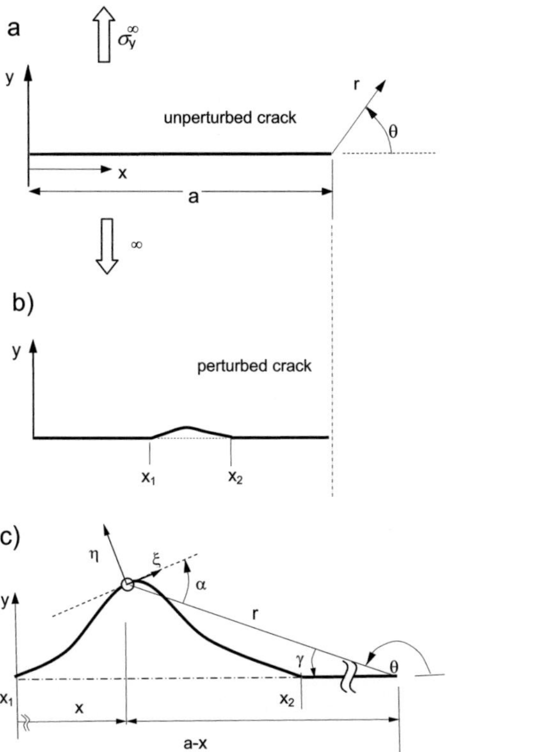

Fig. A5.2 Perturbation in the wake of the crack excluding the tip: a) straight edge crack, b) crack with a small perturbation between x_1 and $x_2 < a$, c) geometric data of the perturbation (y-coordinate exaggerated; origin of the y-axis always in height of the crack tip).

The first order terms of stress intensity factors of the slightly perturbed crack are given by

$$K_{\text{I}}^{(1)} = K_{\text{I}0} + \int_0^a \Delta\sigma_{\eta\eta} h_{11}\, dx \qquad\qquad (A5.2.3)$$

$$K_{\text{II}}^{(1)} = \int_0^a \Delta\tau_{\xi\eta} h_{22}\, dx \qquad\qquad (A5.2.4)$$

The stresses $\Delta\sigma_{\eta\eta}$ and $\Delta\tau_{\xi\eta}$ are the stress fields in the vicinity of the unperturbed crack. They act in the $\xi\eta$-plane along the prospective perturbation and do not fulfil the traction-free boundary conditions at the curved crack. The stress fields yield stress intensity factors which can be calculated in first order by means of the mode-I and mode-II weight functions h_{11} and h_{22} for the unperturbed straight crack with $h_{11}=h_{22}$ for the semi-infinite body.

The normal stresses near a straight edge crack are of second order in y, i.e. $\Delta\sigma_{\eta\eta}\propto y^2$ and, therefore, $\Delta\sigma_{\eta\eta}$ cannot create a K_I contribution in first order. Consequently, the evaluation of (A5.2.3) considering only first-order terms in y, yields

$$K_I^{(1)} = K_{I0} = \sigma_y^\infty F\sqrt{\pi a} \qquad (A5.2.5)$$

i.e. the mode-I stress intensity factor is unaffected by the perturbation.

If in addition remote stresses σ_x^∞ and τ_{xy}^∞ exist, the additional mode-II stress intensity factors in first order are

$$K_{II}^{(2)} = -\sigma_x^\infty \int_0^a h_{22}\, y'(x)\, dx \qquad (A5.2.6)$$

and

$$K_{II}^{(3)} = \tau_{xy}^\infty F\sqrt{\pi a} \qquad (A5.2.7)$$

In the special case of a perturbation ending directly at the crack tip, $x_2=a$, the perturbation can exhibit a certain slope $y'(a)$ as indicated in Fig. A5.3 by the dash-dotted straight line. The perturbed crack and the unperturbed one constitute a kink at $x=a$ defining the angle ω by $\tan(y)=y'=\omega$.

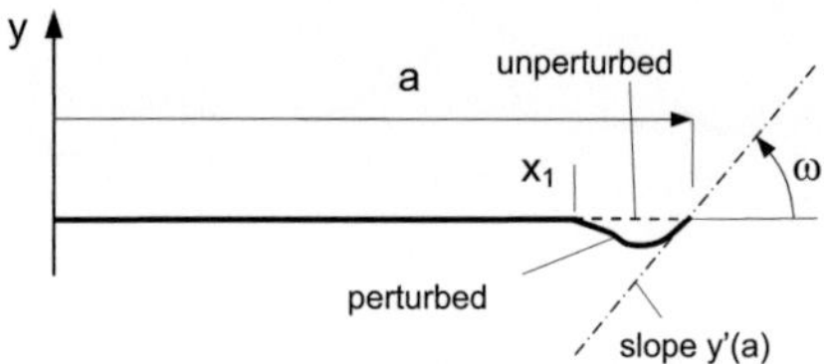

Fig. A5.3 Crack perturbation ending at the crack tip, $x_2=a$.

It was shown by Cotterell and Rice [A5.2] that the mode-II stress intensity factor for such a crack is

$$K_{II}^{(4)} = K_{I0}\, g_{21}(\omega) = F\sigma_y^\infty \sqrt{\pi a}\; g_{21}(\omega) \qquad (A5.2.8)$$

with

$$g_{21} = \sin(\omega/2)\cos^2(\omega/2) \xrightarrow[(\omega\to 0)]{} \tfrac{1}{2}\omega = \tfrac{1}{2}y'(a) \qquad (A5.2.9)$$

For the evaluation of the stress intensity factor K_{II} by use of eq.(A5.2.4), we first have to determine the stress field in the vicinity of the straight crack. The shear stress $\Delta\tau_{\xi\eta}$ results from the usual transformation of the radial stress σ_r and the shear $\tau_{r\theta}$ (for the origin of the coordinates r, θ, see Fig. A5.2a) according to

$$\Delta\tau_{\xi\eta} = -\sigma_r \sin\alpha\cos\alpha + \tau_{r\theta}(\cos^2\alpha - \sin^2\alpha) \rightarrow -\sigma_r\alpha + \tau_{r\theta} \qquad (A5.2.10)$$

where α is the angle between the ξ and r-coordinates.
The angles α and γ in Fig. A5.2c are

$$\alpha = y'(x) - \frac{y(a) - y(x)}{a - x} \qquad (A5.2.11)$$

$$\gamma = \frac{y(a)}{a} - \frac{y(a) - y(x)}{a - x} \qquad (A5.2.12)$$

With the abbreviations

$$\frac{\sigma_r}{\sigma_y^\infty} = f_1(\tfrac{x}{a}), \quad \frac{\tau_{r\theta}}{\sigma_y^\infty} = f_2(\tfrac{x}{a})\gamma \qquad (A5.2.13)$$

it holds for $\tau_{r\theta}$

$$\Delta\tau_{\xi\eta} = -\sigma_y^\infty f_1(\tfrac{x}{a})\left(y'(x) - \frac{y(a) - y(x)}{a - x} \right) + \sigma_y^\infty f_2(\tfrac{x}{a})\left(\frac{y(a)}{a} - \frac{y(a) - y(x)}{a - x} \right) \qquad (A5.2.14)$$

By summing up all the stress intensity factor terms $K_{II}^{(i)}$ we obtain the total K_{II} as

$$K_{II} = \sum_{(i)} K_{II}^{(i)} = \tfrac{1}{2} y'(a) F \sigma_y^\infty \sqrt{\pi a} - \sigma_x^\infty \int_0^a h_{22} y'(x)\,dx + \tau_{xy}^\infty 1.1215\sqrt{\pi a}$$

$$+ \sigma_y^\infty \int_0^a \left[f_1(\tfrac{x}{a})\left(\frac{y(a) - y(x)}{a - x} - y'(x) \right) - f_2(\tfrac{x}{a})\frac{y(a) - y(x)}{a - x} \right] h_{22}\,dx \qquad (A5.2.15)$$

A5.2.2 Numerical data

The stresses in the vicinity of the normal edge crack in the half-space were computed with the analytical procedure proposed by Wigglesworth [A5.5]. For this purpose the first 100 terms of the stress series expansions were determined in [A5.6]. The stresses series were determined by the evaluation of the first 100 terms of the Williams stress expansion. The normalised stresses, expressed by (A5.2.13), were found to be

$$f_1 = \sigma_r / \sigma_y^\infty = \sum_{n=0}^{14} d_n (1 - x/a)^n + R_1 \qquad (A5.2.16)$$

$$R_1 = -(0.259\tfrac{x}{a} + 0.0003364)\exp(-29.792 x/a) \qquad (A5.2.17)$$

$$f_2 = \sum_{n=0}^{14} (n+1) d_n (1 - x/a)^n + R_2 \qquad (A5.2.18)$$

$$R_2 = -0.1551\exp(-\kappa x/a) + 0.1177\exp(-2\kappa x/a) + 0.32627\exp(-4\kappa x/a) \qquad (A5.2.19)$$

with κ=17.173 and the coefficients d_n compiled in Table A5.1.

It should be noted that the first coefficient d_0 of Table A5.1 represents the so-called T-stress, namely,

$$T = \sigma_y^\infty \, d_0 = -0.5259676026 \, \sigma_y^\infty \qquad (A5.2.20)$$

Additional values of d_n can be found in the third column of Table A2.2.

The geometric function F was found in Section A2.1 to be

$$F = 1.12152225523... \qquad (A5.2.21)$$

Evaluation of the integral in (A5.2.15) needs the weight function solution. The computations in [A5.4] were performed with the weight function proposed in [A5.7], namely

$$h_{11} = h_{22} = \sqrt{\frac{2}{\pi a}} \left(\frac{1}{\sqrt{1-x/a}} + \sum_{n=0}^{5} c_n (1-x/a)^{n+1/2} \right) \qquad (A5.2.22)$$

with c_0=0.58852, c_1=0.031854, c_2=0.463397, c_3=0.227211, c_4=-0.828528, c_5=0.351383.

A5.3 Applications

A5.3.1 Computation of K_{II} for a slant edge crack

In this section, it will be demonstrated that the mode-II stress intensity factor of the slant crack with a small angle β (Fig. A5.4) can be computed from the stress intensity factor of a normal edge crack. The slant crack is considered as the perturbed crack with x_1=0 and x_2=a. Having in mind that the slant crack is characterized by a constant slope

$$\frac{y(a) - y(x)}{a - x} = y'(x) = \beta \qquad (A5.3.1)$$

the first term in the second integral of eq.(A5.2.15) disappears. For loading by remote y-tractions, it results with $y(a)$=0

$$K_{II} = \sigma_y^\infty \sqrt{\pi a} F \tfrac{1}{2} \beta - \sigma_y^\infty \beta \int_0^a f_2(\tfrac{x}{a}) \, h_{22} \, dx \qquad (A5.3.2)$$

The numerical evaluation of (A5.3.2) yields

$$F_{\mathrm{II}} = \frac{K_{\mathrm{II}}}{\sigma_y^\infty \sqrt{\pi a}} = 0.6904\,\beta \qquad\qquad (A5.3.3)$$

which is in good agreement with finite element results of [A5.8] expressed by

$$F_{\mathrm{II}} \simeq 0.692\beta + O(\beta^3) \qquad\qquad (A5.3.4)$$

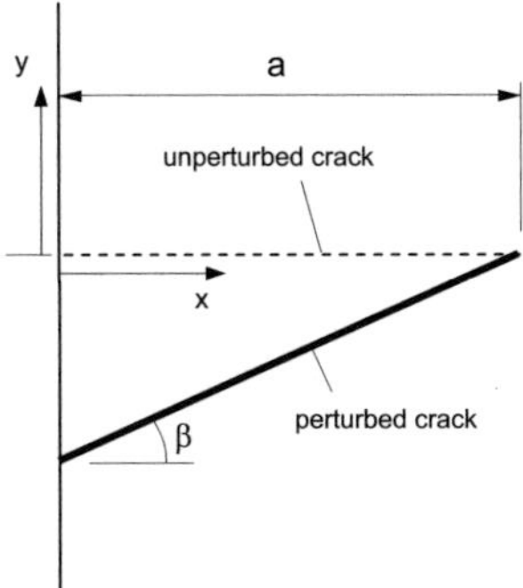

Fig. A5.4 The slant edge crack as the perturbed crack with the kink edge crack normal to the surface and disappearing kink length as the unperturbed crack.

Table A5.1 Coefficients d_n for eqs.(A5.2.16) and (A5.2.18).

$d_0= -0.525968$	$d_5= -0.0193802$	$d_{10}= -0.0158427$
$d_1=0.384983$	$d_6= -0.030546$	$d_{11}= -0.0116784$
$d_2=0.21431$	$d_7= -0.0306629$	$d_{12}= -0.00845447$
$d_3=0.0868761$	$d_8= -0.026327$	$d_{13}= -0.00604695$
$d_4=0.0142438$	$d_9= -0.0208939$	$d_{14}= -0.0042866$

A5.3.2 Mode-II stress intensity factor for a kink crack with finite kink length

A kink crack with finite kink length is represented as a perturbed crack (Fig. A5.5a). The first-order results obtained with eq.(A5.2.15) are plotted in Fig. A5.5b as the curve. Finite element results are entered as the squares and circles. The diamond square represents the limit case $F_{\mathrm{II}}/\beta=1.1215/2$ for $c/a\to0$ by Cotterell and Rice [A5.2]. A good agreement is obvious.

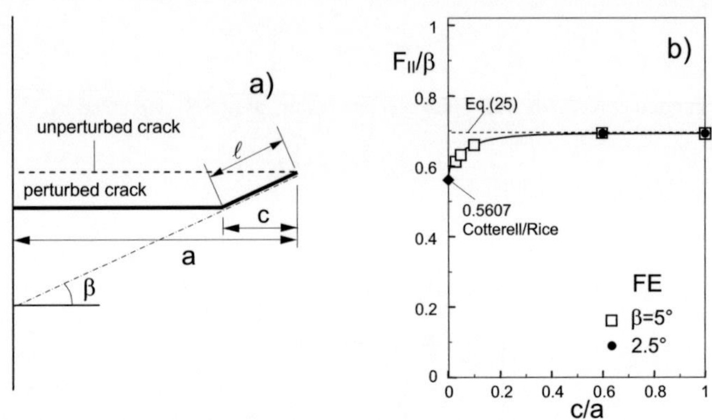

Fig. A5.5 Kink crack as the perturbed crack, a) geometric data, b) results from eq.(A5.2.15) (curve), finite element (FE) computations (squares and circles), and limit case from [A5.2] (diamond square).

References A5

[A5.1] Goldstein, R.V., Salganik, R.L., Brittle fracture of solids with arbitrary cracks, Int. J. Fract. **10**(1974), 507-523.

[A5.2] Cotterell, B. and Rice, J.R., Slightly curved or kinked cracks, International Journal of Fracture **16**(1980), 155-169.

[A5.3] Pham, V.-B., Bahr, H.-A., Bahr, U., Fett, T., Balke, H., Crack paths and the problem of global directional stability, Int. J. Fract. **141**(2006), 513-534.

[A5.4] Fett, T., Rizzi, G., Bahr , H.A., Bahr, U., Pham, V.B., Balke, H., A general weight function approach to compute mode-II stress intensity factors and crack paths for slightly curved or kinked cracks in finite bodies, Engng. Fract. Mech. **75**(2008), 2246-2259.

[A5.5] Wigglesworth, L.A., Stress distribution in a notched plate, Mathematica **4**(1957), 76-96.

[A5.6] Fett, T., Rizzi, G., Bahr, H.A., Bahr, U., Pham, V.-B., Balke, H., Analytical solutions for Stress intensity factor, T-stress and weight function for the edge-cracked half-space, Int. J. Fract., Letters in Fracture and Micromechanics, DOI 10.1007/s10704-007-9152-8.

[A5.7] Fett, T., Munz, D., Stress Intensity Factors and Weight Functions, Computational Mechanics Publications, Southampton, UK, 1997.

[A5.8] Fett, T., Rizzi, G., Weight functions for stress intensity factors and T-stress for oblique cracks in a half-space, Int. J. Fract. **132**(2005), L9-L16.

PART B

EFFECTS OF THE T-STRESS IN BRITTLE MATERIALS

In fracture mechanics interest is focussed mostly on stress intensity factors which describe the singular stress field ahead of a crack tip and govern fracture of a specimen when a critical stress intensity factor is reached. The influence of T on the fracture mechanics behaviour of metals was already mentioned in Section A. Since the author is predominantly working on ceramics and glass, some examples from this field of work will be addressed below.

The T-stress term must have an influence on several fracture mechanics features. Well-known from literature are the effects on

- Path stability [B1-B4]: Local path stability during crack propagation is often discussed in terms of the T-stress [B1]. This aspect shall be discussed here in detail for ceramic materials.

- Size of phase transformation zones [B5]: In materials undergoing stress-induced phase transformations (e.g. transformation-toughened ceramics), the size of phase transformation zones at the crack tip is larger under positive than under compressive T-stress. Consequently, a steeper R-curve has to be expected for positive than for compressive T-stress.

- Size of micro-cracking zones [B5]: A very similar effect has to be expected for micro-cracking zones in polycrystalline ceramics as well as for domain switching zones in piezoelectric materials.

In particular, effects occurring in the crack wake are affected predominantly by the T-term, because the singular stress field caused by the stress intensity factor disappears near the crack faces.

- As an example, the effect of T on crack-face interactions in coarse-grained ceramics is addressed [B6].

[B1] Cotterell, B., Rice, J.R., Slightly curved or kinked cracks, Int. J. Fract., Vol. **16**, pp. 155-169 (1980).

[B2] Melin, S., The influence of the T-stress on the directional stability of cracks, Int. J. Fract., Vol **114**, pp. 259-265 (2002).

[B3] Pham, V.-B., Bahr, H.-A., Bahr, U., Fett, T., Balke, H., Crack paths and the problem of global directional stability, Int. Journal of Fracture **141**(2006), 513-534.

[B4] Fett, T., Rizzi, G., Munz, D., Hoffmann, M.J., Oberacker, R., Wagner, S., Bridging interactions in ceramics and crack path stability, Journal of the Ceramic Society of Japan, **114**(2006), 1038-1043.

[B5] Giannakopoulos, A.E., Olsson, M., Influence of the non-singular stress terms on small-scale supercritical transformation toughness, J. Am. Ceram. Soc., Vol. **75**, pp. 2761-2764 (1992).

[B6] Fett, T., Friction-induced bridging effects caused by the T-stress, Engng. Fract. Mech. **59**(1998), 599-606.

<u>B1</u>

T-stress and path stability

B1.1 Local path stability

The question of crack path stability and instability is important for understanding crack propagation. Most of experimental investigations and numerical computations in this field were carried out on materials exhibiting plasticity effects [B1.1] or on interface problems [B1.2, B1.3].

Local path stability during crack propagation is often discussed in terms of the T-stress [B1.4]. This aspect shall be addressed here for ceramics [B1.5].

Cotterell and Rice [B1.4] investigated the behaviour of the path of an originally straight crack in an infinite body under mode I loading. They applied a perturbation method to obtain the stress intensity factors of a slightly curved or kinked crack and used the solution to examine the directional stability of a straight crack after a disturbance.

Figure B1.1 illustrates a crack kinking situation. A straight crack of initial length a_0 is considered. By application of a disturbing K_{II} stress intensity factor, the crack kinks and grows out of the initial straight plane by an angle of Θ_0 (Fig. B1.1a). The disturbing mode-II loading may be caused e.g. by a small unavoidable misalignment of the loading arrangement. For small K_{II}-values, $K_{II} \ll K_I$, the kink angle is

$$\Theta_0 = -\frac{2K_{II}(a_0)}{K_I(a_0)} \tag{B1.1.1}$$

where $K_I(a_0)$ and $K_{II}(a_0)$ are the stress intensity factors for the initial crack of length a_0, i.e. for the crack situation before kinking.

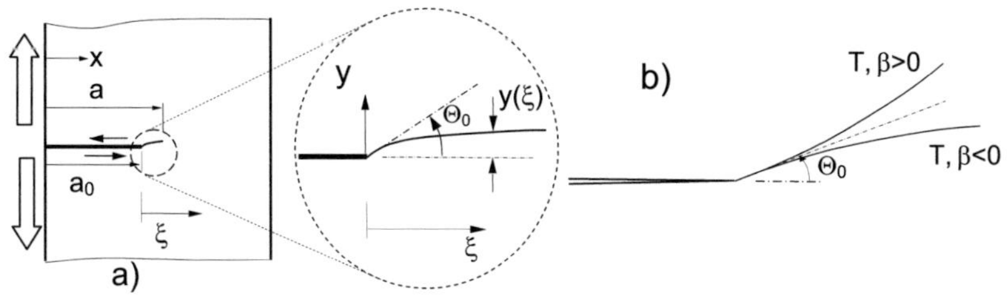

Fig. B1.1 a) Geometrical data of a crack growing under mode-I loading (vertical arrows) with a superimposed small mode-II disturbance (horizontal arrows), b) general influence of the T-stress after crack kinking.

Cotterell and Rice [B1.4] analyzed the crack development for an isotropic material after kinking on the basis of the condition of *local symmetry*, i.e. on the requirement of a disappearing mode-II stress intensity factor at the tip of the actual (grown) crack, K_{II}=0. The total mode-II stress intensity factor K_{II} was given in [B1.4] by the integral equation

$$K_{II}(a) = K_{II}(a_0) + \tfrac{1}{2} y'(a) K_I(a_0) - \sqrt{\frac{2}{\pi}} T \int_0^{a-a_0} \frac{y'(\xi)}{\sqrt{a-a_0-\xi}} \, d\xi \qquad (B1.1.2)$$

valid for small crack extensions a-a_0<<a, small deviations from the initial crack plane, y, and small derivatives y'<<1 of the crack trajectory. From the condition $K_{II}(a)$=0, the solution of (B1.1.2) was derived as

$$y = \frac{\Theta_0}{\beta^2} \frac{a_0 \pi}{8} \left[\exp\!\left(\frac{8\chi}{\pi} \beta^2 \right) \mathrm{erfc}\!\left(-\beta\sqrt{\frac{8\chi}{\pi}} \right) - 1 - \frac{4}{\pi} \beta\sqrt{2\chi} \right] \quad , \; \chi = \frac{a-a_0}{a_0} \qquad (B1.1.3)$$

with the initial biaxiality ratio β [B1.6]

$$\beta = \frac{T\sqrt{\pi a_0}}{K_I(a_0)} \qquad (B1.1.4)$$

Although eq.(B1.1.3) is a solution for short crack extensions only, this relation allows to discuss the effect of local path stability.

Discussions in literature [B1.7, B1.8] deal also with *global path stability* after a longer crack extension, i.e. with the general validity of the T-stress criterion, its application to finite cracks, and loading by non-homogeneous tractions.

B1.1.1 Path stability prediction for ceramics test specimens

The most important conclusion of [B1.4] is illustrated in Fig. B1.1b, namely, increasing deviation from the prescribed kink angle for β>0 and decreasing deviations for β<0.

From eq.(B1.1.3), it has to be expected that crack path stability is only guaranteed for β<0. In nearly all fracture mechanics test specimens, however, the T-stress and, consequently, the biaxiality ratio β are positive, at least in the commonly used range of crack lengths [B1.5].

In Fig. B1.2a the biaxiality ratios are plotted for the compact tension (CT) specimen, the 4-point bending specimen, and the opposite roller test [B1.9]. The solutions for the double cleavage drilled compression (DCDC) specimen and the double cantilever beam (DCB) specimen are plotted in Fig. B1.2b and Fig. B1.2c, respectively. Most of the specimens show positive biaxiality ratios. There are two exceptions for standard test specimens, namely, small cracks in bending bars with a relative crack length a/W<0.35 (specimen width W) and the DCDC specimen that shows strongly negative β in the whole range of possible crack lengths.

In different crack-containing specimens, the path stability may differ strongly even at the same biaxiality ratio β. This may be illustrated for the cases of a CT specimen and a crack in a bar loaded by opposite cylinders ($\beta \approx 0.6$).

Figure B1.3 shows the deviations of the crack from the linear propagation direction under the angle Θ_0 that is prescribed by the disturbing mode-II loading contribution. The results for a crack in a bar under opposite cylinder loading at two crack lengths a_0 are shown by the solid curves. Results for the CT specimen are given by the dashed curve. The curves are plotted for maximum crack extensions of 1/3 of the initial crack length, having in mind that eq.(B1.1.3) is valid for small extensions exclusively.

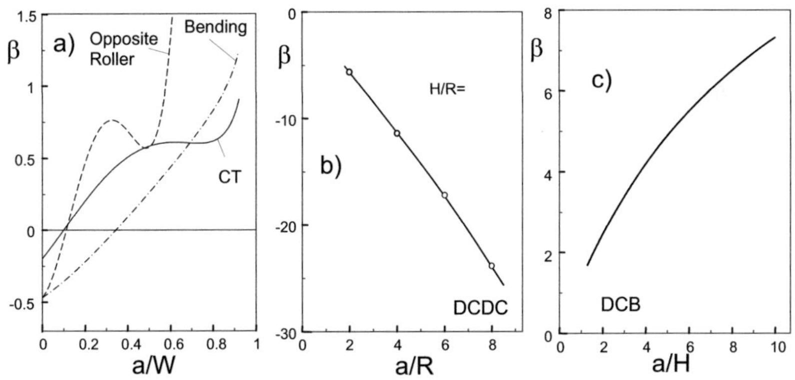

Fig. B1.2 a) Biaxiality ratio for 4-point bending test, CT specimen, and opposite roller test; b) for the DCDC specimen ($2H$=specimen height, R=hole radius), c) for the DCB specimen.

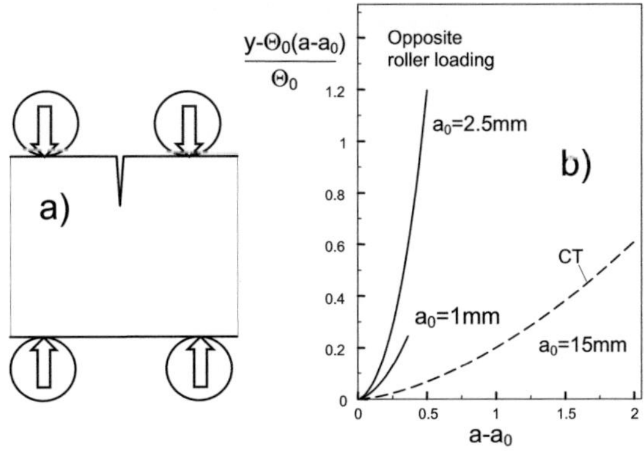

Fig. B1.3 a) Opposite roller fracture toughness tests, b) deviation of the crack path from the straight-line behaviour for a crack of length a_0=1 mm and 2.5 mm in an opposite roller toughness test and in a CT specimen of typical crack length of a_0=15 mm; curves plotted for maximum crack extensions of 1/3 of the initial crack length.

For a better understanding of the effect of the parameters β and a_0, the series expansion of eq.(B1.1.1) may be considered

$$y = (a - a_0)\Theta_0 + \frac{8}{3\pi}\sqrt{\frac{2}{a_0}}\,\beta\Theta_0(a - a_0)^{3/2} + O(a - a_0) \qquad (B1.1.5)$$

It becomes obvious that the effect of the T-stress is proportional to $\beta/\sqrt{a_0}$. At the same biaxiality ratio β, short cracks are significantly more sensitive to path instabilities than long cracks.

In Table B1.1 the parameter $\beta/\sqrt{a_0}$ is compiled for several test specimens and crack lengths. The best path stability is found for the DCDC specimen with negative values for $\beta/\sqrt{a_0}$. In principle, all specimens with positive β and, consequently, positive $\beta/\sqrt{a_0}$ are path-unstable. The instability effect is small for the CT specimen and also for the edge-cracked bending bar with a short crack. Problems in path stability, however, have to be expected for deep cracks in bending bars, DCB specimens, and edge-cracked bars under opposite cylinder loading.

In literature, innumerable experimental results on R-curves are reported. In contrast to the expectation from eq.(B1.1.3), however, no crack-path instability worth mentioning was detected. This is not astonishing for DCDC tests, CT tests, and bending bars with short cracks because of their negative or only moderately positive parameter $\beta/\sqrt{a_0}$ (Table B1.1). Steinbrech et al. [B1.10], for example, measured R-curves on coarse-grained alumina in bending up to relative crack lengths of about a/W=0.9, where strong path instability has to be expected.

The same holds for tests with opposite roller loading and DCB tests [B1.5], although only positive biaxiality ratios are involved in these tests.

Table B1.1 Ranking of path stability for different test specimens with typical crack lengths.

Test	Crack length a_0	Biaxiality ratio β	Path instability parameter $\beta/\sqrt{a_0}$ (mm$^{-1/2}$)
DCDC (R=0.5 mm)	2 mm	-12	-8.5
	4 mm	-24	-12
CT (W=30 mm)	15 mm	0.6	0.155
Bending (W= 4mm)	1 mm	-0.14	-0.14
	2 mm	0.26	0.18
	3.5 mm	1.05	0.56
DCB (H=12.5 mm)	30 mm	2.85	0.52
Opposite cylinder loading (W= 4mm)	1 mm	0.67	0.67
	2.5 mm	1.7	1.07

B1.1.2 Influence of bridging-induced mode-II R-curve on path stability

In [B1.5], the surprising effect of path stability under positive T-stresses was interpreted as a consequence of the crack resistance curve.

For coarse-grained materials, a mode-I shielding stress intensity factor term $K_{I,sh}$ exists that shields the crack tip partially from the applied loads. This term is caused by crack bridging in the wake of a growing crack. It has to be expected that crack-face interactions will also affect crack extension under pure or superimposed mode-II loading.

The interlocking of the two opposite crack faces must result in shear tractions which suppress the shear displacements in the crack-tip region which would be caused by the externally applied mode-II loading. These shear bridging tractions, τ_{br}, are illustrated in Fig. B1.4 together with the mode-I bridging tractions σ_{br}.

In [B1.5] it was outlined that the shear tractions generated under small mode-II load contributions can result in a disappearing effective crack-tip stress intensity factor $K_{II,tip}$.

The mode-II shielding stress intensity factor can be computed from the distribution of the shear tractions over the crack. It holds by use of the mode-II weight function h_{II}

$$K_{II,sh} = \int_{a_0}^{a} \tau(x)\, h_{II}(x)\, dx \qquad \text{(B1.1.6)}$$

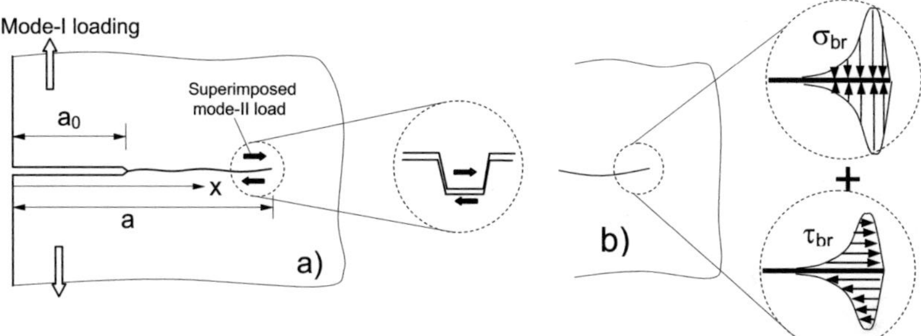

Fig. B1.4 Crack grown from a notch, a) geometrical data, loading, and shielding against shear deformation by bridging interactions, b) modelling of crack-face interactions by friction under loading with bridging stresses.

The actual mode-II crack tip stress intensity factor $K_{II,tip}$ then results from

$$K_{II,tip} = \begin{cases} 0 & \text{for} \quad K_{II,appl} + K_{II,sh} \leq 0 \\ K_{II,appl} + K_{II,sh} & \text{else} \end{cases} \qquad \text{(B1.1.7)}$$

It governs the local stability of crack paths. Crack paths are stable for any disturbing K_{II}-contribution that fulfils the upper part of eq.(B1.1.7).

If the value $K_{II,tip}$ does not disappear, the crack must kink by an angle of Θ out of the initial crack plane and will propagate then under $K_{I,tip}=K_{I0}$ and $K_{II,tip}=0$. For small values of $K_{II,tip}/K_{I0}$, the crack kink angle Θ can be expressed by a modified form of eq.(B1.1.1)

$$\Theta = -2\frac{K_{II,tip}}{K_{I0}} \qquad\qquad (B1.1.8)$$

B1.2 Global path stability

In the first-order approximation by Cotterell and Rice [B1.4] the special case of local stability (or instability) was considered for small crack extensions.

A simple procedure was proposed in [B1.11] that allows computing the stress intensity factors for slightly curved and kinked cracks even at large crack extensions. If a body is loaded by remote stresses σ_y^∞, σ_x^∞, and τ_{xy}^∞, the mode-II stress intensity factor reads (see Section A5)

$$K_{II} = F\sigma_y^\infty\sqrt{\pi a}\,\frac{1}{2}\,y'(a) - \sigma_x^\infty\int_0^a h_{II}\,y'(x)\,dx + \tau_{xy}^\infty F\sqrt{\pi a}$$

$$+\sigma_y^\infty\int_0^a\left[f_1(\tfrac{x}{a})\left(\frac{y(a)-y(x)}{a-x}-y'(x)\right)-f_2(\tfrac{x}{a})\frac{y(a)-y(x)}{a-x}\right]h_{II}\,dx \qquad (B1.2.1)$$

with the geometric function F=1.1215, the mode-II weight function h_{II}, and the functions f_1 and f_2 given by eqs.(A5.2.16)-(A5.2.19).

Here, the case of an edge-cracked semi-infinite body may be considered with the initial crack of length a_0 oriented in the x-direction (Fig. B1.5).

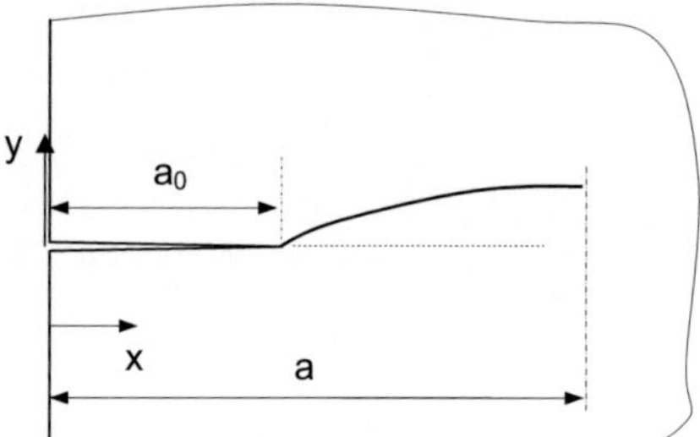

Fig. B1.5 Crack in a semi-infinite body.

Equation (B1.2.1) can be applied for the computation of crack paths y=f(x) by setting K_{II}=0. This relation is not restricted to small crack extensions as is required for (B1.1.2). The only remaining conditions for the validity of the first-order analysis are the assumptions of small values of y<<a and y'<<1. The influence of the free surface at x=0 is considered by the geometric function F and the weight function h_{II} for the edge-cracked half-space.

As an example of application, the edge-cracked half-space under remote stresses in x and y direction (Fig. B1.6a) was analyzed in [B1.11]. As the disturbance a pair of forces P at the crack mouth was chosen. The curved edge crack paths in Fig. B1.6b were calculated from K_{II}=0.

In Fig. B1.6b, the crack paths calculated by solving the integro-differential equation (B1.2.1) are compared with the results from FE simulation in [B1.8, B1.11]. A very good agreement of both methods can be concluded. The maximum differences between the two solutions are 0.9% at $a/a_0=3$ and $\sigma_x^\infty / \sigma_y^\infty =0.8$.

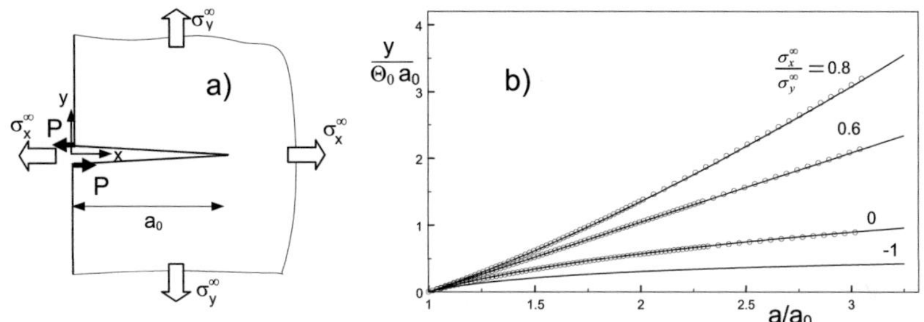

Fig. B1.6 a) Biaxially stressed half-space with a disturbance force P, b) crack paths of the original edge crack after disturbance by P, solid lines: numerical integration of the first-order integro-differential equation (B1.2.1), hollow circles: Finite element simulation.

References B1:

B1.1 Tvergaard, V., Hutchinson, J.W., Effect of T-stress on mode I crack growth resistance in a ductile solid, Int. J. of Solids and Structures **31**(1994), 823-833

B1.2 He, M.Y., Hutchinson, J.W., Kinking of a crack out of an interface, J. Appl. Mech. **56**(1989), 270-278.

B1.3 He, M.Y. Bartlett, A., Evans, A.G., Hutchinson, J.W., Kinking of a crack out of an interface: Role of in-plane stress, J. Am. Ceram. Soc. **74**(1991), 767-771.

B1.4 Cotterell, B., Rice, J.R., Slightly curved or kinked cracks, Int. J. Fract., **16**(1980), 155-169.

B1.5 T. Fett, G. Rizzi, D. Munz, M. Hoffmann, R. Oberacker, S. Wagner, Bridging interactions in ceramics and consequences on crack path stability, Journal of the Ceramic Society of Japan, **114**(2006), 1038-1043.

B1.6 Leevers, P.S., Radon, J.C., Inherent stress biaxiality in various fracture specimen geometries, Int. J. Fract. **19**(1982), 311-325.

B1.7 Melin, S., The influence of the T-stress on the directional stability of cracks, Int. J. Fract. (2002)**114**, 259-265.

B1.8 Pham, V.-B., Bahr, H.-A., Bahr, U., Fett, T., Balke, H., Prediction of crack paths and the problem of directional stability, Int. J. Fract. **141**(2006), 513-534.

B1.9 Fett, T., Munz, D., Thun, A toughness test device with opposite roller loading, Engng. Fract. Mech., Vol. **68**, pp. 29-38 (2001).

B1.10 Steinbrech, R., Reichl, A., Schaarwächter, W., R-curve behaviour of long cracks in alumina, J. Am. Ceram. Soc.. **73**(1990), 2009–2015.

B1.11 Fett , T., Rizzi, G., Bahr , H.-A., Bahr, U., Pham, V.-B., Herbert Balke, H., A general weight function approach to compute mode-II stress intensity factors and crack paths for slightly curved or kinked cracks in finite bodies, Engng. Fract. Mech. **75**(2008), 2246-2259.

B2

Effect of T-stress on phase transformation zones

B2.1 Phase transformation in zirconia ceramics

Due to the singular stress field near a crack tip in transformation-toughened zirconia, the material undergoes a stress-induced martensitic transformation and the tetragonal phase changes to the monoclinic phase (t- to m-ZrO$_2$). This transformation occurs when the characteristic local stress σ_{char} reaches a critical value $\sigma_{char,c}$

$$\sigma_{char} = \sigma_{char,c} \qquad (B2.1.1)$$

The result is a crack-tip transformation zone. Several stress criteria for the onset of phase transformation were applied in literature.

In one of the earliest attempts [B2.1], it was assumed that volume strains of the phase transformation only are playing a part in the transformation criterion, because the transformation shear strains are nearly annihilated by twinning. Since volume strains are proportional to the hydrostatic stress σ_{hyd}, a hydrostatic transformation criterion was proposed [B2.1]

$$\sigma_{hyd} = \tfrac{1}{3}(\sigma_x + \sigma_y + \sigma_z) = \sigma_{hyd,c} \qquad (B2.1.2)$$

In compressive experiments over a wide range of multiaxial stress states, it was found that also the shear strains have to be included in the transformation criterion [B2.2, B2.3]. This was done by adding a von-Mises stress contribution σ_{VM} to the hydrostatic term [B2.2, B2.3]

$$\frac{\sigma_{hyd}}{\sigma_{hyd,c}} + \frac{\sigma_{VM}}{\sigma_{VM,c}} = 1 \qquad (B2.1.3)$$

and later backed by theoretical considerations [B2.4].

In addition, other criteria were used, such as for instance the maximum normal stress criterion

$$\sigma_1 = \sigma_{1,c} \qquad (B2.1.4)$$

in which σ_1 indicates the first principal stress.

In this Chapter let us concentrate on the simple model of the hydrostatic stress criterion. For the special case of small-scale transformation conditions (transformation zone size negligible compared to crack size and component dimensions), McMeeking and Evans [B2.1] and Budiansky et al. [B2.5] computed the transformation zone, neglecting the perturbation of the stress field due to transformations.

B2.2 Phase transformation zone and R-curve in presence of T-stress

B2.2.1 Phase transformation zone

To the knowledge of the author, the first theoretical study of the effect of T-stress on phase transformation zones was published by Giannakopoulos and Olsson [B2.6]. This investigation is the basis of the following considerations.

In the presence of a T-stress contribution, the *hydrostatic stress* near the tip of the crack under plane strain conditions reads

$$\sigma_{hyd} = \tfrac{1+\nu}{3}\left(\frac{2K_{\mathrm{I}}}{\sqrt{2r\pi}}\cos(\varphi/2) + T \right) \tag{B2.2.1}$$

From (B2.1.2) and (B2.2.1), the shape $r(\varphi)$ of the phase transformation zone for plane strain results as

$$r = \frac{8}{3\sqrt{3}}\,\omega\cos^2(\theta/2) \tag{B2.2.2}$$

with the height ω of the zone

$$\omega = \frac{(1+\nu)^2}{4\sqrt{3}\pi}\left(\frac{K_{\mathrm{I}}}{\sigma_{\mathrm{hyd,c}} - (1+\nu)T/3} \right)^2 \tag{B2.2.3}$$

(for r, φ, and ω see Fig. B2.1). Figure B2.1a illustrates the transformation zone for a non-extending crack, Fig. B2.1b the shape after a crack extension of Δa. Due to the martensitic transformation, a volumetric expansion strain of about 4.5% occurs. These strains cause tensile stresses at a certain distance ahead of the crack tip and compressive stresses along the length Δa at the crack line. The compressive stresses lead to a shielding stress intensity factor which has to be overcome during crack propagation, i.e. the applied stress intensity factor must be increased to maintain stable crack growth.

In later, more complicated numerical studies (e.g. [B2.7]) the influence of shear stresses and strains on the transformation criterion and on the zone calculation also was taken into consideration.

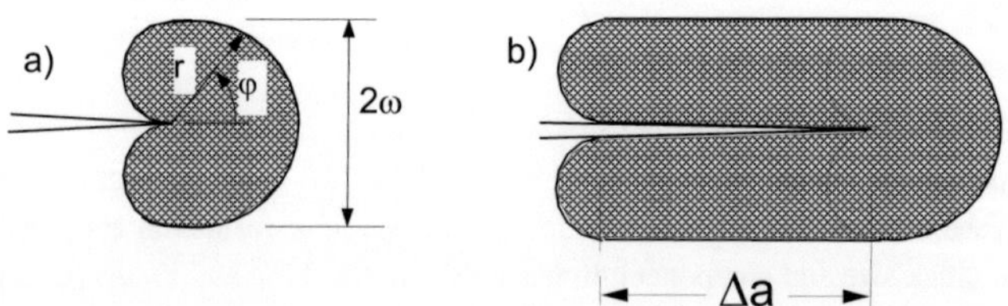

Fig. B2.1 a) Phase transformation zone ahead of a crack tip, b) zone after crack extension.

B2.2.2 Influence of T on the R-curve

McMeeking and Evans [B2.1] computed the crack resistance curve under small-scale transformation conditions assuming weak transformations. This means that the singular stresses caused by the stress intensity factor only are considered, whereas the stresses caused by the phase transformations were neglected. From the analysis in [B2.1], the surface tractions result in a shielding stress intensity factor K_{sh}

$$K_{sh} = p \oint_{\Gamma} \mathbf{n} \cdot \mathbf{h} \, dS \qquad (B2.2.4)$$

where Γ is the contour line of the transformation zone and dS is a line length increment. The vector $\mathbf{h}$ represents the weight function $\mathbf{h} = (h_y, h_x)^{\mathrm{T}}$ with the components h_y and h_x. In the special case of a pure dilatational transformation, the surface tractions are given by the normal pressure p defined by

$$p = \frac{\varepsilon^{\mathrm{T}} f E}{3(1 - 2v)}, \qquad (B2.2.5)$$

where ε^{T} is the volumetric phase transformation strain, f the volume fracture of transformed material, and v Poisson's ratio.

In Fig. B2.2 the shielding (residual) stress intensity factor for the case $T=0$, denoted as $K_{sh,0}$, is plotted for the phase transformation zone shown in Fig. B2.1b. The shielding stress intensity factor tends asymptotically to a value of 0.22.

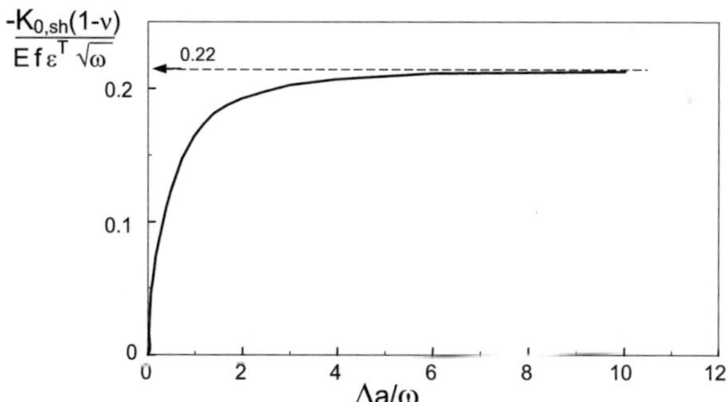

Fig. B2.2 Normalised shielding stress intensity factor $K_{0,sh}$ in the absence of a T-term computed with the method proposed by McMeeking and Evans [B2.1].

As a consequence of (B2.2.4), it may be concluded that the shielding stress intensity factor K_{sh} must be proportional to the square root of the zone height

$$K_{sh} \propto \sqrt{\omega} \qquad (B2.2.6)$$

with the factor of proportionality depending on the zone length Δa, the elastic constants E and v, and the transformation strain ε^{T}.

In the presence of a T-stress, the shielding stress intensity factor is

$$K_{sh} = \frac{K_{0,sh}}{1 - \frac{1+v}{3}\,T/\sigma_{hyd,c}} \tag{B2.2.7}$$

Introducing the biaxiality ratio β results in

$$K_{sh} = \frac{K_{0,sh}}{1 - \lambda\,\dfrac{\beta}{\sqrt{\pi\,a/W}}} \tag{B2.2.8}$$

with the dimensionless parameter

$$\lambda = \frac{1+v}{3}\,\frac{K_{I0}}{\sqrt{W}\,\sigma_{hyd,c}} \tag{B2.2.9}$$

Figure B2.3 shows the shielding stress intensity factor for an edge-cracked bending bar under weak transformation conditions.

For very small initial cracks with $a/W\to0$, the quantity $\beta\sqrt{(a/W)}$ tends to $-\infty$ and, consequently, the zone height tends to zero, i.e. the shielding stress intensity factor disappears. In the case of long cracks with $a/W\to1$, the zone height tends to $+\infty$, since $\beta\to+\infty$.

At least for an extreme relative crack length of $a/W\to1$, the conditions of small-scale behaviour and weak transformation assumed are violated, because not only the zone height must be small compared to the crack ($\omega\!\ll\!a$) but also to the crack ligament $\omega\!\ll\!(W\text{-}a)$. This is indicated by the dashed parts of the curves in Fig. B2.3.

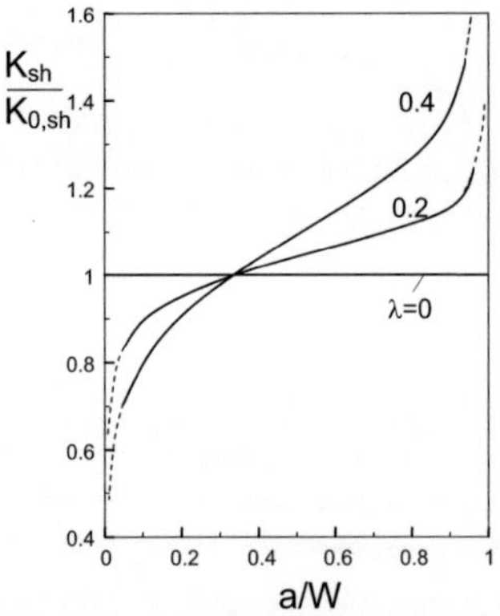

Fig. B2.3 Influence of the T-stress on the shielding stress intensity factor (bending).

The preceding results are approximations, because rather rigid restrictions had to be chosen to allow for a simple analysis. For a more exact derivation, the following points have to be considered by the analysis:

- The phase transformation causes a stress field that has to be added to the singular stresses. This requires an iterative procedure.

- Instead of the near-tip weight function, the weight function for the finite test specimen has to be used in (B2.2.4).

- The transformation zones cause an additional T-stress-term of about

$$T_{zone} \cong (0.435 - \tfrac{1}{6}\arctan(\Delta a / b))\frac{\varepsilon^{\mathrm{T}} E}{1-\nu} \qquad (B2.2.10)$$

(see Section C22 and [B2.8]).

References B2

B2.1 McMeeking, R.M., Evans, A.G. Mechanics of transformation-toughening in brittle materials, J. Am. Ceram. Soc. **65**(1982), 242–246.

B2.2 Reyes, P.E., Cherng, J.S., Chen, I.W., Transformation plasticity of CeO2-stabilized tetragonal zirconia polycrystals: I, Stress assistance and autocatalysis, J. Am. Ceram. Soc. 71(1988), 343-353.

B2.3 Chen, I.W., Model of transformation toughening in brittle material, J. Am. Ceram. Soc. 74(1991), 2564-2572.

B2.4 Sun, Q., Hwang, K.C., Yu, S.W., Micromechanics constitutive model of transformation plasticity with shear and dilatation effect, J. Mech. Phys. Sol. 39(1991), 507-524.

B2.5 Budiansky, B., Hutchinson, J.W., Lambropoulos, J.C., Continuum theory of dilatant transformation toughening in ceramics, Int. J. Solids Struct. **19**(1983), 337–355.

B2.6 Giannakopoulos, A.E., Olsson, M., Influence of the non-singular stress terms on small-scale supercritical transformation toughness, J. Am. Ceram. Soc. **75**, pp. 2761-2764 (1992).

B2.7 Stam, G.T.M., van der Giessen, E., Meijers, P., Effect of transformation-induced shear strains on crack growth in zirconia-containing ceramics, Int. J. Solids Struct. **31**(1994), 1923–1948.

B2.8 Fett, T., Rizzi, G., Fracture mechanics parameters of crack surface zones under volumetric strains, Int. J. Fract. **127**(2004), L117-L124.

B3

Effect of T-stress on micro-cracking zones

B3.1 Micro-cracking zones

In a poly-crystalline material, e. g. ceramics, the high stresses ahead of a crack result in the fracture of favourably oriented grain boundaries (Fig. B3.1). This micro-cracking at grain boundaries is caused by internal and superimposed externally applied stresses. The internal stresses are a consequence of thermal expansion mismatch in differently oriented grains. Different stress criteria for micro-cracking were used in literature.

In this section we will use a critical value of the first invariant of the stress tensor (hydrostatic stress), as proposed by Evans and Faber [B3.1], and to a minor extent an effective stress used in the study of Charalambides and McMeeking [B3.2].

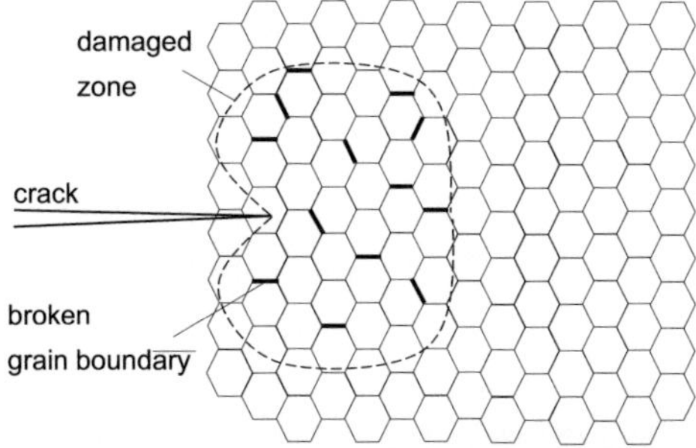

Fig. B3.1 Broken grain boundaries in a region ahead of a crack tip defining the micro-cracking zone.

B3.1.1 Stress criteria for micro-cracking

In [B3.1] the shape and size of the micro-crack zone is assumed to be governed by the condition of a critical value of the hydrostatic stress being responsible for cracking, i.e.

$$\sigma_{hyd} = \tfrac{1}{3}(\sigma_r + \sigma_\varphi + \sigma_z) = \sigma_{hyd,c} \tag{B3.1.1}$$

In eq.(B3.1.1), σ_r and σ_φ are the stress components in polar coordinates and σ_z is the stress in thickness direction. It holds for σ_z

$$\sigma_z = \begin{cases} 0 & \textit{for plane stress} \\ \nu(\sigma_r + \sigma_\varphi) & \textit{for plane strain} \end{cases} \tag{B3.1.2}$$

where ν is Poisson's ratio.

In [B3.2] the cracking condition is expressed by an effective stress

$$\sigma_{eff} = \sigma_{eff,c} \tag{B3.1.3}$$

with σ_{eff} defined by the principal stresses σ_1, σ_2, σ_3

$$\sigma_{eff} = \sqrt{\sigma_1^2 + \sigma_2^2 + \sigma_3^2} \qquad \text{(B3.1.4)}$$

The principal stresses are given by the three σ solutions of

$$\sigma^3 - (\sigma_x + \sigma_y + \sigma_z)\sigma^2 + (\sigma_x\sigma_y + \sigma_y\sigma_z + \sigma_x\sigma_z - \tau_{xy}^2)\sigma - (\sigma_x\sigma_y\sigma_z - \sigma_z\tau_{xy}^2) = 0 \qquad \text{(B3.1.5)}$$

B3.1.2 Micro-cracking ahead of a crack

The stresses ahead of a mode-I loaded crack are

$$\sigma_\varphi = \frac{K_I}{\sqrt{2r\pi}}\cos^3(\varphi/2) + T\sin^2\varphi + O(r^{1/2}) \qquad \text{(B3.1.6a)}$$

$$\sigma_r = \frac{K_I}{\sqrt{2r\pi}}\cos(\varphi/2)(1 + \sin^2(\varphi/2)) + T\cos^2\varphi + O(r^{1/2}) \qquad \text{(B3.1.6b)}$$

with the origin of the polar coordinates r and φ located at the crack tip. The higher-order stress terms $O(r^{1/2})$ may be neglected in the following considerations.

It is not the aim of this Section to compute the "true" zone size and shape, but rather to show the principal influence of the constant stress term on the zone size and its consequence on the crack tip toughness. In order to simplify the analysis, two assumptions will be made:

I.   The crack length a and the ligament W-a (W=specimen width) are assumed to be large compared to the length of the micro-cracking zone

II.   The case of "weak micro-cracking" is considered, i.e. the near-tip stress field is assumed to be unaffected by the micro-cracks. This behaviour is similar to the case of "weak transformation" zones at the tip of ceramics undergoing phase transformations under high stresses, where the effect of volume change during transformation is ignored [B3.3].

B3.1.3 Size and shape of the micro-cracking zone

The *hydrostatic stress* near the tip of the crack under plane strain conditions reads

$$\sigma_{hyd} = \tfrac{1+v}{3}\left(\frac{2K_I}{\sqrt{2r\pi}}\cos(\varphi/2) + T\right) \qquad \text{(B3.1.7)}$$

This case allows for an analysis that is identical to that of the phase transformation zones in zirconia (see Section B2). The zone contour resulting from condition (B3.1.1) is given by

$$r_c = \tfrac{2}{\pi}\left(\frac{K_{I0}\cos(\varphi/2)}{\tfrac{3}{1+v}\sigma_{hyd,c} - T}\right)^2 \qquad \text{(B3.1.8)}$$

The shape of this zone is plotted in Fig. B3.2 for several values of *T*. The results for the two criteria are very similar. Therefore, the hydrostatic stress criterion will be applied exclusively in the further numerical analyses.

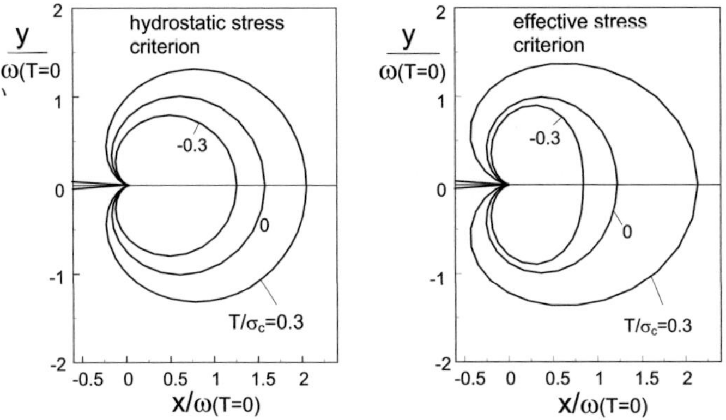

Fig. B3.2 Micro-cracking zones for the two cracking criteria.

Below, the influence of T-stress will be discussed for the case of an edge-cracked bending bar. The biaxiality ratio β [B3.4] for such a crack of length a is given by [B3.5]

$$\beta = \frac{-0.469 + 1.2825\alpha + 0.6543\alpha^2 - 1.2415\alpha^3 + 0.07568\alpha^4}{\sqrt{1-\alpha}}, \quad \alpha = a/W \quad (B3.1.9)$$

and plotted in Fig. B3.3 versus the relative crack length a/W.

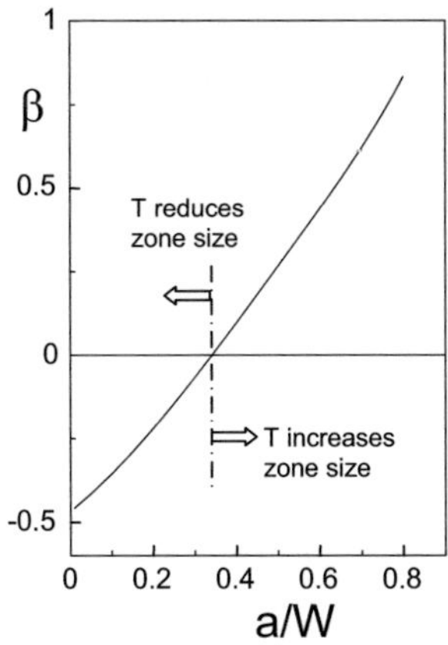

Fig. B3.3 Biaxiality ratio β for an edge-cracked bending bar (4-point bending).

B3.2 R-curve for a hydrostatic stress criterion

According to the analyses of McMeeking and Evans [B3.3] and Giannakopoulos and Olsson [B3.6], the shielding stress intensity factor under weak and small-scale micro-cracking (in presence of a T-stress term) is given by

$$K_{sh} = \frac{K_{0,sh}}{1 - \frac{1+\nu}{3} T / \sigma_{hyd,c}}$$ (B3.2.1)

where $K_{0,sh}$ denotes the shielding stress intensity factor for the case of $T=0$. Introducing the biaxiality ratio β results in

$$K_{sh} = \frac{K_{0,sh}}{1 - \lambda \dfrac{\beta}{\sqrt{\pi a / W}}}$$ (B3.2.2)

with the dimensionless parameter

$$\lambda = \frac{1+\nu}{3} \frac{K_{10}}{\sqrt{W} \sigma_{hyd,c}}$$ (B3.2.3)

For negative T and β, i.e. for relative crack lengths $a/W<0.37$, the micro-cracking zone size r_c decreases. Consequently, the shielding stress intensity factor is reduced as indicated in Fig. B3.3. In case of longer cracks with $a/W \geq 036$, the inverse effect occurs. For details, see Section B2.2.2.

References B3:

[B3.1] Evans, A.G., Faber, K.T., Crack-growth resistance of microcracking brittle materials, J. Am. Ceram. Soc. **67**(1984), 255-260.

[B3.2] Charalambides, P.G., McMeeking, R.M., Near-tip mechanics of stress-induced microcracking in brittle materials, J. Am. Ceram. Soc. **71**(1988), 465-472.

[B3.3] McMeeking, R.M., Evans, A.G. Mechanics of transformation-toughening in brittle materials, J. Am. Ceram. Soc. **65**(1982), 242–246.

[B3.4] Leevers, P.S., Radon, J.C., Inherent stress biaxiality in various fracture specimen geometries, Int. J. Fract. **19**(1982), 311-325.

[B3.5] Fett, T., T-stress and stress intensity factor solutions for 2-dimensional cracks, VDI-Verlag, 2002.

[B3.6] Giannakopoulos, A.E., Olsson, M., Influence of the non-singular stress terms on small-scale supercritical transformation toughness, J. Am. Ceram. Soc. **75**, 2761-2764 (1992).

B4

T-stress and crack-face bridging

B4.1 T-stress contribution generated by crack-face bridging

For coarse-grained ceramic materials, a shielding stress intensity factor term exists, that shields the crack tip partially from the applied loads. This term is often caused by crack bridging in the wake of a growing crack.

In cases of materials without a crack resistance curve (R-curve), the externally applied load exclusively is responsible for the stress intensity factor K_I and the T-stress. In ceramics with R-curve effects caused by crack-face bridging, an additional crack loading by the bridging stresses will occur. In this case, the question arises: What are the T-term and the biaxiality ratio in the presence of bridging stresses?

The T-stress term caused by the externally applied mechanical load is denoted here as T_{appl}. In a material with an R-curve effect due to bridging stresses $\sigma_{br}(x)<0$ acting in the crack wake (Fig. B4.1), a T-stress portion T_{br} is created that can be computed by the Green's function (or weight function) technique. For bridging stresses disappearing at $x=a$, it results

$$T_{br} = \int_0^a t(x,a)\sigma_{br}(x)dx , \quad \sigma_{br} < 0 \tag{B4.1.1}$$

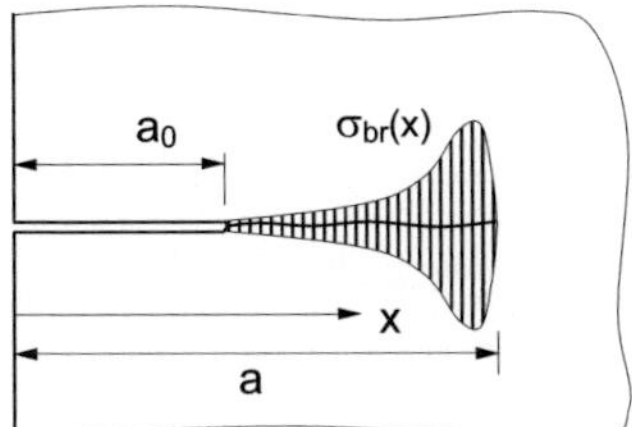

Fig. B4.1 Bridging stresses in the wake of a crack grown from a notch of depth a_0.

The Green's function t in (B4.1.1) can be approximated by a two-term expression [B4.1, B4.2] (see also Section A4.3.3). For the edge-cracked rectangular bar of width W, a three-term Green's function was determined in [B4.1] as

$$t = \frac{1}{a}(C_0\sqrt{1-x/a} + C_1(1-x/a)^{3/2}) \tag{B4.1.2}$$

with the coefficients C_0 and C_1 expressed by

$$C_0 = \frac{15(-0.3889 + 1.8706\,\alpha - 2.0012\,\alpha^2 - 1.0544\,\alpha^3 + 2.283\,\alpha^4 - 0.3932\,\alpha^5)}{8(1-\alpha)^2} \tag{B4.1.3}$$

$$C_1 = \frac{35(0.5487 - 2.1127\,\alpha + 2.1180\,\alpha^2 + 1.1845\,\alpha^3 - 2.0864\,\alpha^4 + 0.3932\,\alpha^5)}{8(1-\alpha)^2} \qquad (B4.1.4)$$

($\alpha = a/W$). Apart from the T-stress term T_{br}, also the bridging stress intensity factor, K_{br}, can be calculated from the bridging stresses. It holds

$$K_{br} = \int_0^a h(x,a)\sigma_{br}(x)\,dx \qquad (B4.1.5)$$

with the weight function $h(x,a)$, which is available in literature for most crack types (see Part A).

The total T-stress term, T_{total}, really present in the near-tip region is given by the sum of the applied and the bridging contributions, i.e. by

$$T_{total} = T_{appl} + T_{br} \qquad (B4.1.6)$$

If K_{appl} denotes the stress intensity factor caused by the externally applied load, the total stress intensity factor K_{total} representing the singular stresses near the crack tip is given by

$$K_{total} = K_{appl} + K_{br} \qquad (B4.1.7)$$

On the basis of the externally applied load, the (applied) biaxiality ratio β_{appl} is given as

$$\beta_{appl} = \frac{T_{appl}\sqrt{\pi a}}{K_{appl}} \qquad (B4.1.8)$$

The real (total) biaxiality ratio can be written as

$$\beta_{total} = \frac{T_{total}\sqrt{\pi a}}{K_{total}} = \frac{(T_{appl} + T_{br})\sqrt{\pi a}}{K_{appl} + K_{br}} \neq \beta_{appl} \qquad (B4.1.9)$$

Consequently, β_{total} is different from β_{appl}, and path stability for cracks in materials with bridging effects must deviate from path stability for materials without a bridging behaviour. Since the total stress intensity factor during stable crack extension equals the crack-tip toughness, $K_{total}=K_{I0}$, eq.(B4.1.9) may be expressed as

$$\beta_{total} = \frac{(T_{appl} + T_{br})\sqrt{\pi a}}{K_{I0}} = \beta_{appl}\frac{K_{appl}}{K_{I0}} + \frac{T_{br}}{K_{I0}}\sqrt{\pi a} \qquad (B4.1.10)$$

Computation of T_{br} for a sintered reaction-bonded silicon nitride

As an example of application, let us consider a result from literature. In [B4.3] the R-curve of a commercial, sintered reaction-bonded silicon nitride (SRBSN) was studied. A narrow notch was introduced in a bending bar by using the razor blade procedure [B4.4, B4.5]. The notch was extended in a stiff loading device. After stable crack extension, crack opening displacement (COD) measurements were carried out under load. From these results, bridging stresses were derived [B4.3].

Figure B4.2 shows the bridging stress distribution in the crack wake, $\sigma_{br}(a\text{-}x)$. From this, T_{br} results by inserting the stress distribution into eq.(B4.1.5) and β_{total} from eq.(B4.1.9).

The biaxiality ratio β_{appl} caused by the applied bending load can be expressed as

$$\beta_{appl} = \frac{-0.469 + 1.2825\alpha + 0.6543\alpha^2 - 1.2415\alpha^3 + 0.07568\alpha^4}{\sqrt{1-\alpha}} \qquad (B4.1.11)$$

In Fig. B4.3a the applied and the total biaxiality ratios are plotted versus the relative crack length a/W. The main effect of an R-curve behaviour on the biaxiality ratio is represented by the ratio $K_{appl}/K_{I0}>1$. The steepness of the function $\beta_{total}=f(a/W)$ is strongly increased compared to the function $\beta_{appl}=f(a/W)$. The direct influence of the bridging tractions on the T-stress is of minor importance, as shown by the difference of the circles and squares in Fig. B4.3b. The region of negative biaxiality ratios is slightly extended from $0 \leq a/W \leq 0.34$ to $0 \leq a/W \leq 0.36$ by the contribution of bridging tractions. Roughly, it can be concluded that in the region of $\beta_{appl}<0$ (for the edge-cracked bending bar at $a/W<0.34$) the effective biaxiality ratio becomes stronger negative and crack path stability is promoted. In the region with $\beta_{appl}>0$, an increased crack path instability has to be expected.

Based on these results, eq.(B4.1.10) can be approximated by

$$\beta_{total} \cong \beta_{appl} \frac{K_{appl}}{K_{I0}} \qquad (B4.1.12)$$

From the considerations made above, path stability can be concluded for a slightly extended range of crack lengths.

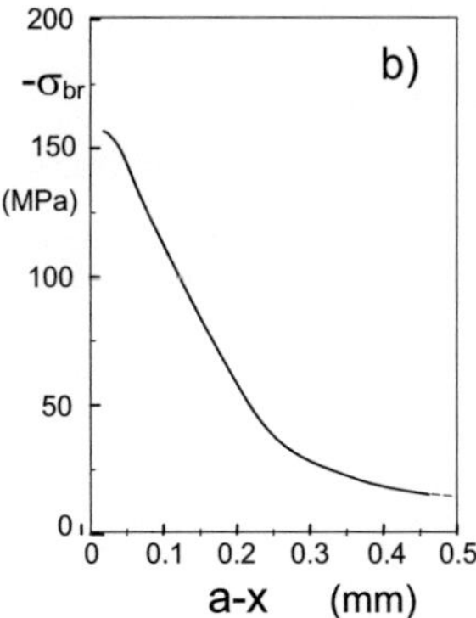

Fig. B4.2 Bridging stress distribution in the wake of a crack in sintered reaction-bonded silicon nitride [B4.3] after Δa=0.52 mm crack extension.

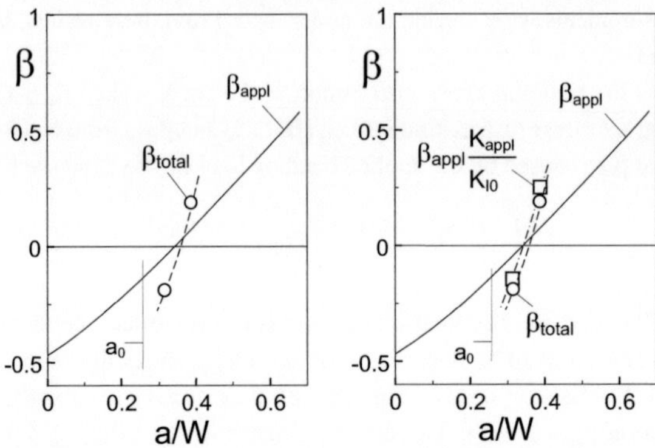

Fig. B4.3 a) Applied biaxiality ratio for an edge-cracked bending bar (solid curve) and total biaxiality ratios for the two bending results from [B4.3] (circles), b) contribution of the ratio K_{appl}/K_{I0} and of the bridging stresses.

B4.2 Bridging stress contribution caused by the T-stress term

In the bridging model by Mai and Lawn [B4.6], tractions are transmitted between the upper and lower crack faces by friction. Large grains with the lattice orientation different from the surrounding matrix undergo local residual stresses by thermal mismatch. Figure B4.4 shows the simple case of a large grain with the thermal expansion coefficients α_1 in the c-axis direction and α_2 normal to the c-axis. The "matrix" in which this grain is embedded is assumed to have average material parameters, for instance an average expansion coefficient of $\overline{\alpha}$ approximated as

$$\overline{\alpha} = \frac{\alpha_1 + 2\alpha_2}{3} \tag{B4.2.1}$$

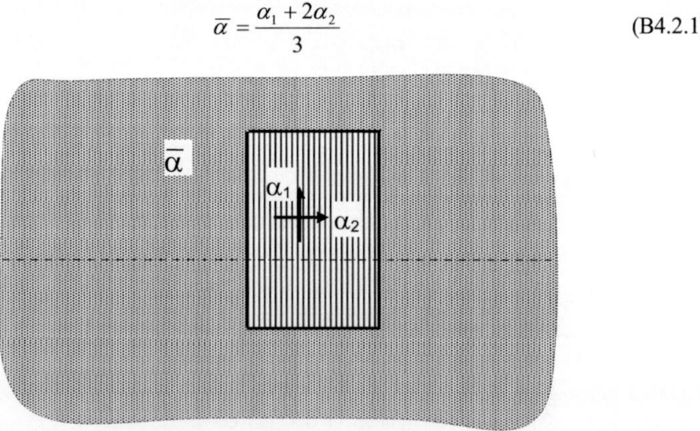

Fig. B4.4 A single anisotropic grain embedded in a matrix of average material parameters (dash-dotted line: prospective crack plane).

As an approximation the different elastic constants in the different lattice directions are neglected and assumed to be sufficiently identical with the average values.

After cooling down from the sintering temperature, the temperature change ΔT gives rise for
thermal strains in the grain with respect to the matrix

$$\Delta\varepsilon_1^{(th)} = -\Delta T(\alpha_1 - \overline{\alpha}) \tag{B4.2.2}$$

$$\Delta\varepsilon_2^{(th)} = -\Delta T(\alpha_2 - \overline{\alpha}) \tag{B4.2.3}$$

The thermal stresses caused by these strains are

$$\sigma_1^{(th)} = -CE\Delta T(\alpha_1 - \overline{\alpha}) \tag{B4.2.4}$$

$$\sigma_2^{(th)} = -CE\Delta T(\alpha_2 - \overline{\alpha}) \tag{B4.2.5}$$

with a coefficient C depending on Poisson's ratio v. Generally, the quantity C is in the range
of $1<C<1/(1-2v)$ depending on the specially chosen boundary conditions (plane strain, gener-
alized plane strain, plane stress). If $\alpha_1>\alpha_2$, it results $\alpha_1>\overline{\alpha}$ and $\alpha_2<\overline{\alpha}$. Consequently, we
have different stress signs in the two lattice directions with tension in 1-direction and com-
pression in 2-direction. In the case of Al_2O_3 it is $\alpha_1-\alpha_2\cong0.55\times10^{-6}/°$, E=360 GPa, and ΔT= -
1000°. For an average value of C=1.5, the typical mismatch stress is $\sigma_1^{(th)}\cong130$ MPa, $\sigma_2^{(th)}\cong$ -
65 MPa, i.e. stresses in the order of about $|\sigma^{(th)}|\approx 100$ MPa result. The mismatch stresses will
decrease when a crack passes in the vicinity creating a new free surface. Whereas the stress
component normal on the new surface disappears completely, the stress parallel to the crack
face is slightly reduced.

In Fig. B4.5, a large grain is shown, acting as a crack bridging event. For reasons of simplici-
ty, a 2-dimensional bridging contact may be assumed ($L/D\rightarrow\infty$). The x-component of the
thermal mismatch tractions is indicated. During crack-face separation resulting in an increasing displacement, δ, a friction stress σ_{fr} occurs which is proportional to the x-stress component.

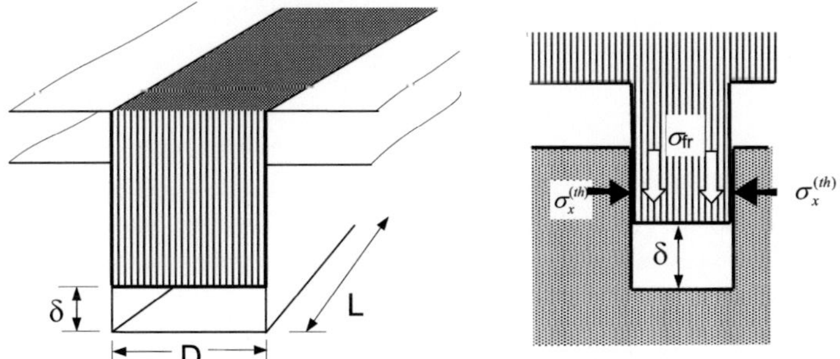

Fig. B4.5 Crack surface interactions due to a local frictional bridging event.

The loads transferred by crack face interactions are localized at single grains. They can be modelled in a more homogeneous way by so-called bridging stresses σ_{br} which average the localized interactions over a large number of grains. If $\sigma_x^{(th)}$ denotes the x-stress component due to thermal mismatch, the bridging stress σ_{br} can be expressed by

$$\sigma_{br} \cong \mu\sigma_x^{(1)} \tag{B4.2.6}$$

defining an effective friction coefficient μ.

In the preceding considerations the x-stress clamping the large grains was assumed to be generated by thermal mismatch exclusively. A second source for the occurrence of an x-stress component is the existence of a non-disappearing T-stress [B4.7]. Whereas the influence of thermal mismatch is an intrinsic effect independent on the special test specimen, the influence of the T-stress reflects an influence of the chosen specimen. Since T can be negative and positive, the bridging stresses can be increased and decreased by this stress term. Frictional bridging is possible only if the total x-stress component is negative resulting in "clamping" effects. Therefore, only those bridging events are of importance.

If $\sigma_{0,br}$ denotes the bridging stresses in the absence of a T-stress term, the bridging stresses in presence of T result simply from (B4.2.6) as

$$\sigma_{br} = \sigma_{0,br}\,\frac{T+\sigma_x^{(th)}}{\sigma_x^{(th)}} \tag{B4.2.7}$$

By use of the biaxiality ratio β this reads

$$\sigma_{br} = \sigma_{0,br}\left(1+\frac{\beta}{\sigma_x^{(th)}\sqrt{\pi a}}K_I\right) \tag{B4.2.8}$$

Having in mind that during stable crack extension the crack-tip stress intensity factor must equal the so-called crack-tip toughness K_{I0}, it follows

$$\sigma_{br} = \sigma_{0,br}\left(1+\frac{\beta}{\sigma_x^{(th)}\sqrt{\pi a}}K_{I0}\right) \tag{B4.2.9}$$

The bridging stresses shield the crack tip from the external loads. The related bridging stress intensity factor K_{br} can be computed from the distribution of the bridging stresses along the crack wake by use of the weight function technique.

Figure B4.6a shows the geometric data of the most commonly used bending test specimen and Fig. B4.1 illustrates the bridging stress distribution (here for the case of a bar with an initial notch of depth a_0).

$$K_{br} = \left(1+\frac{\beta}{\sigma_x^{(th)}\sqrt{\pi a}}K_{I0}\right)\int_0^a h(x,a)\,\sigma_{0,br}(x)\,dx = \left(1+\frac{\beta}{\sigma_x^{(th)}\sqrt{\pi a}}K_{I0}\right)K_{0,br} \tag{B4.2.10}$$

with the bridging stress intensity factor $K_{0,br}$ at $T=0$. In Fig. B4.6a, the biaxiality ratio for a 4-point bending bar is plotted. The bridging stress intensity factor is represented by the ratio

$K_{br}/K_{0,br}$ in Fig. B4.7a. While for long cracks, $0.2<a/W<0.8$, the influence of the T-stress is moderate with $K_{br}/K_{0,br} \cong 1\pm0.2$, it becomes significant in the case of very short cracks (size comparable with natural cracks), Fig. B4.7b.

Finally, the crack resistance curve ("R-curve") results as

$$K'_R = K_{I0} - K_{br} = K_{I0} - \left(1 + \frac{\beta}{\sigma_x^{(th)}\sqrt{\pi a}} K_{I0}\right) K_{0,br} \tag{B4.2.11}$$

with $K_{0,br}<0$.

In early investigations on R-curve behaviour of ceramics it was often assumed that the R-curve would be a material specific property. In the actual literature there is common agreement that the bridging stresses are the real material specific quantity. The influence of the T-stress again gives rise for an influence of the specially chosen test specimens on bridging properties. Having this in mind we have to consider the bridging stress relation for $T=0$ as the true material property.

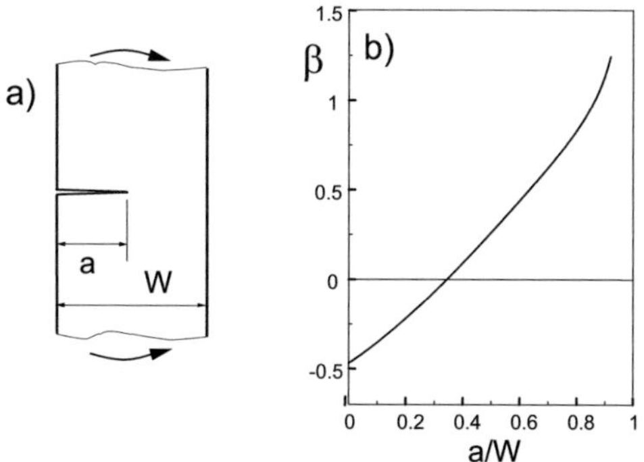

Fig. B4.6 a) Geometry of a bending bar, b) biaxiality ratio for 4-point bending tests.

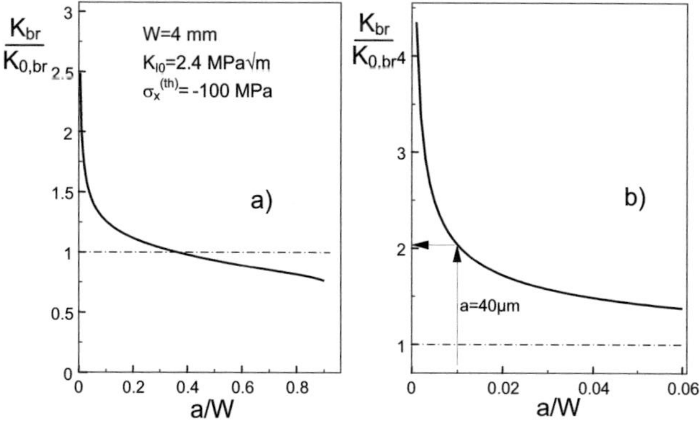

Fig. B4.7 a) Influence of the T-stress on the bridging stress intensity factor, b) details for small cracks.

In this context, it must be confessed that there were no systematic experimental investigations so far to establish proof of a T-stress effect.

The R-curve effects mentioned before for crack bridging by large grains should be also relevant for crack bridging by fibres or by whiskers in reinforced ceramics. Measurements on such model materials might simplify the experimental evidence.

References B4

[B4.1] Fett, T., Munz, D. Stress Intensity Factors and Weight Functions, Computational Mechanics Publications, Southampton, UK, (1997).

[B4.2] Wang, X., Elastic T-stress solutions for semi-elliptical surface cracks in finite thickness plates, Engng. Fract. Mech., Vol. **70**, pp. 731–756 (2003).

[B4.3] Fett, T., Munz, D., Kounga Njiwa, A.B, Rödel, J., Quinn, G.D., Bridging stresses in sintered reaction-bonded Si_3N_4 from COD measurements, J. Europ. Ceram. Soc., **25**(2005), 29-36.

[B4.4] Nishida, T., Pezzotti, G., Mangialardi, T., Paolini, A.E., "Fracture mechanics evaluation of ceramics by stable crack propagation in bend bar specimens", *Fract. Mech. Ceram.* **11** (Eds. R.C. Bradt, D.P.H. Hasselman, D. Munz, M. Sakai, V.Y. Shevchenko), 107–114 (1996).

[B4.5] Kübler, J., Fracture toughness using the SEVNB method: Preliminary results, Ceramic Engineering & Science Proceedings, Vol. **18**, pp. 155-162 (1997).

[B4.6] Mai, Y., Lawn, B.R., Crack-interface grain bridging as a fracture resistance mechanism in ceramics: II. Theoretical fracture mechanics model, J. Am. Ceram. Soc. **70**(1987), 289.

[B4.7] Fett, T., Friction-induced bridging effects caused by the T-stress, Engng. Fract. Mech. **59** (1998), 599-606.

PART C

COMPENDIUM OF STRESS INTENSITY FACTOR AND T-STRESS SOLUTIONS

The content of this section is divided in the following items

1) Cracks in infinite bodies

   Section C1 (Internal crack)

2) Cracks in the semi-infinite body (half-space)

   Section C2 (straight, oblique, and kink edge crack)

3) Semi-infinite cracks

   Sections C3 (kink crack) and C4 (fork crack)

4) Cracks in finite bodies

   Internally cracked components: Sections C5 and C6

   Edge-cracked components: Sections C7 to C10

   Double-edge-cracked components: Sections C11 and C12

5) Fracture mechanics test specimens

6) Miscellaneous crack problems

<u>C1</u>

Crack in an infinite body

C1.1   Couples of forces

The T-stress term resulting from a couple of symmetric point forces (see Fig. C1.1) can be derived from the Westergaard stress function [C1.1], which for this special case reads

$$Z = \frac{2P}{\pi} \frac{\sqrt{a^2 - x^2}}{(z^2 - x^2)\sqrt{1-(a/z)^2}} \quad , \quad z = y + ix \tag{C1.1.1}$$

The real part of eq.(C1.1.1) gives the x-stress component for $y=0$

$$\sigma_x\big|_{y=0} = \mathrm{Re}\{Z\} = \frac{2P}{\pi} \frac{\sqrt{a^2 - x^2}\, x'}{(x'^2 - x^2)\sqrt{x'^2 - a^2}} \tag{C1.1.2}$$

Its singular part

$$\sigma_{x,sing}\big|_{y=0} = \frac{2P}{\pi} \frac{\sqrt{a/2}}{\sqrt{a^2 - x^2}\,\sqrt{x'-a}} \tag{C1.1.3}$$

provides the well-known stress intensity factor solution

$$K = \lim_{x'\to a} \sqrt{2\pi(x'-a)}\,\sigma_x = \sqrt{\frac{a}{\pi}}\,\frac{2P}{\sqrt{a^2 - x^2}} \tag{C1.1.4}$$

Then, the regular stress term reads

$$\sigma_{x,reg}\big|_{y=0} = \frac{2P}{\pi} \frac{(a^2 - x^2)x' - \sqrt{a/2}\,(x'^2 - x^2)\sqrt{x'+a}}{(x'^2 - x^2)\sqrt{x'^2 - a^2}\,\sqrt{a^2 - x^2}} \tag{C1.1.5}$$

and for the T-stress term it results

$$T = \lim_{x'\to a}\sigma_{x,reg} = \begin{cases} 0 & \text{for } x < a \\ \text{undefined } (\to \infty) & \text{for } x = a \end{cases} \tag{C1.1.6}$$

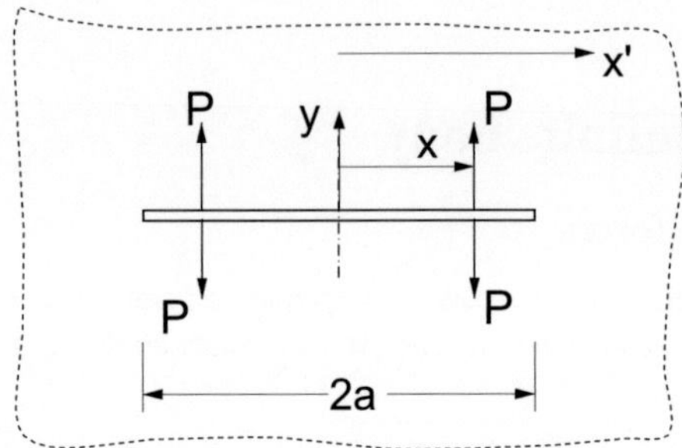

Fig. C1.1 Crack in an infinite body loaded by symmetric couples of forces.

C1.2 Constant crack-face loading

In the case of a constant crack-face pressure $p = \text{const.}$ (Fig. C1.2a), the stress function reads

$$Z = p\left[\frac{z}{\sqrt{z^2 - a^2}} - 1\right] \tag{C1.2.1}$$

resulting in the x-stress of

$$\sigma_x\big|_{y=0} = p\left[\frac{x'}{\sqrt{x'^2 - a^2}} - 1\right] \tag{C1.2.2}$$

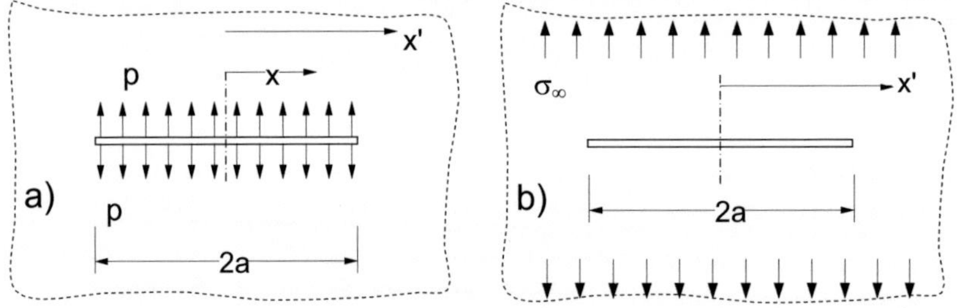

Fig. C1.2 Crack in an infinite body under a) constant crack-face pressure, b) remote tension.

The stress intensity factor results from eq.(C1.2.2) as

$$K_{\mathrm{I}} = p\sqrt{\pi a} \tag{C1.2.3}$$

and the T-stress term as

$$T = 0 \quad . \tag{C1.2.4}$$

C1.3　　Remote tension

In the case of the crack under the remote tensile stress σ_∞, Fig. C1.2b, the stress function reads

$$Z = \sigma_\infty \frac{z}{\sqrt{z^2 - a^2}} \qquad\qquad (C1.2.5)$$

yielding the x-stress of

$$\sigma_x\big|_{y=0} = \sigma_\infty \frac{x'}{\sqrt{x'^2 - a^2}} \qquad\qquad (C1.2.6)$$

with the stress intensity factor resulting as

$$K_I = \sigma_\infty \sqrt{\pi a} \qquad\qquad (C1.2.7)$$

and the T-stress term as

$$T = -\sigma_\infty \ . \qquad\qquad (C1.2.8)$$

Reference C1

[C1.1] Irwin, G.R., Analysis of stresses and strains near the end of a crack transversing a plate, J. Appl. Mech. **24** (1957), 361-364.

C2

Crack in a semi-infinite body

C2.1 Edge crack normal to the surface

The *stress intensity factor* of an edge-cracked semi-infinite body (Fig. C2.1) is under the remote stress $\sigma = \sigma_\infty$ is (see Section A2.1)

$$K = \sigma_\infty F \sqrt{\pi a} \qquad\qquad (C2.1.1)$$

$$F = 1.12152225523... \qquad\qquad (C2.1.2)$$

The *weight function* can be described by

$$h = \sqrt{\frac{2}{\pi a}} \left(\frac{1}{\sqrt{1-x/a}} + \sum_{n=0}^{N} D_n (a - x/a)^{n+\frac{1}{2}} \right) \qquad\qquad (C2.1.3)$$

An approximate representation of h was given in [C2.1] with the coefficients

$$D_0 = 0.58852, \quad D_1 = 0.031854, \quad D_2 = 0.463397$$

$$D_3 = 0.227211, \quad D_4 = -0.828528, \quad D_5 = 0.351383$$

On the basis of the Wigglesworth analysis [C2.2], the crack opening displacements v can be determined and the weight function h results simply from the Rice [C2.3] equation

$$h = \frac{E'}{K} \frac{\partial v}{\partial a} \qquad\qquad (C2.1.4)$$

A weight function was given in [C2.1] determined from the first 12 coefficients of [C2.2] together with an extended solution with coefficients up to $n=22$. By application of Mathematica [C2.4] one can easily increase the accuracy of these coefficients [C2.5]. In Table C2.1 the first coefficients are compiled.

The *T-stress* and the *biaxiality ratio* β are

$$T = -0.5259676026\, \sigma_\infty \qquad\qquad (C2.1.5)$$

$$\beta = -0.46897652 \qquad\qquad (C2.1.6)$$

The *Green's function for T-stresses* reads

$$T = \sigma_x\big|_{x=a} - \sigma_y\big|_{x=a} + \int_0^a t(x,a)\,\sigma(x)\,dx \qquad (C2.1.7)$$

with

$$t = \tfrac{1}{a}\sum_{n=1}^{\infty} C_n (1 - x/a)^{(2n-1)/2} \qquad (C2.1.8)$$

An approximate solution for t is

$$t = \tfrac{1}{a}[0.345(1 - x/a)^{1/2} + 0.087(1 - x/a)^{3/2} + 0.733(1 - x/a)^{5/2}] \qquad (C2.1.9)$$

or with reduced accuracy (see Fig. A4.6d) by the linear relation

$$t \approx \frac{0.948}{a}\int_0^a \sigma_y(x)(1 - x/a)\,dx \qquad (C2.1.10)$$

Table C2.1 Coefficients D_n for eq.(C2.1.3).

n	D_n	n	D_n	n	D_n
0	0.568846	10	-0.00177336	20	0.00017319
1	0.243546	11	-0.00077382	21	0.00014924
2	0.077759	12	-0.00021180	22	0.00012834
3	0.0083769	13	0.00008136	23	0.00011041
4	-0.014199	14	0.00021773	24	0.00009516
5	-0.0173687	15	0.00026719	25	0.00008223
6	-0.0140855	16	0.00027134	26	0.00007127
7	-0.00970142	17	0.00025389	27	0.00006197
8	-0.00603396	18	0.00022780	28	0.000054065
9	-0.00344057	19	0.00019983		

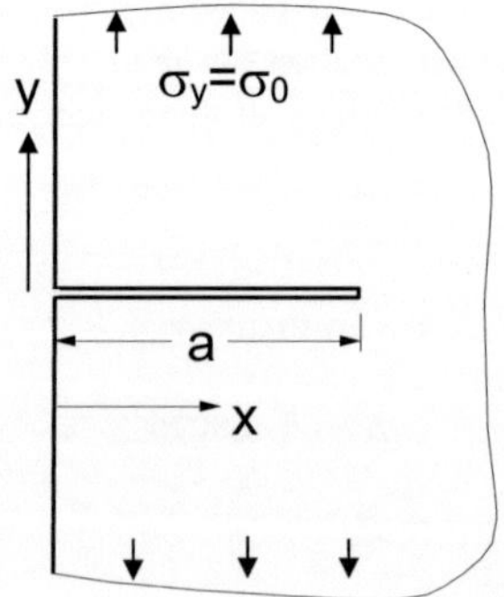

Fig. C2.1 Edge crack in a semi-infinite body loaded by remote y-stresses σ_0.

C2.2 Oblique crack in the half-space

A slant edge crack in a semi-infinite body under an angle φ to the x-axis is illustrated in Fig.
C2.2. This crack is loaded either by pairs of normal forces P or shear forces Q at a distance of
x from the crack mouth (Fig. C2.2a) or constant tractions (Fig. C2.2b) in ξ and y direction, σ_ξ
and σ_y.

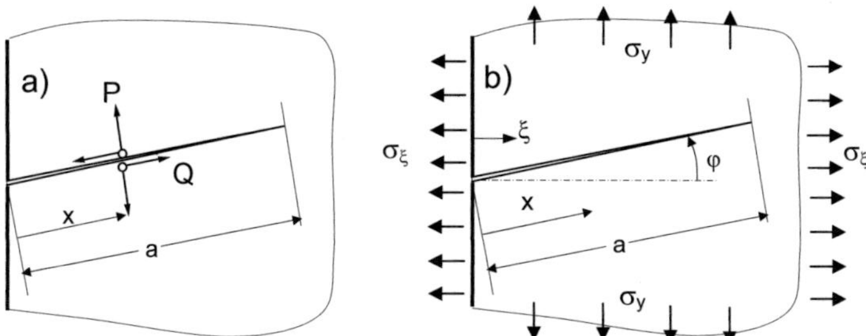

Fig. C2.2 Oblique edge crack in a half space, a) loaded with concentrated forces acting on the crack
faces, b) loaded by constant tractions.

C2.2.1 Stress intensity factors and weight functions

As outlined in Section A3.1, the stress intensity factors under a combined crack-face loading
can be superimposed resulting in

$$K_\mathrm{I} = \int\limits_0^a [h_{11}(x,a)\sigma_n(x) + h_{12}(x,a)\tau(x)]dx \tag{C2.2.1}$$

$$K_\mathrm{II} = \int\limits_0^a [h_{21}(x,a)\sigma_n(x) + h_{22}(x,a)\tau(x)]dx \tag{C2.2.2}$$

The subscripts of the weight functions indicate the type of the resulting stress intensity factor,
the superscripts are related to the type of the tractions. In relations (C2.2.1) and (C2.2.2), the
stress σ_n is the stress normal to the crack and τ is the shear stress acting in the crack plane.
The weight function contributions can be expressed by power series expansions as shown in
[C2.1]

$$h_{ij} = \sqrt{\frac{2}{\pi a}} \sum_{n=0}^{\infty} D_n^{(ij)} (1-x/a)^{n-1/2} \tag{C2.2.3}$$

with
$$D_0^{(11)} = D_0^{(22)} = 1, \quad D_0^{(12)} = D_0^{(21)} = 0 \ . \tag{C2.2.4}$$

The coefficients of the weight functions can be obtained from the stress intensity factors for pairs of normal or shear forces, Fig. C2.2a, as described by eq(A3.1.7-A3.1.9) [C6].

The finite element results for concentrated normal and shear forces were fitted according to eq.(C2.2.3) using terms with $n=1$ and 2, exclusively. The coefficients are compiled in Table C2.2. At small angles of $\varphi \leq 30°$, the mixed weight functions can be approximated by

$$h_{12} \cong -\sqrt{\frac{2}{\pi a}}\left(0.942\sqrt{1-x/a} - 0.411(1-x/a)^{3/2}\right)\varphi \qquad (C2.2.5)$$

$$h_{21} \cong -\sqrt{\frac{2}{\pi a}}\left(0.689\sqrt{1-x/a} + 1.2501(1-x/a)^{3/2}\right)\varphi \qquad (C2.2.6)$$

with φ in radian.

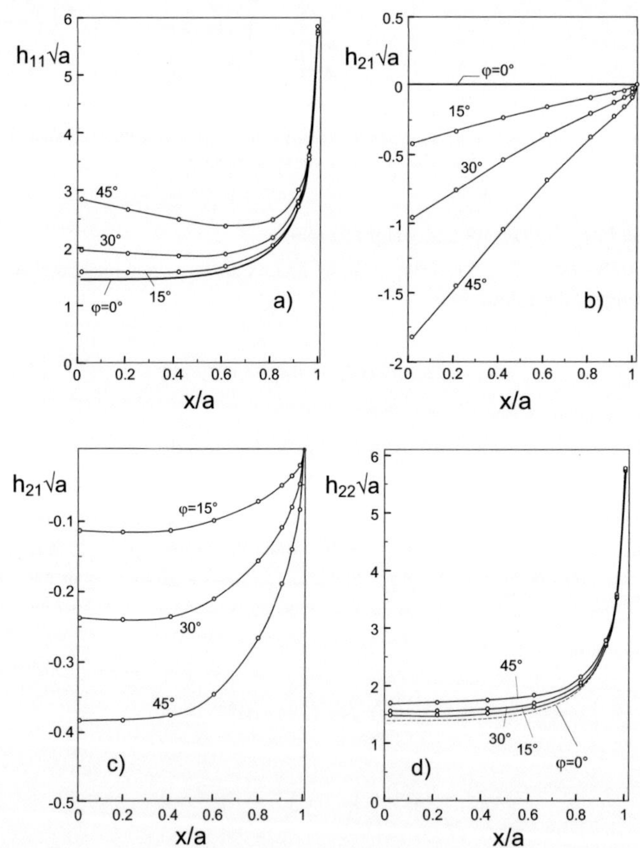

Fig. C2.3 Weight functions from stress intensity factors for pairs of concentrated forces, a) and b) for normal forces P, c) and d) for shear Q.

Table C2.2 Coefficients for approximate weight functions of stress intensity factors, eq.(C2.2.3).

$\varphi\,(°)$	$D_1^{(11)}$	$D_2^{(11)}$	$D_1^{(12)}$	$D_2^{(12)}$	$D_1^{(21)}$	$D_2^{(21)}$	$D_1^{(22)}$	$D_2^{(22)}$
0	0.568	0.283	0	0	0	0	0.568	0.283
15	0.678	0.306	-0.233	0.094	-0.179	-0.351	0.607	0.276
30	1.051	0.413	-0.500	0.222	-0.399	-0.807	0.756	0.210
45	1.880	0.679	-0.846	0.421	-0.737	-1.576	1.061	0.022
60	3.787	1.600	-1.383	0.783	-1.461	-3.308	1.727	-0.253
75	10.32	6.574	-2.557	1.702	-4.270	-10.03	3.600	-1.382

C2.2.2 T-stress and Green's functions

For the most general case, the integral representation must read

$$T = \int_0^a t^{(1)}(x,a)\,\sigma_n\,dx + \int_0^a t^{(2)}(x,a)\,\tau_{xy}\,dx - \sigma_y\big|_{x=a} + \sigma_x\big|_{x=a} \tag{C2.2.7}$$

Green's functions $t^{(1)}$ and $t^{(2)}$ can be determined as the T-term for a pair of concentrated forces P and Q acting normal and parallel to the crack face. They can be expressed by

$$t^{(1)} = \frac{1}{a}\sum_{n=1}^{\infty} C_n^{(1)}(1 - x/a)^{(2n-1)/2} \tag{C2.2.8}$$

$$t^{(2)} = \frac{1}{a}\sum_{n=1}^{\infty} C_n^{(2)}(1 - x/a)^{(2n-1)/2} \tag{C2.2.9}$$

Finite element results are shown in Fig. C2.4. From the T-stresses for the concentrated forces P and Q, the Green's functions were obtained. A 3-terms fit of the data with respect to eqs.(C2.2.8) and (C2.2.9) yields the coefficients compiled in Table C2.3.

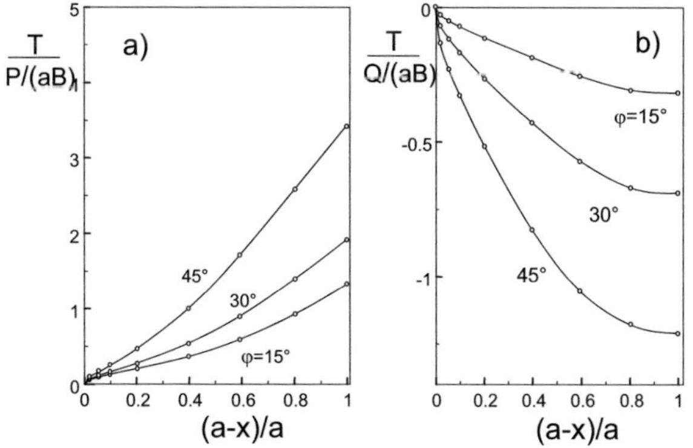

Fig. C2.4 Green's functions for T-stress under concentrated forces P and Q.

Table C2.3 Coefficients for weight functions of T-stress, eqs.(C2.2.8) and (C2.2.9).

φ (°)	$C_1^{(1)}$	$C_2^{(1)}$	$C_3^{(1)}$	$C_1^{(2)}$	$C_2^{(2)}$	$C_3^{(2)}$
0	0.345	0.087	0.733	0	0	0
15	0.346	0.279	0.701	-0.185	-0.413	0.277
30	0.381	0.956	0.582	-0.446	-0.861	0.613
45	0.461	2.882	0.085	-0.911	-1.422	1.126

C2.2.3 Stress intensity factor for remote stresses

By applying of the weight functions given before, the stress intensity factors and the T-stress term were computed for a remote stress in η-direction, σ_η, and for a constant stress in ξ-direction, σ_ξ. The normal and shear tractions to be used in eqs.(C2.2.1), (C2.2.2), and (C2.2.7) are

$$\sigma_n = \sigma_\xi \sin^2 \varphi + \sigma_y \cos^2 \varphi \tag{C2.2.10}$$

$$\tau = (\sigma_y - \sigma_\xi)\sin \varphi \cos \varphi \tag{C2.2.11}$$

Figure C2.5 shows the geometric functions for the stress intensity factors defined by

$$K_\mathrm{I} = \sigma F_\mathrm{I}\sqrt{\pi a}, \quad K_\mathrm{II} = \sigma F_\mathrm{II}\sqrt{\pi a} \tag{C2.2.12}$$

In Fig. C2.5 the squares represent results obtained with the weight function method. The circles indicate FE results.

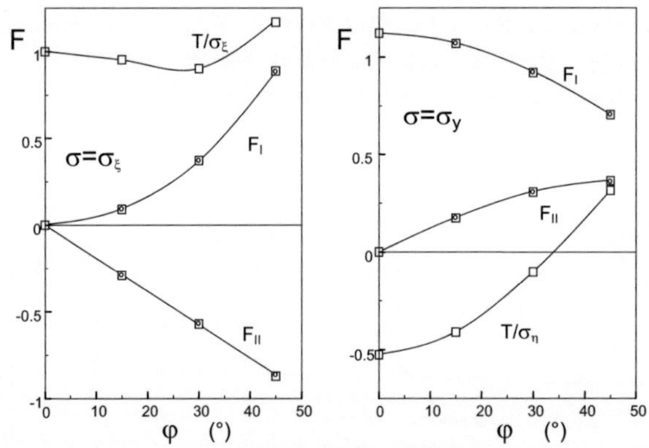

Fig. C2.5 Mixed-mode stress intensity factors and T stress; a) constant stresses in ξ-direction, b) remote stresses in η-direction. Squares: Weight function method, circles: Finite element results.

For small angles φ, it holds

$$F_\mathrm{I} = 1.12152 - 0.768\varphi^2 + O(\varphi^4) \tag{C2.2.13}$$

$$F_{\mathrm{II}} = 0.6905\,\varphi + O(\varphi^3) \qquad\qquad (\mathrm{C2.2.14})$$

resulting in the stress intensity factor ratio

$$\frac{K_{\mathrm{II}}}{K_{\mathrm{I}}} = \frac{F_{\mathrm{II}}}{F_{\mathrm{I}}} \cong 0.6156\,\varphi \qquad\qquad (\mathrm{C2.2.15})$$

with φ in radian.

C2.3 Kink edge crack

The edge crack with a kink is illustrated in Fig. C2.6. Stress intensity factors and T-stress were computed for constant stresses in y- and x- direction and for a constant pressure p on the crack faces [C2.7]. The results are compiled in Tables C2.4-C2.6 and Fig. C2.7. The results of Table C2.6 also reflect the well-known feature that the fracture mechanics parameters for crack-face pressure are identical with the sum of parameters under $\sigma_x=\sigma_y$ loading.

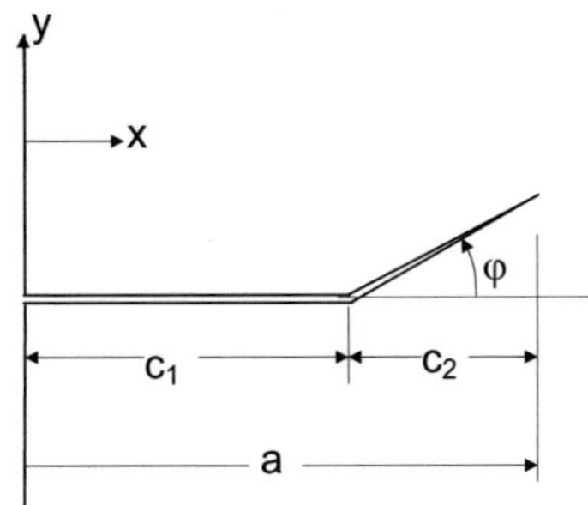

Fig. C2.6 Kinked edge crack.

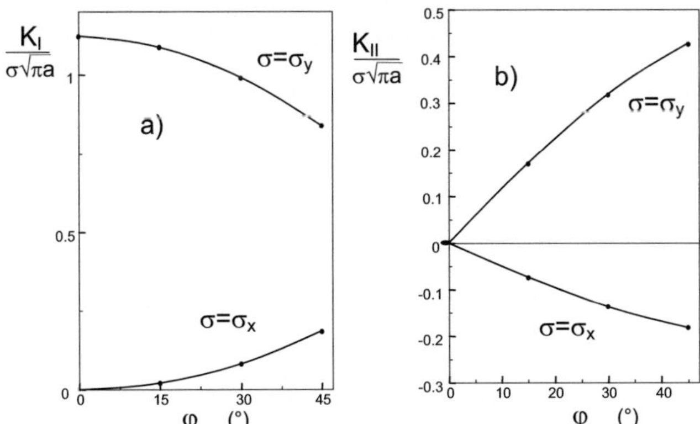

Fig. C2.7 Stress intensity factors for $c_1/a=0.9$ under loading in x and y- direction.

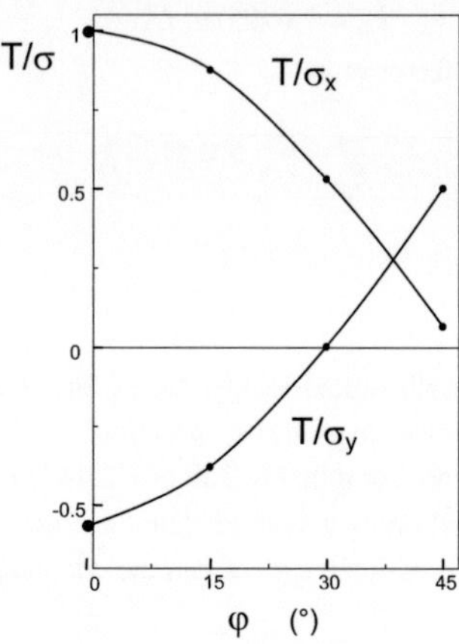

Fig. C2.8 T-stress for $c_1/a=0.9$ under loading in x and y- direction.

Table C2.4 Stress intensity factors and T-stress under constant x-stresses from [C2.7].

c_1/a	c_2/a	φ (°)	$K_I/\sigma_x\sqrt{a\pi}$	$K_{II}/\sigma_x\sqrt{a\pi}$	T/σ_x
		0	0	0	1
0	1	15	0.0928	-0.296	0.9545
0	1	30	0.400	-0.613	0.9026
0	1	45	1.056	-1.036	1.169
0.9	0.1	15	0.0203	-0.0742	0.8732
0.9	0.1	30	0.0810	-0.1380	0.5284
0.9	0.1	45	0.1838	-0.1831	0.0636
0.95	0.05	15	0.0140	-0.0519	0.8730
0.95	0.05	30	0.0559	-0.0964	0.5284
0.95	0.05	45	0.1258	-0.1273	0.0628
0.97	0.03	15	0.0108	-0.0401	0.8731
0.97	0.03	30	0.0428	-0.0744	0.5288
0.97	0.03	45	0.0961	-0.0980	0.0643
1	0	15	0	0	
1	0	30	0	0	
1	0	45	0	0	

In order to check the accuracy, the finite element results were compared with the highly precise results of Noda and Oda [C2.8] obtained by using the body force method. In Table C2.7, the results from [C2.7] are compared with those of [C2.8] for c_1/a=0.1 and loading in y-direction. Maximum deviations are less than 0.25%.

Table C2.5 Stress intensity factors and T-stress under remote y-stresses.

c_1/a	c_2/a	φ (°)	$K_I / \sigma_y \sqrt{a\pi}$	$K_{II} / \sigma_y \sqrt{a\pi}$	T / σ_y
		0	1.1215	0	-0.526
0	1	15	1.088	0.177	-0.526
0	1	30	0.989	0.329	-0.411
0	1	45	0.838	0.434	-0.1013
0.9	0.1	15	1.087	0.1696	-0.3805
0.9	0.1	30	0.989	0.3172	0.0002
0.9	0.1	45	0.838	0.4255	0.4985
0.95	0.05	15	1.061	0.1625	-0.3506
0.95	0.05	30	0.967	0.304	0.1132
0.95	0.05	45	0.824	0.4044	0.6999
0.97	0.03	15	1.050	0.159	-0.4415
0.97	0.03	30	0.960	0.294	0.2309
0.97	0.03	45	0.820	0.3918	0.9231
1	0	15	1.093	0.1437	
1	0	30	1.011	0.2695	
1	0	45	0.887	0.3626	

Table C2.6 Stress intensity factors and T-stress under constant internal pressure p.

c_1/a	c_2/a	φ (°)	$K_I / p\sqrt{a\pi}$	$K_{II} / p\sqrt{a\pi}$	T / p
		0	1.1215	0	0.474
0	1	15	1.179	-0.119	0.544
0	1	30	1.387	-0.285	0.804
0	1	45	1.893	-0.602	1.484
0.9	0.1	15	1.108	0.0951	0.4926
0.9	0.1	30	1.070	0.1792	0.5309
0.9	0.1	45	1.022	0.2424	0.5616
0.95	0.05	15	1.075	0.1086	0.5226
0.95	0.05	30	1.023	0.2045	0.6418
0.95	0.05	45	0.950	0.2771	0.7630
1	0	15	1.093	0.1437	
1	0	30	1.011	0.2695	
1	0	45	0.887	0.3626	

Table C2.7 Comparison of stress intensity factors from [C2.7] with data of Noda and Oda [C2.8] for $c_2/a=0.1$.

φ (°)	$K_I/\sigma_y\sqrt{a\pi}$ [C2.7]	$K_{II}/\sigma_y\sqrt{a\pi}$ [C2.7]	$K_I/\sigma_y\sqrt{a\pi}$ [C2.8]	$K_{II}/\sigma_y\sqrt{a\pi}$ [C2.8]
15	1.087	0.1693	1.087	0.170
30	0.989	0.3172	0.990	0.317
45	0.839	0.4255	0.841	0.426

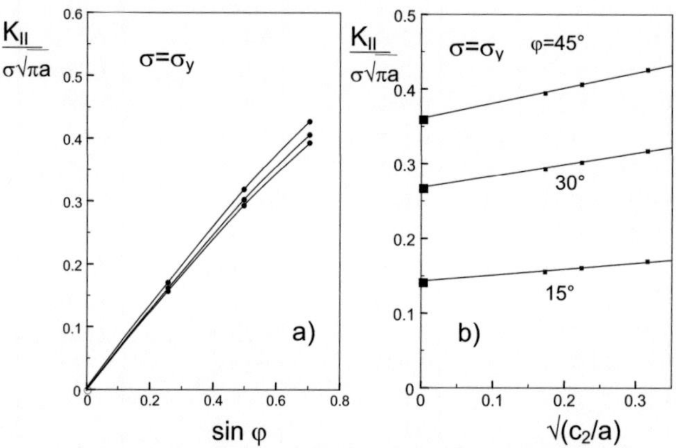

Fig. C2.9 Influence of kink length and angle on the mode-II stress intensity factor for $\sigma=\sigma_y$.

In Fig. C2.9 the mode-II stress intensity factors for remote stress in y-direction are plotted as a function of the sine of the angle φ and the square root of the parameter c_2/a. Figure C2.10 represents the data for constant stress in x-direction and Fig. C2.11 for constant crack-face pressure p. The values for $c_2/a=0$ in Fig. C2.9 were obtained from the limit case of a small kink crack ahead of a semi-infinite crack in an infinite body that is loaded by a mode-I contribution, exclusively [C2.9].

For small angles φ and $c_2 \ll a$, the data shown in Figs. C2.9-C2.11 can be expressed by the following approximations:

$\sigma=\sigma_y$:

$$\frac{K_{II}}{\sigma_y\sqrt{\pi a}} \cong 1.1215\sin(\varphi/2)\cos^2(\varphi/2)+0.2168\sin\tfrac{3}{2}\varphi\sqrt{\frac{c_2}{a}} \qquad (C2.3.1)$$

$\sigma=\sigma_x$:

$$\frac{K_{II}}{\sigma_x\sqrt{\pi a}} \cong -\frac{\sqrt{8}}{\pi}\sin\varphi\cos^{1/2}\varphi\sqrt{c_2/a}-0.204\frac{c_2}{a}\varphi^2 \qquad (C2.3.2)$$

crack-face pressure p:

$$\frac{K_{II}}{p\sqrt{\pi a}} \cong 1.1215\sin(\varphi/2)\cos^2(\varphi/2) - 0.399\sin\tfrac{3}{2}\varphi\sqrt{\frac{c_2}{a}} \qquad (C2.3.3)$$

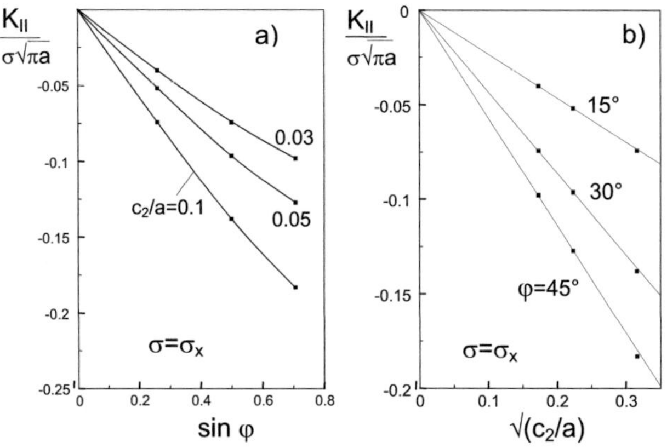

Fig. C2.10 Influence of kink length and angle on the mode-II stress intensity factor for $\sigma=\sigma_x$.

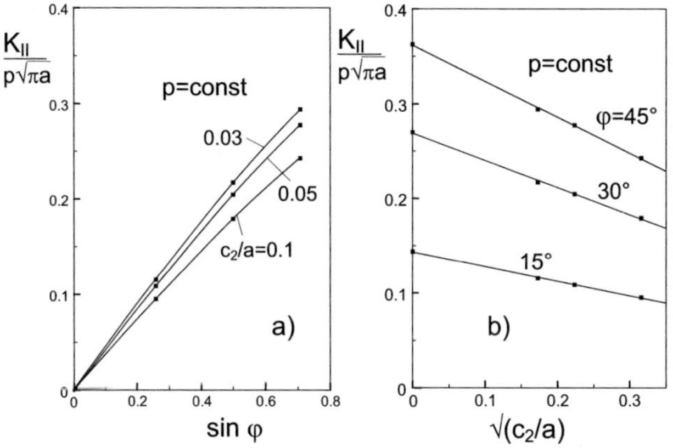

Fig. C2.11 Influence of kink length and angle on the mode-II stress intensity factor at crack-face pressure p.

References C2

[C2.1] Fett, T., Munz, D., Stress intensity factors and weight functions, Computational Mechanics Publications, Southampton, 1997.

[C2.2] Wigglesworth, L.A., Stress distribution in a notched plate, Mathematica 4(1957), 76-96.

[C2.3] Rice, J.R., Some remarks on elastic crack-tip stress fields, Int. J. Solids and Structures **8**(1972), 751-758.

[C2.4] Mathematica 3.0, Wolfram Research Inc., USA.

[C2.5] Fett, T., Rizzi, G., Bahr , H.A., Bahr, U., Pham, V.B., Balke, H., A general weight function approach to compute mode-II stress intensity factors and crack paths for slightly curved or kinked cracks in finite bodies, to be published in Engng. Fract. Mech.

[C2.6] Fett, T., Rizzi, G., Weight functions for stress intensity factors and T-stress for oblique cracks in a half-space, Int. J. Fract. **132**(2005), L9-L16.

[C2.7] Fett, T., Rizzi, G., T-stress solutions determined by finite element computations, Report FZKA 6937, Forschungszentrum Karlsruhe, 2004, Karlsruhe.

[C2.8] Noda, N.A., Oda, K., Numerical solutions of the singular integral equations in the crack analysis using the body force method, Int. J. Fract. **58**(1992), 285-304.

[C2.9] Isida, M., Nishino, T., Formulae of stress intensity factor of bent cracks in plane problems, Trans. Japan. Soc. Mech. Engrs. **48/430**(1982), 729-738.

C3

Semi-infinite kink crack

C3.1 Stress intensity factors and weight functions

C3.1.1 Approximate stress intensity factors from the Cotterell and Rice analysis

During spontaneous failure or during subcritical crack growth under mixed-mode loading, an abrupt change of the initial crack plane occurs. The behaviour of such kinked cracks was discussed early in terms of the crack-tip stress field by Lawn and Wilshaw [C3.1] and in terms of estimated stress intensity factors and T-stress by Cotterell and Rice [C3.2].

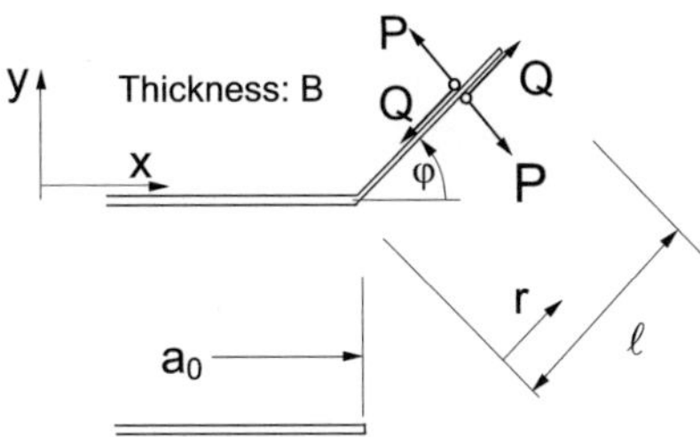

Fig. C3.1 Kink crack with kink length ℓ, and kink angle φ, loaded by point forces P and Q.

A straight crack of length a_0 is shown in Fig. C3.1. A kink of length ℓ with a sudden change of the original crack direction by an angle φ is assumed at its end. Following the analysis by Cotterell and Rice [C3.2], the local mixed-mode stress intensity factors $K_I(\ell)$ and $K_{II}(\ell)$ representing the singular stresses ahead of the kink can be computed from the singular stress field produced ahead of the original (unkinked) crack. Taking into consideration the singular stress term and the first regular term, the near-tip stress field caused by the original crack of length a can be described by

$$\sigma_\varphi = \frac{K_I(a_0)}{\sqrt{2r\pi}}g_{11} + \frac{K_{II}(a_0)}{\sqrt{2r\pi}}g_{12} + T\sin^2\varphi \tag{C3.1.1}$$

$$\tau_{r\varphi} = \frac{K_I(a_0)}{\sqrt{2r\pi}}g_{21} + \frac{K_{II}(a_0)}{\sqrt{2r\pi}}g_{22} - T\sin\varphi\cos\varphi \tag{C3.1.2}$$

with the distance r from the tip of the original crack and the angular functions

$$g_{11} = \cos^3(\varphi/2) \tag{C3.1.3}$$

$$g_{12} = -3\sin(\varphi/2)\cos^2(\varphi/2) \tag{C3.1.4}$$

$$g_{21} = \sin(\varphi/2)\cos^2(\varphi/2) \qquad (C3.1.5)$$

$$g_{22} = \cos(\varphi/2)(1 - 3\sin^2(\varphi/2)) \qquad (C3.1.6)$$

In (C3.1.1) and (C3.1.2) $K_I(a_0)$ and $K_{II}(a_0)$ are the stress intensity factors of the original (un-kinked) crack. By considering these stresses on the projective plane of the kink as the stresses in the "uncracked component" and regarding the kink of length ℓ as the "crack", the weight function procedure provides the stress intensity factors $K_I(\ell)$ and $K_{II}(\ell)$ related to the tip of the kink.

This technique was applied by Cotterell and Rice [C3.2] for the case of the simplified asymptotic weight function

$$h_{11} = h_{22} = \sqrt{\frac{2}{\pi r}} \ . \qquad (C3.1.7)$$

ignoring mixed weight function terms, i.e. for $h_{12} = h_{21} = 0$.

In this approximation, the stress intensity factors at the tip of the kink can be written as

$$K_I(\ell) = K_I(a_0)\,g_{11} + K_{II}(a_0)\,g_{12} + b_1 T\sqrt{\ell} \qquad (C3.1.8)$$

$$K_{II}(\ell) = K_I(a_0)\,g_{21} + K_{II}(a_0)\,g_{22} + b_2 T\sqrt{\ell} \qquad (C3.1.9)$$

with the angular functions

$$b_1 = \sqrt{\frac{8}{\pi}}\sin^2\varphi \qquad (C3.1.10)$$

$$b_2 = -\sqrt{\frac{8}{\pi}}\sin\varphi\cos\varphi \qquad (C3.1.11)$$

Highly accurate stress intensity factors for kinked cracks were reported by Bilby et al. [C3.3], Hayashi and Nemath-Nasser [C3.4], Lo [C3.5], and Isida and Nishino [C3.6]. These results showed deviations from eqs.(C3.1.8) and (C3.1.9), especially for the terms with g_{21} and g_{22}.

C3.1.2 Weight function procedure including higher-order weight function terms

The stress intensity factors $K(\ell)$ for the kinked cracks were computed in [C3.7] by the weight function method as

$$K_I(\ell) = \int_0^\ell \sigma_\varphi h_{11}(r,\ell)\,dr + \int_0^\ell \tau_{r\varphi} h_{12}(r,\ell)\,dr \qquad (C3.1.12)$$

$$K_{II}(\ell) = \int_0^\ell \sigma_\varphi h_{21}(r,\ell)\,dr + \int_0^\ell \tau_{r\varphi} h_{22}(r,\ell)\,dr \qquad (C3.1.13)$$

with the weight functions $h(r,\ell)$ as defined in [C3.8]

$$h_{ij} = \sqrt{\frac{2}{\pi\ell}} \sum_{n=0}^{\infty} D_n^{(ij)} (1 - r/\ell)^{n-1/2} \tag{C3.1.14}$$

with
$$D_0^{(11)} = D_0^{(22)} = 1, \quad D_0^{(12)} = D_0^{(21)} = 0 \tag{C3.1.15}$$

The stress intensity factors at the tip of the kink can be written as

$$K_{\mathrm{I}}(\ell) = K_{\mathrm{I}}(a_0)C_{11} + K_{\mathrm{II}}(a_0)C_{12} \tag{C3.1.16}$$

$$K_{\mathrm{II}}(\ell) = K_{\mathrm{I}}(a_0)C_{21} + K_{\mathrm{II}}(a_0)C_{22} \tag{C3.1.17}$$

with the coefficients

$$C_{11} = \frac{g_{11}}{\sqrt{\pi}} \sum_{n=0}^{\infty} D_n^{(11)} \frac{\Gamma(n+1/2)}{\Gamma(n+1)} + \frac{g_{21}}{\sqrt{\pi}} \sum_{n=0}^{\infty} D_n^{(12)} \frac{\Gamma(n+1/2)}{\Gamma(n+1)} \tag{C3.1.18}$$

$$C_{12} = \frac{g_{12}}{\sqrt{\pi}} \sum_{n=0}^{\infty} D_n^{(11)} \frac{\Gamma(n+1/2)}{\Gamma(n+1)} + \frac{g_{22}}{\sqrt{\pi}} \sum_{n=0}^{\infty} D_n^{(12)} \frac{\Gamma(n+1/2)}{\Gamma(n+1)} \tag{C3.1.19}$$

$$C_{21} = \frac{g_{11}}{\sqrt{\pi}} \sum_{n=0}^{\infty} D_n^{(21)} \frac{\Gamma(n+1/2)}{\Gamma(n+1)} + \frac{g_{21}}{\sqrt{\pi}} \sum_{n=0}^{\infty} D_n^{(22)} \frac{\Gamma(n+1/2)}{\Gamma(n+1)} \tag{C3.1.20}$$

$$C_{22} = \frac{g_{12}}{\sqrt{\pi}} \sum_{n=0}^{\infty} D_n^{(21)} \frac{\Gamma(n+1/2)}{\Gamma(n+1)} + \frac{g_{22}}{\sqrt{\pi}} \sum_{n=0}^{\infty} D_n^{(22)} \frac{\Gamma(n+1/2)}{\Gamma(n+1)} \tag{C3.1.21}$$

(Γ=Gamma function). The application of these improved relations of course needs knowledge of the coefficients $D_n^{(ij)}$. A number of coefficients were reported in [C3.7].

For this purpose, finite element computations were performed for point forces P and Q at the kink at variable distance r/ℓ from the tip of the kink crack (Fig. C3.1). Figure C3.2 shows the so obtained weight functions as the symbols.

The coefficients $D_n^{(ij)}$ were determined by application of a fit procedure. They are compiled in Tables C3.1 and C3.2.

Table C3.1 Coefficients for the weight functions h_{11} and h_{22}.

β	$D_1^{(11)}$	$D_2^{(11)}$	$D_3^{(11)}$	$D_1^{(22)}$	$D_2^{(22)}$	$D_3^{(22)}$
30°	0.002757	0.001539	-0.000088	0.03657	0.00724	0.01350
60°	0.045588	0.026197	-0.004604	0.13621	0.04521	0.02746
90°	0.236603	0.167142	-0.059586	0.28997	0.18191	-0.04287

Table C3.2 Coefficients for the mixed weight function terms h_{12} and h_{21}.

β	$D_1^{(12)}$	$D_2^{(12)}$	$D_3^{(12)}$	$D_1^{(21)}$	$D_2^{(21)}$	$D_3^{(21)}$
30°	-0.00893	-0.00347	0.001890	-0.009522	-0.00222	-0.00714
60°	-0.06765	-0.02787	0.021586	-0.07466	-0.01029	-0.06851
90°	-0.21696	-0.07370	0.084911	-0.22917	-0.11355	-0.18260

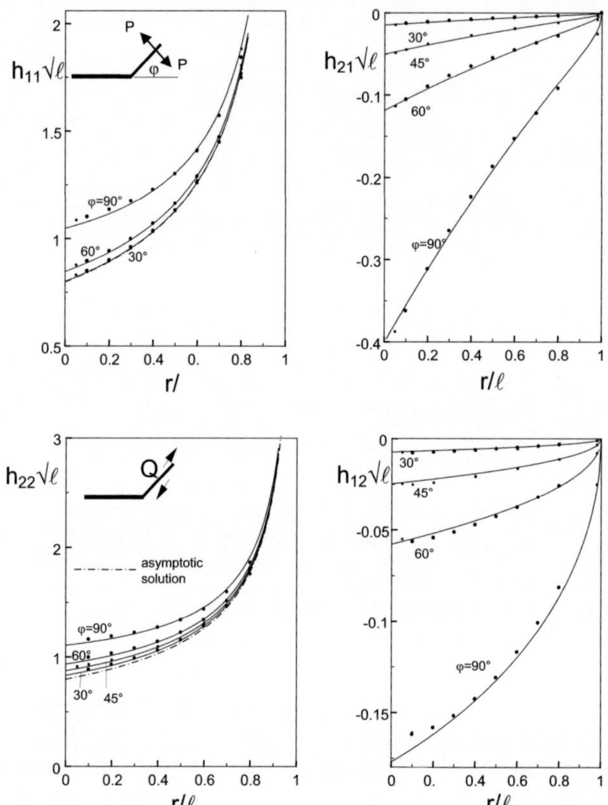

Fig. C3.2 Approximate weight functions eqs.(C3.1.22-C3.1.25) (curves) compared with finite element results (circles: data from [C3.7], squares: unpublished finite element results).

Approximate weight functions can be sufficiently expressed by use of the first series terms exclusively as given by eqs.(C3.1.8) and (C3.1.9). The next higher approximation is

$$h_{11} = \sqrt{\frac{2}{\pi\ell}}\left(\frac{1}{\sqrt{1-r/\ell}} + D_1^{(11)}\sqrt{1-r/\ell}\right) \qquad (C3.1.22)$$

$$h_{22} = \sqrt{\frac{2}{\pi\ell}}\left(\frac{1}{\sqrt{1-r/\ell}} + D_1^{(22)}\sqrt{1-r/\ell}\right) \qquad (C3.1.23)$$

$$h_{21} = \sqrt{\frac{2}{\pi\ell}}\left(D_1^{(21)}\sqrt{1-r/\ell} + D_2^{(21)}(1-r/\ell)^{3/2}\right) \qquad (C3.1.24)$$

$$h_{12} = \sqrt{\frac{2}{\pi\ell}}\left(D_1^{(12)}\sqrt{1-r/\ell} + D_2^{(12)}(1-r/\ell)^{3/2}\right) \qquad (C3.1.25)$$

The coefficients $D_n^{(ij)}$ are functions of the kink angle exclusively and can be fitted from data reported in [C3.7] as

$$D_1^{(11)} \cong 0.05135\,\varphi^4 \,, \quad D_1^{(22)} \cong 0.1562\,\varphi^2$$

$$D_1^{(21)} \cong -0.04523\,\varphi^3 \,, \quad D_2^{(21)} \cong -0.08449\,\varphi^3 \qquad (C3.1.26)$$

$$D_1^{(12)} \cong -0.064\,\varphi^3 \,, \quad D_2^{(12)} \cong 0.000705\,\varphi^8$$

with φ to be inserted in radian. These weight function approximations are shown in Fig. C3.2 as the solid curves together with the numerical results from [C3.7] (circles) and additional finite element results (squares).

C3.1.3 Computation of the coefficients C_{ij}

Using the weight functions, the mixed-mode stress intensity factors $K_I(\ell)$ and $K_{II}(\ell)$ were determined. From eqs.(C3.1.18-C3.1.21), the coefficients C_{11}, C_{12}, C_{21}, and C_{22} were obtained. Table C3.3 shows these coefficients for different kink angles. These data are introduced in Fig. C3.3 as circles. An excellent agreement with the numerical results from literature is evident. These results indicate that the higher-order weight function terms missing in [C3.2] are responsible for the differences between the Cotterell-Rice approximation and the exact solution.

For $0\leq\varphi\leq90°$ the dashed curves in Fig. C3.3 can be described by the simple relation of

$$C_{12} \cong g_{12} - \tfrac{1}{3}\sin^3(\tfrac{1}{2}\varphi), \quad C_{22} \cong g_{22} + \tfrac{1}{6}\sin^2(\tfrac{4}{5}\varphi) \qquad (C3.1.27)$$

A more increased accuracy can be reached by use of the solutions reported in [C3.9].

Table C3.3 Coefficients for the stress intensity factors $K_I(\ell)$ and $K_{II}(\ell)$.

φ	C_{11}	C_{22}	C_{21}	C_{21}
0°	1	0	1	0
30°	0.901708	-0.72983	0.24052	0.79694
60°	0.655697	-1.16820	0.36946	0.30721
90°	0.372209	-1.19417	0.34845	-0.19696

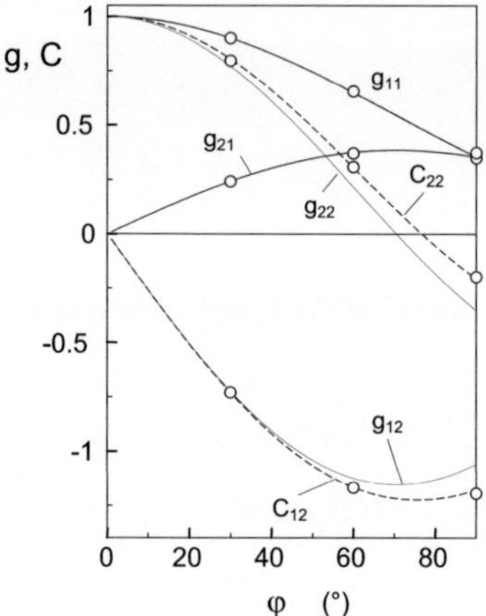

Fig. C3.3 Stress intensity factor for a kink crack, as obtained from the weight function (circles) and compared with the approximation by Cotterell and Rice [C3.2] (solid curves). Dashed curves: representation by eq.(C3.1.27).

C3.1.4 Stress intensity factors caused by the T-stress

Knowledge of the weight function allows to compute the stress intensity factor contributions caused by the T-stress term, T, the first regular stress term. By use of the normal and shear stress components

$$\sigma_n = T \sin^2 \varphi , \quad \tau_{r\varphi} = -T \sin \varphi \cos \varphi \tag{C3.1.28}$$

the stress intensity factors result from (C3.1.12) and (C3.1.13) as

$$K_{\mathrm{I}}(\ell) = b_1 T \sqrt{\ell} \ , \quad K_{\mathrm{II}}(\ell) = b_2 T \sqrt{\ell} \tag{C3.1.29}$$

with the coefficients defined by

$$b_1 = \sqrt{\frac{2}{\pi}} \left[\sin^2 \varphi \sum_{n=0}^{\infty} \frac{D_n^{(11)}}{n+\frac{1}{2}} - \sin \varphi \cos \varphi \sum_{n=0}^{\infty} \frac{D_n^{(12)}}{n+\frac{1}{2}} \right] \tag{C3.1.30}$$

$$b_2 = \sqrt{\frac{2}{\pi}} \left[- \sin \varphi \cos \varphi \sum_{n=0}^{\infty} \frac{D_n^{(22)}}{n+\frac{1}{2}} + \sin^2 \varphi \sum_{n=0}^{\infty} \frac{D_n^{(21)}}{n+\frac{1}{2}} \right] \tag{C3.1.31}$$

Neglecting all coefficients with $n>0$ leads to the approximations reported by Cotterell and Rice [C3.2]

$$b_1 \cong \sqrt{\frac{8}{\pi}} \sin^2 \varphi \qquad (C3.1.32)$$

$$b_2 \cong -\sqrt{\frac{8}{\pi}} \sin \varphi \cos \varphi \qquad (C3.1.33)$$

Evaluation of eqs.(C3.1.30) and (C3.1.31) yields the coefficients b_1 and b_2 compiled in the second and third columns of Table C3.4. Columns 4 and 5 contain data as taken from a diagram in [C3.10] (where b_1 and b_2 are computed for a negative kink direction, i.e. for $-\varphi$) and columns 6 and 7 present the approximate solution according to eqs.(C3.1.32) and (C3.1.33). Taking into account the limited accuracy for extracting the data from the plot in [C3.10], agreement of these data with the results from (C3.1.30) and (C3.1.31) is good.

Table C3.4 Coefficients for the stress intensity factors $K_I(\ell)$ and $K_{II}(\ell)$ due to the T-stress, eq. (C3.1.29).

φ	b_1	b_2	b_1	b_2	b_1	b_2
30°	0.402	-0.704	0.46	-0.70	0.399	-0.691
60°	1.238	-0.737	1.30	-0.73	1.197	-0.691
90°	1.761	-0.200	1.80	-0.17	1.596	0

C3.2 T-stress and Green's function

A kink crack ahead of a semi-infinite crack was modelled in [C3.11] by a crack of length $a=450\times\ell$ in a plate of height $900\times\ell$ and width $900\times\ell$. The T-stresses for concentrated forces (see Fig. C3.1) are plotted in Fig. C3.4. The data were fitted according to

$$t^{(1)} = \frac{1}{\ell}\sum_{n=1}^{N} D_n^{(1)}(1-r/\ell)^{(2n-1)/2} \ , \quad t^{(2)} = \frac{1}{\ell}\sum_{n=1}^{N} D_n^{(2)}(1-r/\ell)^{(2n-1)/2} \qquad (C3.2.1)$$

for $N=5$ and 6 terms with the coefficients compiled in Tables C3.5 and C3.6.

Table C3.5 Coefficients for eq.(C3.2.1) as obtained under normal forces P (forked crack under symmetrical load).

φ	$D_1^{(1)}$	$D_2^{(1)}$	$D_3^{(1)}$	$D_4^{(1)}$	$D_5^{(1)}$
30°	0.1096	-0.5694	-4.4385	7.9583	-3.5310
45°	-0.0662	-0.1610	-1.177	1.0680	-0.0130
60°	-0.057	-0.190	0.3400	-0.9839	0.7213
90°	0.0982	0.2950	-0.5274	1.1234	-0.4927

Table C3.6 Coefficients for eq.(C3.2.1) as obtained under shear forces Q (forked crack under symmetrical load).

φ	$D_1^{(2)}$	$D_2^{(2)}$	$D_3^{(2)}$	$D_4^{(2)}$	$D_5^{(2)}$	$D_6^{(2)}$
15°	-0.1355	-6.5790	45.026	-89.694	75.588	-23.430
30°	-0.1301	-1.3333	3.9832	-2.8202	0.4441	-
45°	-0.2310	-0.0820	-1.0491	2.4794	-1.2822	-
60°	-0.2877	-0.1457	-0.2574	0.2991	0.0004	-
90°	-0.3821	-0.2418	-0.0300	-0.2447	0.1720	-

Considering the conditions of symmetry and anti-symmetry, a fit procedure for $N=3$ terms yields the coefficients of

$$D_n^{(1)} = A_n^{(1)}\varphi^2 + B_n^{(1)}\varphi^4 \tag{C3.2.2}$$

$$D_n^{(2)} = A_n^{(2)}\varphi + B_n^{(2)}\varphi^3 \tag{C3.2.3}$$

with the numbers $A_n^{(i)}, B_n^{(i)}$ compiled in Table C3.7 (for φ in radian).

Table C3.7 Coefficients for eqs.(C3.2.2) and (C3.2.3).

n	$A_n^{(1)}$	$B_n^{(1)}$	$A_n^{(2)}$	$B_n^{(2)}$
1	0.0742	-0.00462	-0.1223	-0.02335
2	-0.1277	0.0531	0.1144	-0.0995
3	0.1763	-0.00651	-0.2288	0.07505

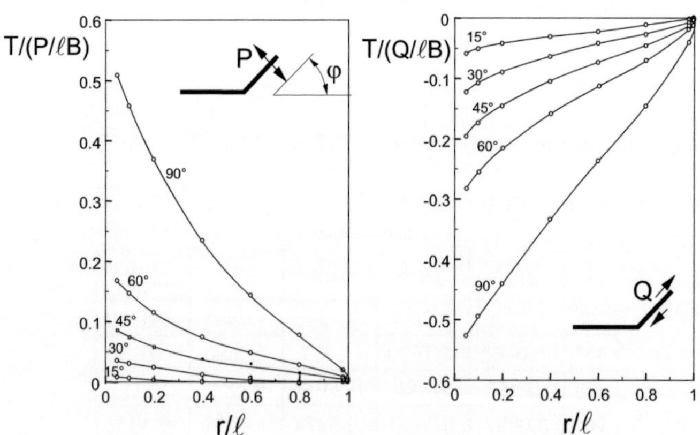

Fig. C3.4 Green's functions for the T-stress of kinked cracks.

References C3:

[C3.1] Lawn, B. and Wilshaw, T.R., Fracture of Brittle Solids, Cambridge University Press (1975).

[C3.2] Cotterell, B. and Rice, J.R., Some remarks on elastic crack-tip stress fields, Int. J. Fract. **16**(1980), 155-169.

[C3.3] Bilby, B.A., Cardew, G.E., and Howard, I.C., Stress intensity factors at the tip of kinked and forked cracks, in Fracture 1977, Vol.3, University of Waterloo Press, 197.

[C3.4] Hayashi, K., Nemath-Nasser, S., Energy-release rate and crack kinking under combined loading, J. Appl. Mech. **48**(1981), 520-524.

[C3.5] Lo, K.K., Analysis of branched cracks, J. Appl. Mech. **45**(1978), 797-802.

[C3.6] Isida, M., Nishino, T., Formulae of stress intensity factors of bent cracks in plane problems, Trans. Japan Soc. Mech. Engrs. **48-430**(1982), 729-738.

[C3.7] Fett, T., Pham, V.B., Bahr, H.A., Weight functions for kinked semi-infinite cracks, Engng. Fract. Mech., **71**(2004), 1987-1995.

[C3.8] Fett, T., Munz, D. Stress Intensity Factors and Weight Functions, Computational Mechanics Publications, Southampton, UK, 1997

[C3.9] Kageyama, K., Okamura, H., Elastic analysis of infinitesimally kinked crack under tension and transverse shear, Trans. Japan. Soc. Mech. Engnrs., **48**-430(1982), 783-791.

[C3.10] He, M.Y., Bartlett, A., Evans, A.G., Hutchinson, J.W., Kinking of a crack out of an interface: role of in-plane stress, J. Am. Ceram. Soc. **74**(1991), 767-771.

[C3.11] Fett, T., Rizzi, G., Bahr, H.A., Green's functions for the T-stress of small kink and fork cracks, Engng. Fract. Mech. 73(2006), 1426-1435.

C4

Semi-infinite fork cracks

C4.1 Stress intensity factors and weight functions

Forked cracks occur for instance in thermal shock problems at the moment of crack arrest after an extended phase of spontaneous crack extension. This is the case in the centre region of thermally shocked circular disks. Also in fatigue experiments on metals branching takes place (see e.g. [C4.1]). Such a crack is illustrated in Fig. C4.1a together with the initially straight crack of length a_0 present at the moment before forking. Highly precise stress intensity factor solutions are available for the special case of remote tractions (e.g. [C4.2, C4.3]). For locally varying stress distributions along the kink (e.g. in presence of crack bridging effects), stress intensity factor computation needs knowledge of the weight function. The following sections provide FE solutions.

C4.1.1 Loading on one branch

Figure C4.1 illustrates the case of the upper branch loaded by pairs of concentrated forces. The weight functions under normal forces P are plotted in Fig. C4.2 for point A. The "asymptotic solution" entered as the dashed curve is given by

$$h_{11,asympt} = \sqrt{\frac{2}{\pi(\ell - r)}} \tag{C4.1.1}$$

The weight function for point B on the lower branch is shown in Fig. C4.3.

Loading the upper branch by a pair of shear forces yields the weight functions of Fig. C4.4 for point A. The weight functions for point B are given in Fig. C4.5.

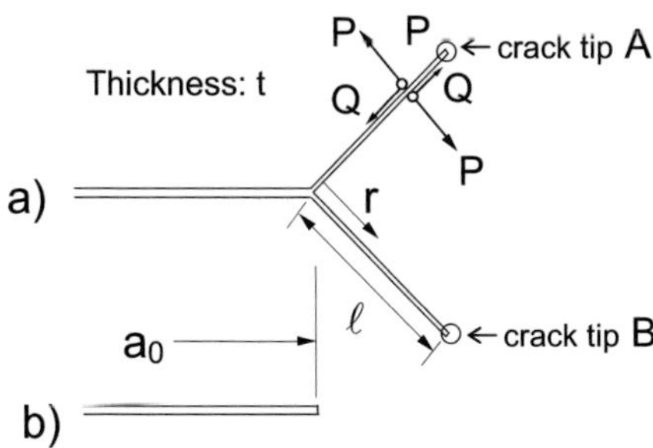

Fig. C4.1 a) Geometrical data of a semi-infinite forked crack in an infinite body, b) reference crack of length a_0.

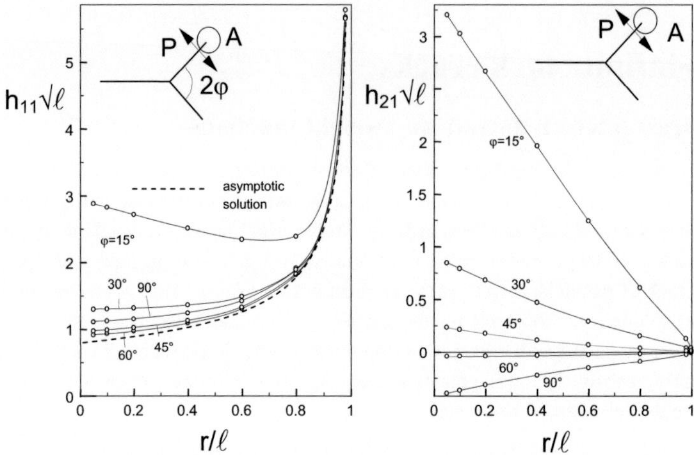

Fig. C4.2 Weight functions for a fork crack under loading by a normal force *P* on the upper branch, stress intensity factor also for the upper branch (A).

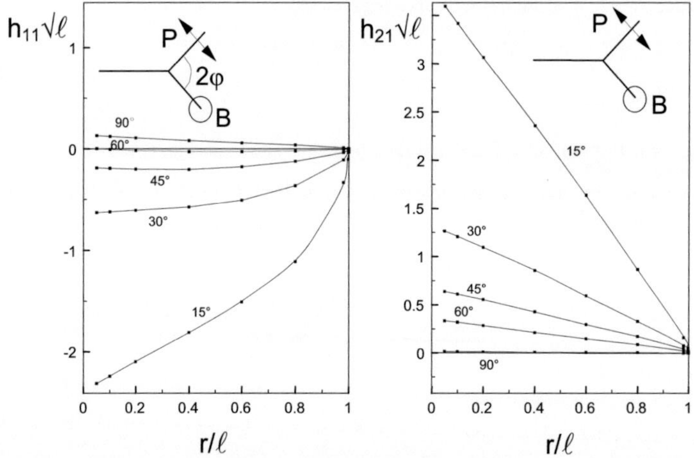

Fig. C4.3 Weight functions for a fork crack under loading by a normal force *P* on the upper branch, stress intensity factor for the lower branch (B).

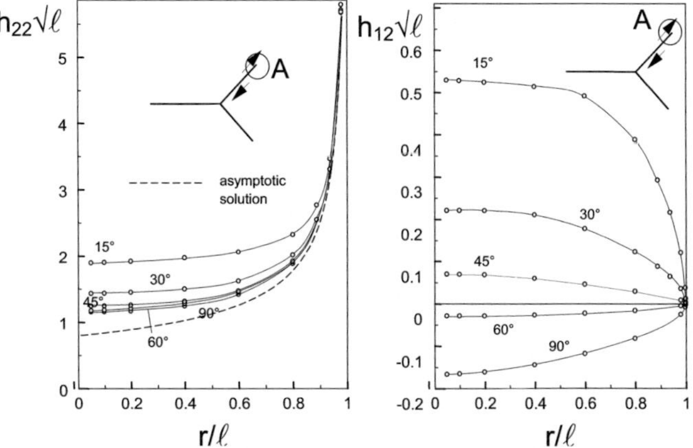

Fig. C4.4 Weight functions for a fork crack under shear loading by a shear force Q on the upper branch, stress intensity factor also for the upper branch (A).

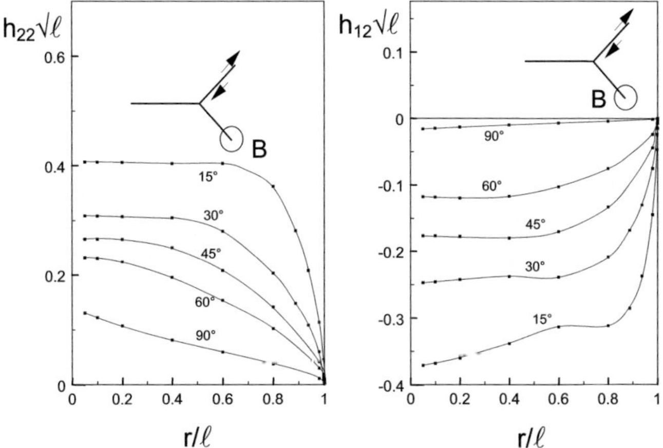

Fig. C4.5 Weight functions for a fork crack under shear loading by a shear force Q on the upper branch, stress intensity factor for the lower branch (B).

C4.1.2 Loading on both branches

The weight functions for symmetric loading are shown in Fig. C4.6 for normal forces P and in Fig. C4.7 for shear forces Q. These results can also be obtained by adding and subtracting the weight functions of Figs. C4.2-C4.5. From the results of Figs. C4.6 and C4.7, the coefficients for the weight function representations

$$h_{ij} = \sqrt{\frac{2}{\pi\ell}} \sum_{n=0}^{\infty} D_n^{(ij)} (1 - r/\ell)^{n-1/2} \tag{C4.1.1}$$

with
$$D_0^{(11)} = D_0^{(22)} = 1, \quad D_0^{(12)} = D_0^{(21)} = 0 \tag{C4.1.2}$$

can be determined (see also [C4.4] and Section A3).

Table 4.1 Coefficients for a 4-terms weight function h_{11} (symmetric loading).

φ	$D_1^{(11)}$	$D_2^{(11)}$	$D_3^{(11)}$	$D_4^{(11)}$
15°	-2.3693	5.7222	-5.6661	2.0038
30°	-0.6906	-0.1901	1.7514	-1.0583
45°	-0.1869	-0.4855	1.1566	-0.5172
60°	0.0353	-0.1374	0.3450	-0.1126
90°	0.3750	0.0358	0.3053	-0.1736

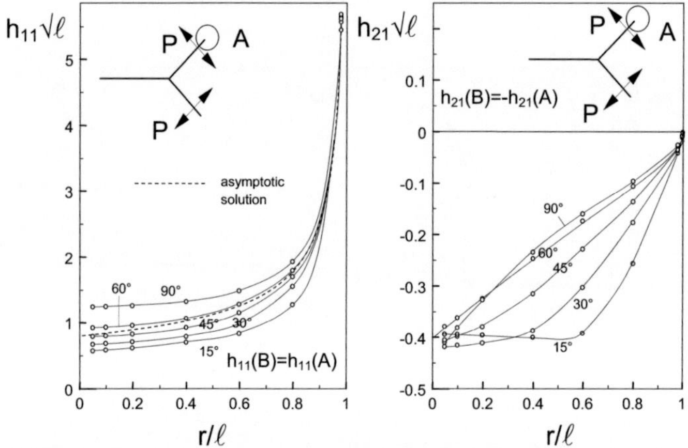

Fig. C4.6 Weight functions for a fork crack under normal force P on both branches.

Table 4.2 Coefficients for a 4-terms weight function h_{22} (symmetric loading)

φ	$D_1^{(22)}$	$D_2^{(22)}$	$D_3^{(22)}$	$D_4^{(22)}$
15°	-0.0691	3.5919	-4.5084	1.8392
30°	-0.0351	0.3243	0.5140	-0.4062
45°	-0.0078	-0.3362	0.9521	-0.3929
60°	-0.0224	0.0093	-0.0012	0.1653
90°	0.2168	0.0563	0.1441	-0.1389

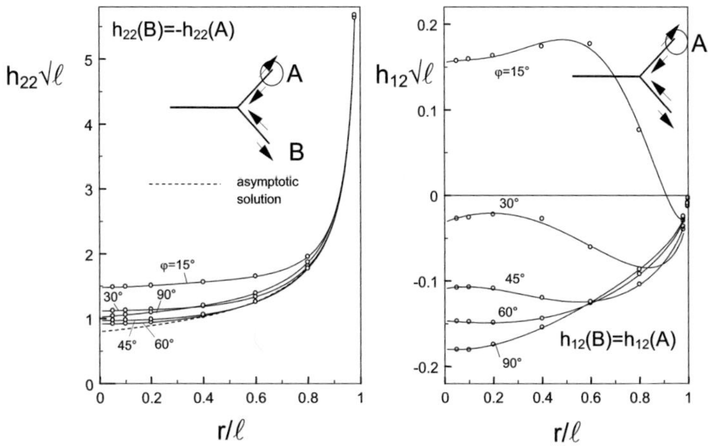

Fig. C4.7 Weight functions for a fork crack under symmetrical shear loading on both branches.

Table 4.3 Coefficients for a 5-terms weight function h_{21} (symmetric loading)

φ	$D_1^{(21)}$	$D_2^{(21)}$	$D_3^{(21)}$	$D_4^{(21)}$	$D_5^{(21)}$
15°	-0.2243	-4.3809	11.2858	-10.8232	3.6477
30°	-0.3036	-1.3556	1.7585	-0.6234	-
45°	-0.3139	-0.3497	-0.1272	0.2778	-
60°	-0.2638	-0.1313	-0.2522	0.1469	-
90°	-0.2529	-0.0022	-0.4668	0.1785	-

Table 4.4 Coefficients for a 5-terms weight function h_{12} (symmetric loading)

φ	$D_1^{(12)}$	$D_2^{(12)}$	$D_3^{(12)}$	$D_4^{(12)}$	$D_5^{(12)}$
15°	-0.3233	4.4178	-10.0998	9.1825	-2.9823
30°	-0.4122	1.0553	-0.8749	0.1926	-
45°	-0.3215	0.0754	0.4026	-0.2933	-
60°	-0.2551	-0.0593	0.2103	-0.0796	-
90°	-0.2401	0.0052	-0.1085	0.1181	-

C4.2 T-stress and Green's function

T-stress results for symmetrically loaded forked cracks are represented in Fig. C4.8. The data plotted refer to the upper part of the crack (i.e. for point A). Under P-load, the T-stress in the lower crack part (point B) is identical with the results at A. Under Q-load, the T-stress at point B has the same value as at point A, but an opposite sign. The coefficients for the Green's function representation according to eq.(A4.4.1) and

$$t^{(1)} = \tfrac{1}{\ell}\sum_{n=1}^{N} D_n^{(1)}(1 - r/\ell)^{(2n-1)/2} \quad , \quad t^{(2)} = \tfrac{1}{\ell}\sum_{n=1}^{N} D_n^{(2)}(1 - r/\ell)^{(2n-1)/2} \qquad (C4.2.1)$$

are compiled in Tables C4.5 and C4.6 up to N=5 and 6.

In Fig. C4.9, the T-stresses are shown for the case of the upper crack part only being loaded. Also in this case the symmetric loading case can be obtained by superposition of these results.

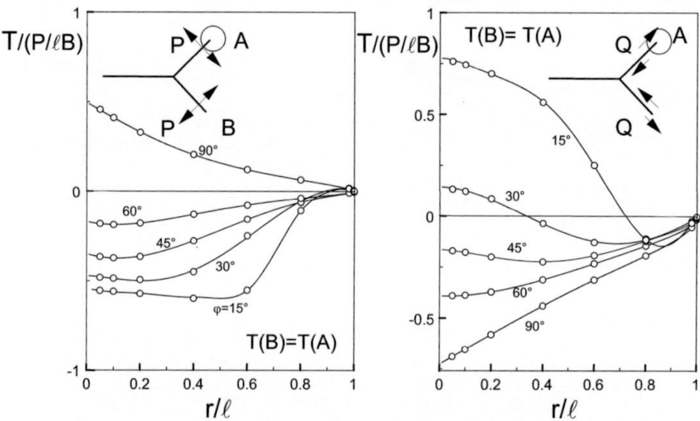

Fig. C4.8 Green's functions for the T-stress of symmetrically loaded forked cracks.

Table C4.5 Coefficients for eq.(C4.2.1) as obtained under normal forces P (symmetrical load).

φ	$D_1^{(1)}$	$D_2^{(1)}$	$D_3^{(1)}$	$D_4^{(1)}$	$D_5^{(1)}$
30°	0.1096	-0.5694	-4.4385	7.9583	-3.5310
45°	-0.0662	-0.1610	-1.177	1.0680	-0.0130
60°	-0.057	-0.190	0.3400	-0.9839	0.7213
90°	0.0982	0.2950	-0.5274	1.1234	-0.4927

Table C4.6 Coefficients for eq.(C4.2.1) as obtained under shear forces Q (symmetrical load).

φ	$D_1^{(2)}$	$D_2^{(2)}$	$D_3^{(2)}$	$D_4^{(2)}$	$D_5^{(2)}$	$D_6^{(2)}$
15°	-0.1355	-6.5790	45.026	-89.694	75.588	-23.430
30°	-0.1301	-1.3333	3.9832	-2.8202	0.4441	-
45°	-0.2310	-0.0820	-1.0491	2.4794	-1.2822	-
60°	-0.2877	-0.1457	-0.2574	0.2991	0.0004	-
90°	-0.3821	-0.2418	-0.0300	-0.2447	0.1720	-

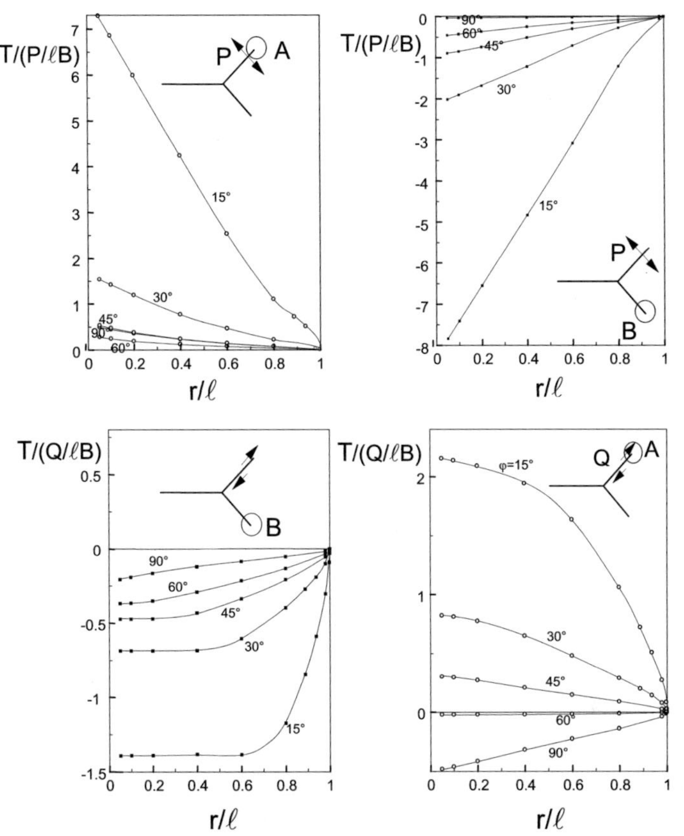

Fig. C4.9 Green's functions for the T-stress of non-symmetrically loaded forked cracks.

References C4:

[C4.1] Suresh, S., Shih, C.F., Plastic near-tip fields for branched cracks, Int. J. Fract. 30(1986), 237-259.

[C4.2] Kitagawa, H., Yuuki, R., Analysis of branched cracks under biaxial stresses, in: Fracture 1977, Taplin D. M. R. ed., University of Waterloo Press Canada, Vol. 3, 201-211.

[C4.3] Kitagawa, H., Yuuki, R., Ohira, T., Crack-morphological aspects in fracture mechanics, Engng. Fract. Mech. 7(1975), 515-529.

[C4.4] Fett, T., Munz, D. Stress Intensity Factors and Weight Functions, Computational Mechanics Publications, Southampton, UK, 1997

C5

Circular disk with internal crack

C5.1 Disk under constant radial load

The circular disk with a symmetrical internal crack is shown in Fig C5.1. This configuration under constant circumferential traction has been analysed with the boundary collocation method (BCM).

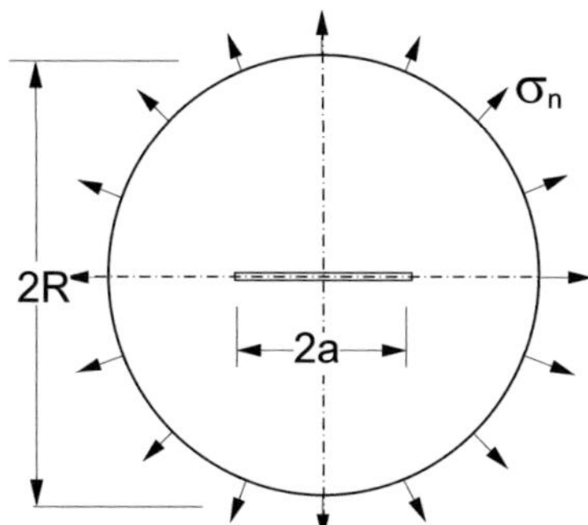

Fig. C5.1 Circular disk with internal crack under constant normal traction at the circumference.

The stress intensity factor solution is given by [C5.1]

$$F = \frac{K}{\sigma_n\sqrt{\pi a}} = \frac{1 - 0.5\alpha + 1.6873\alpha^2 - 2.671\alpha^3 + 3.2027\alpha^4 - 1.8935\alpha^5}{\sqrt{1-\alpha}}. \qquad (C5.1.1)$$

with $\alpha = a/R$. Figure C5.2a shows the BCM results as the circles and the relation (C5.1.1) as the curve (se also Table C5.1).

The T-stress is plotted in Fig. C5.2. The BCM results are introduced by the circles. The curve can be approximated by

$$T/\sigma_n = \frac{-2.34\alpha^2 + 4.27\alpha^3 - 3.326\alpha^4 + 0.9824\alpha^5}{1-\alpha} \qquad (C5.1.2)$$

The T-values compiled in Table C5.1 were extrapolated to $\alpha = 1$. Within the numerical accuracy of extrapolation, the limit values are

$$\lim_{\alpha \to 1} T/\sigma * (1-\alpha) \cong -0.413 = -\frac{1}{\sqrt{\pi^2 - 4}} \qquad (C5.1.3)$$

and for the biaxiality ratio

$$\lim_{\alpha \to 1} \beta \sqrt{1-\alpha} \cong \frac{1}{2} \qquad\qquad (C5.1.4)$$

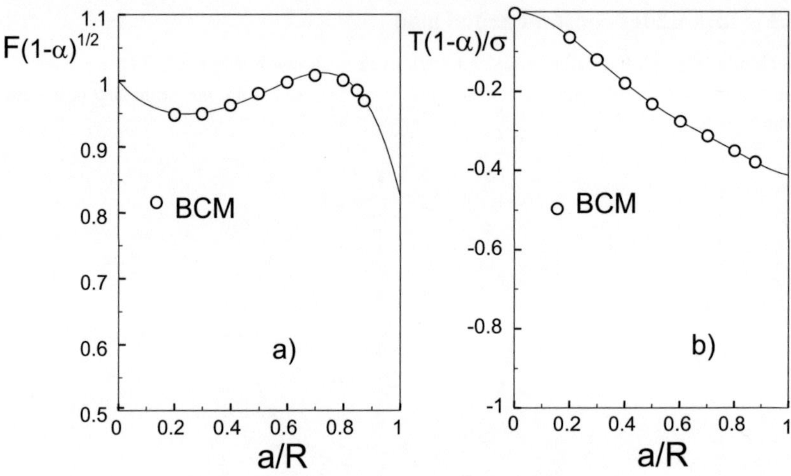

Fig. C5.2 T-stress for an internal crack in a circular disk.

Table C5.1 T-stress, stress intensity factor, and biaxiality ratio for an internally cracked circular disk with constant tensile traction at the circumference (values for $\alpha = 1$ extrapolated).

$\alpha = a/R$	$F \cdot (1-\alpha)^{1/2}$	$T/\sigma \cdot (1-\alpha)$	$\beta \cdot (1-\alpha)^{1/2}$
0	1.000	0.000	0.00
0.1	0.965	-0.019	-0.020
0.2	0.951	-0.064	-0.067
0.3	0.951	-0.120	-0.126
0.4	0.962	-0.176	-0.183
0.5	0.979	-0.228	-0.233
0.6	0.998	-0.275	-0.275
0.7	1.011	-0.315	-0.311
0.8	1.004	-0.352	-0.351
0.9	0.953	-0.385	-0.404
1.0	0.8255	-0.413	-0.50

Figure C5.3 represents the displacement at the crack centre $x=0$ for constant normal tractions σ_n in the form of

$$\delta = \frac{2a\sigma_n}{E'}\frac{1}{\alpha}\ln\frac{1}{1-\alpha}\lambda(\alpha) \ , \quad \alpha = a/R \tag{C5.1.5}$$

The open circles result from boundary collocation computations and the solid curve is the result obtained by application of the procedure proposed by Paris (see e.g. Appendix B in Tada's handbook [C5.2]). The solid circles are analytical values resulting from limit case considerations. The dashed curve in Fig. C5.3 is the solution for the endless parallel strip with an internal crack, as reported by Tada [C5.2].

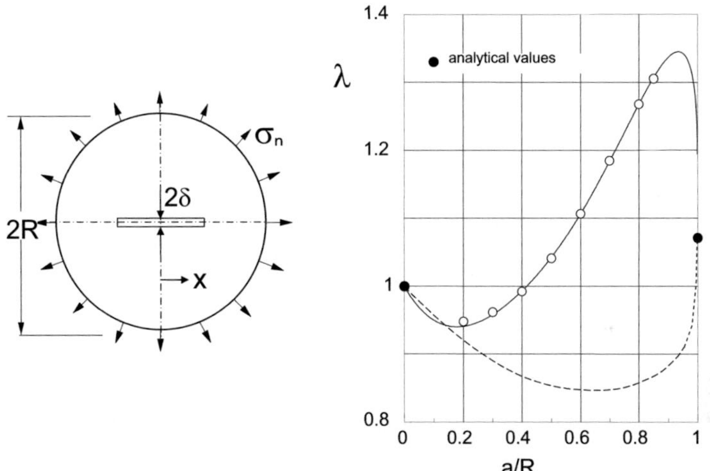

Fig. C5.3 Crack opening displacements for the internally cracked disk according to eq.(C5.1.5) (open circles: BCM results, solid curve: procedure proposed by Paris). Dashed line: data for an internally cracked endless strip [C5.2].

C5.2 Disk partially loaded by normal traction at the circumference

A partially loaded disk is shown in Fig. C5.4a. Constant normal tractions σ_n are applied at the circumference within an angle of 2γ.

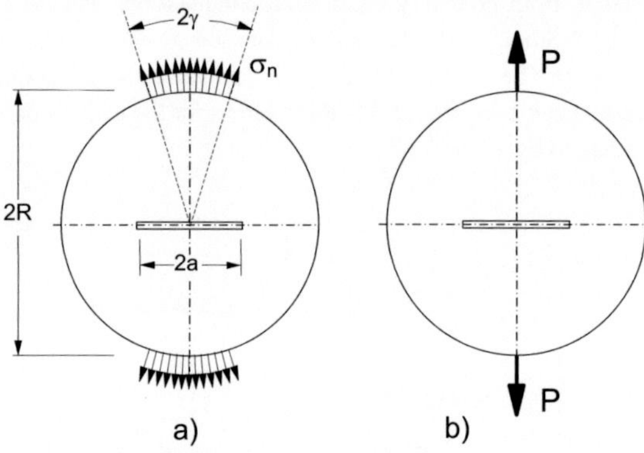

Fig. C5.4 a) Partially loaded disk, b) diametrical loading by a couple of forces (disk thickness: B).

The total force in y-direction results from

$$P_y = 2B\sigma_n \int_0^\gamma R\cos\gamma'd\gamma' = 2B\sigma_n R\sin\gamma \tag{C5.2.1}$$

The geometric function F, defined by

$$K_{\mathrm{I}} = \sigma*\sqrt{\pi a}\,F(a/R) , \tag{C5.2.2}$$

is plotted in Fig. C5.5, with the characteristic stress $\sigma*$ defined as

$$\sigma* = \frac{P_y}{\pi RB} . \tag{C5.2.3}$$

From the limit case $\gamma\to 0$, the solutions for concentrated forces (see Fig. C5.4b) are obtained as represented in Fig. C5.6. Comparison with the results from literature [C5.3-C5.5] reveals good agreement of stress intensity factors. The solution given by Tada et al. [C5.2] (dashed curve in Fig. C5.6) deviates by about 20% near $a/R = 0.8$. The results obtained here can be expressed by

$$K_{\mathrm{I}} = \sigma*\sqrt{\pi a}\,F_P$$
$$F_P = \frac{3 - 1.254\alpha - 1.7013\alpha^2 + 4.0597\alpha^3 - 2.8059\alpha^4}{\sqrt{1-\alpha}} \tag{C5.2.4}$$

with $\alpha=a/R$ and σ^* given by eq.(C5.2.3).

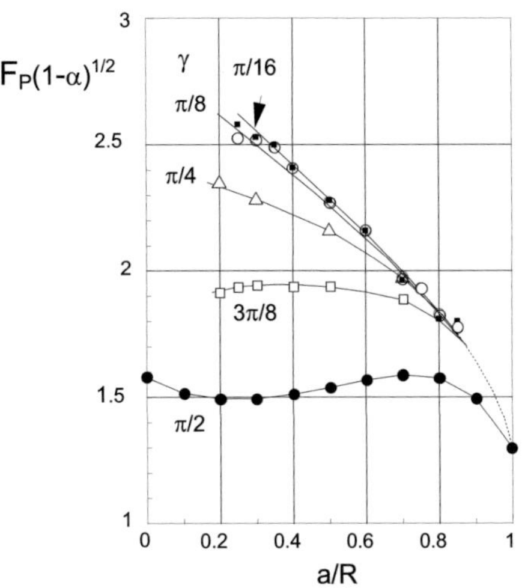

Fig. C5.5 Stress intensity factors for a circular disk, partially loaded over an angle of 2γ (see Fig. C5.4a).

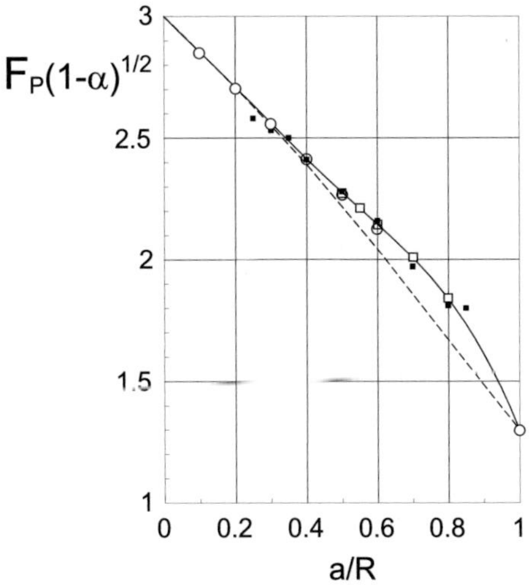

Fig. C5.6 Stress intensity factor and T-stress for a circular disk loaded diametrically by concentrated forces (Fig. C5.4b). Comparison of stress intensity factors; solid squares: partially distributed stresses with an angle of $\gamma=\pi/16$, circles: results by Atkinson et al. [C5.3] and Awaji and Sato [C5.4], open squares: results obtained with the weight function technique [C5.5], dashed line: solution proposed by Tada et al.[C5.2].

The x-stress term T is shown in Fig. C5.7. From the limit case $\gamma \to 0$, the solutions for concentrated forces (see Fig. C5.4b) are obtained as represented in Fig. C5.8.

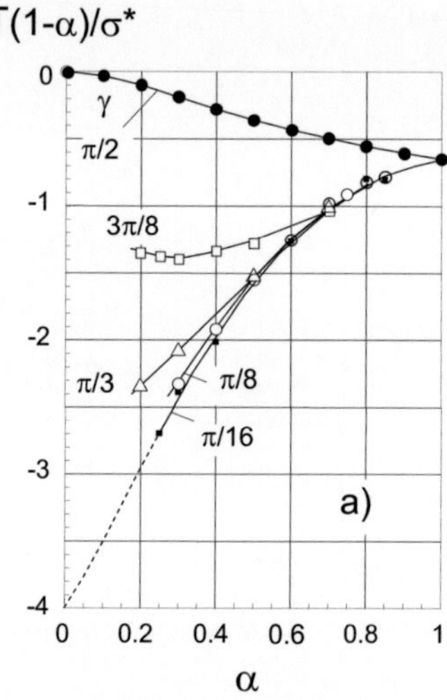

Fig. C5.7 T-stress for a circular disk partially loaded over an angle of 2γ (see Fig. C5.4a).

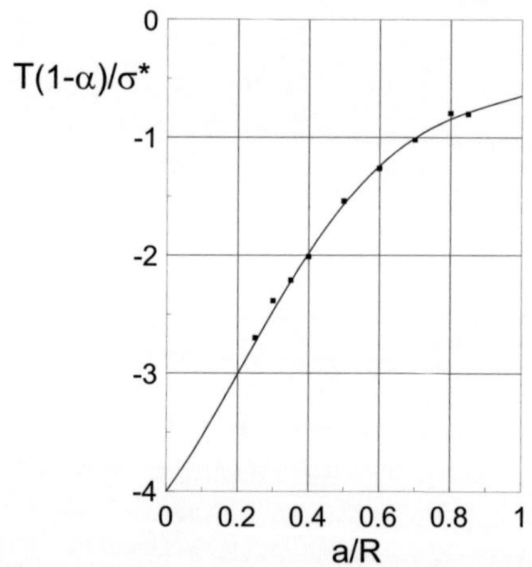

Fig. C5.8 T-stress for a circular disk loaded diametrically by concentrated forces (Fig. C5.4b). T-stress results including partially distributed stresses with an angle of $\gamma = \pi/16$ (squares) and exact limit cases for $\alpha = 0$.

The T-stress can be fitted by

$$\frac{T}{\sigma^*} = \frac{-4(1-\alpha) + 7.6777\,\alpha^2 - 16.0169\,\alpha^3 + 8.7994\,\alpha^4 - 1.10849\,\alpha^5}{1-\alpha} \qquad \text{(C5.2.5)}$$

In this case, the limit value is (at least in very good approximation)

$$\lim_{\alpha \to 1} T/\sigma^*(1-\alpha) \cong -0.648 \cong -\frac{\pi}{2\sqrt{\pi^2 - 4}} \qquad \text{(C5.2.6)}$$

C5.3 Central point forces acting on the crack face

A centrally cracked circular disk loaded by a couple of forces at the crack centre is shown in Fig. C5.9. The corresponding stress intensity factor and T-stress were calculated by boundary collocation computations.

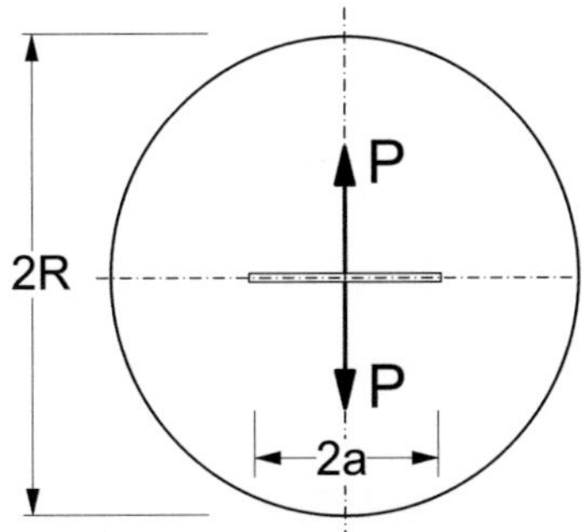

Fig. C5.9 Circular disk with a couple of forces acting on the crack faces.

The stress intensity factor for central point forces is

$$K_{\mathrm{I}} = \frac{P}{\sqrt{\pi a}} F_P \qquad \text{(C5.3.1)}$$

with the geometric function being

$$F_P = \frac{1 - 1.07884\alpha + 8.24956\alpha^2 - 17.9026\alpha^3 + 20.3339\alpha^4 - 9.305\alpha^5}{\sqrt{1-\alpha}}$$

Figure C5.10 gives a comparison of the BCM results with results obtained by Tada et al. [C5.2] using an asymptotic extrapolation technique. Maximum differences are in the order of about 10%.

The T-stress data obtained with the BCM method are plotted in Fig. C5.11 by the squares. Together with the limit value, eq.(C5.2.6), the numerically found T-values were fitted by the polynomial of

$$\frac{T}{\sigma^*} = \frac{-4.1971\alpha + 5.4661\,\alpha^2 - 1.1497\,\alpha^3 - 0.7677\,\alpha^4}{1-\alpha} \tag{C5.3.2}$$

This relation is introduced into Fig. C5.11 as the solid line.

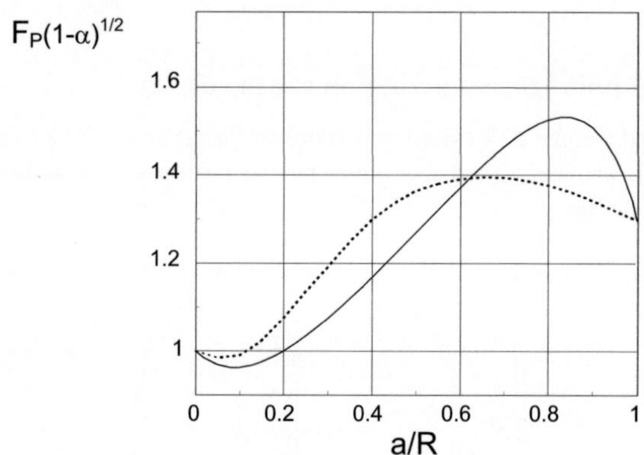

Fig. C5.10 a) Geometric function for a couple of forces *P* at the crack centre. Solid curve: [C5.6], dashed curve: [C5.2].

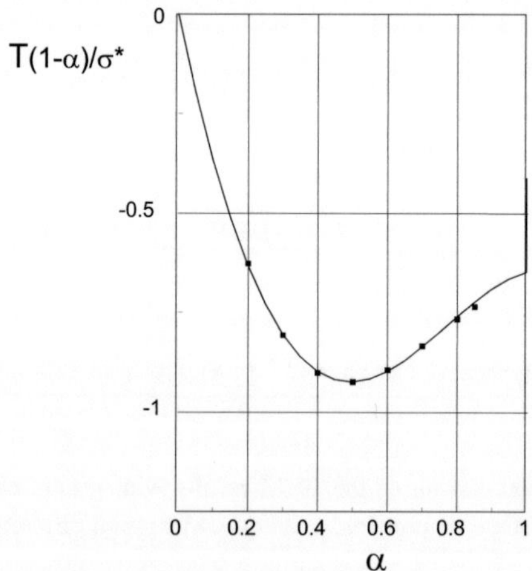

Fig. C5.11 T-stress for an internally cracked circular disk with a couple of forces acting in the crack centre on the crack faces [C5.7]. Symbols: numerical results, solid line: fitting curve.

Mode-I weight function [C5.1] for symmetrical loading $\sigma(x) = \sigma(-x)$:

$$h_1 = \frac{2}{\sqrt{\pi a}}\left[\frac{1}{\sqrt{1-\rho^2}} + D_0\sqrt{1-\rho^2} + D_1(1-\rho^2)^{3/2}\right], \quad \rho = x/a \qquad (C5.3.3)$$

$$D_0 = \frac{8 - 4\alpha + 3.8612\alpha^2 - 15.9344\alpha^3 + 24.6076\alpha^4 - 13.234\alpha^5}{\sqrt{1-\alpha}} - 8 \qquad (C5.3.4)$$

$$D_1 = -\frac{8 - 4\alpha + 0.6488\alpha^2 - 14.1232\alpha^3 + 24.2696\alpha^4 - 12.596\alpha^5}{\sqrt{1-\alpha}} + 8 \qquad (C5.3.5)$$

A two-terms Green's function for the T-stress term reads

$$t = \tfrac{1}{a}C_0(1 - x^2/a^2)^{1/2} + \tfrac{1}{a}C_1(1 - x^2/a^2)^{3/2} \qquad (C5.3.6)$$

with

$$C_0 = \frac{-3.902\alpha^2 + 11.307\alpha^3 - 14.743\alpha^4 + 6.469\alpha^5}{1-\alpha} \qquad (C5.3.7)$$

$$C_1 = \frac{1.2297\alpha^2 - 7.828\alpha^3 + 14.012\alpha^4 - 6.958\alpha^5}{1-\alpha} \qquad (C5.3.8)$$

or in the form

$$t = \tfrac{1}{a}E_0(1 - x^2/a^2) + \tfrac{1}{a}E_1(1 - x^2/a^2)^2 \qquad (C5.3.9)$$

$$E_0 = \frac{-6.8622\alpha + 18.1057\alpha^2 - 22.0173\alpha^3 + 9.3229\alpha^4}{1-\alpha} \qquad (C5.3.10)$$

$$E_1 = \frac{4.1902\alpha - 14.626\alpha^2 + 21.2854\alpha^3 - 9.8117\alpha^4}{1-\alpha} \qquad (C5.3.11)$$

With the Green's function, the stress intensity factor for the diametrical tension specimen (Fig. C5.4b) was computed by using the stress distribution

$$\frac{\sigma_y}{\sigma^*} = \frac{4}{(1+\xi^2)^2} - 1 \quad , \quad \xi = x/R \qquad (C5.3.12)$$

$$\frac{\sigma_x}{\sigma^*} = -1 + \frac{4\xi^2}{(1+\xi^2)^2} \qquad (C5.3.13)$$

The result is plotted in Fig. C5.12. It becomes obvious that in this approximation small deviations between BCM and Green's function results are visible for large α only.

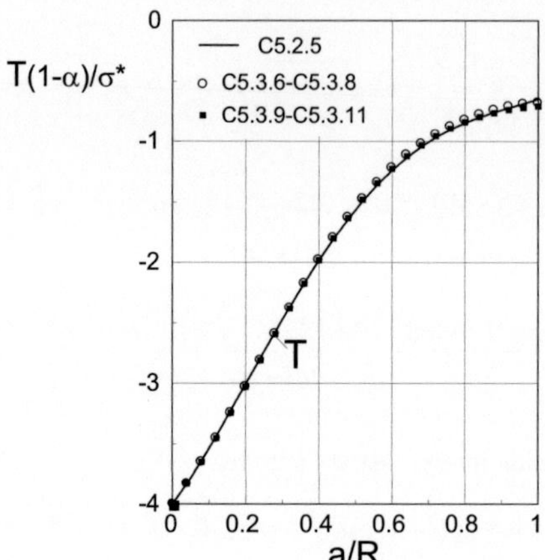

Fig. C5.12 T-stresses for an internally cracked circular disk loaded by a couple of diametric forces at the free boundary (see Fig. C5.4b). Results from 2-terms Green's functions (symbols) compared with results from boundary collocation (BCM) computations (curve represents eq.(C5.2.5)).

C5.4　Mode-II loading

Figure C5.13 shows the crack-face loading by a constant shear stress τ and a pair of concentrated tangential forces Q.

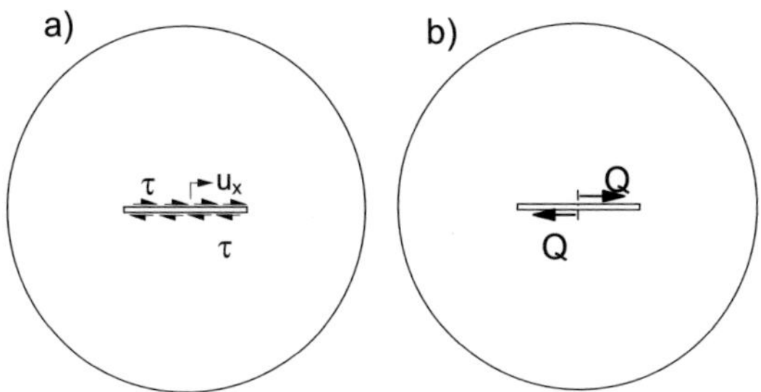

Fig. C5.13 Internal radial crack loaded by shear traction, a) constant shear stress τ, b) pair of concentrated shear forces Q.

The stress intensity factor under constant shear traction τ is [C5.5]

$$K_{II} = \tau F_{II}\sqrt{\pi a} \; , \quad F_{II} = \frac{1 - 0.5\alpha + 0.9274\alpha^2 - 0.88414\alpha^3 + 0.28226\alpha^4}{\sqrt{1-\alpha}} \quad (C5.4.1)$$

with $\alpha = a/R$.

The stress intensity factor for a point load Q (line load over plate thickness B) in the crack centre is [C5.5]

$$K_{II,Q} = \frac{2Q}{B\sqrt{\pi a}} F_{II,Q} \; , \quad F_{II} = \frac{1 - 0.5\alpha + 1.977\alpha^2 - 1.5655\alpha^3 + 0.3851\alpha^4}{\sqrt{1-\alpha}} . \quad (C5.4.2)$$

A mode-II weight function for symmetric loading $\tau(x) = \tau(-x)$ is [C5.5]

$$h_{II} = \frac{2}{\sqrt{\pi a}}\left[\frac{1}{\sqrt{1-\rho^2}} + D_0\sqrt{1-\rho^2} + D_1(1-\rho^2)^{3/2} \right] \quad (C5.4.3)$$

$$D_0 = \frac{5 - 2.5\alpha + 1.4882\alpha^2 - 2.3766\alpha^3 + 1.1028\alpha^4}{\sqrt{1-\alpha}} - 5 \quad (C5.4.4)$$

$$D_1 = \frac{-4 + 2\alpha + 0.4888\alpha^2 + 0.81112\alpha^3 - 0.7177\alpha^4}{\sqrt{1-\alpha}} + 4 \quad (C5.4.5)$$

C5.5 Brazilian disk with internal crack

The mixed-mode loading situation for a disk under diametrically applied concentrated forces is shown in Fig. C5.14. The angle between crack plane and loading line deviates from $\Theta=90°$. This configuration is called Brazilian disk test.

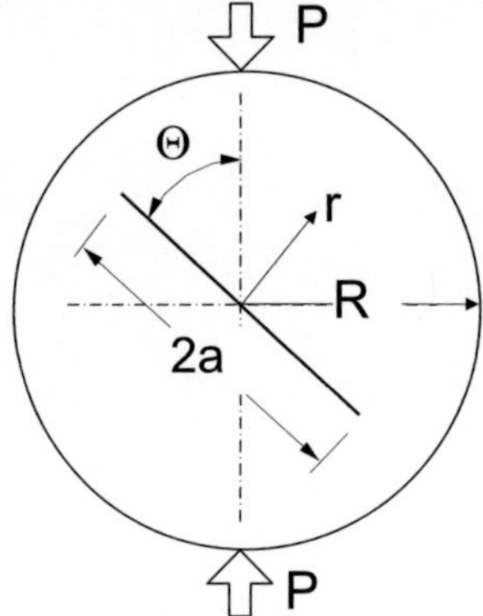

Fig. C5.14 Diametrical compression test with internal crack (disk thickness: B).

The mixed-mode stress intensity factors K_I, K_{II} and related geometric functions F_I, F_{II} read

$$K_I = \sigma * F_I \sqrt{\pi a} = \int_0^a \sigma(x) h_I(x,a)\,dx \qquad (C5.5.1)$$

$$K_{II} = \sigma * F_{II} \sqrt{\pi a} = \int_0^a \tau(x) h_{II}(x,a)\,dx \qquad (C5.5.2)$$

with the characteristic stress $\sigma*$ defined by

$$\sigma* = \frac{P}{\pi R B} \; , \qquad (C5.5.3)$$

(identical with the maximum tensile stress in the centre of the disk).

The tangential, radial, and shear stress components (σ_φ, σ_r, and $\tau_{r\varphi}$) in an uncracked Brazilian disk were given by Erdlac (quoted in [C5.3]) as

$$\sigma_\varphi = 2\sigma * \left[\frac{1}{2} - \frac{(1-\rho\cos\Theta)\sin^2\Theta}{(1+\rho^2-2\rho\cos\Theta)^2} - \frac{(1+\rho\cos\Theta)\sin^2\Theta}{(1+\rho^2+2\rho\cos\Theta)^2} \right] \qquad (C5.5.4)$$

$$\sigma_r = 2\sigma * \left[\frac{1}{2} - \frac{(1-\rho\cos\Theta)(\cos\Theta-\rho)^2}{(1+\rho^2-2\rho\cos\Theta)^2} - \frac{(1+\rho\cos\Theta)(\cos\Theta+\rho)^2}{(1+\rho^2+2\rho\cos\Theta)^2} \right] \qquad (C5.5.5)$$

$$\sigma_{r\varphi} = \frac{2P}{\pi BR}\left[\frac{(1-\rho\cos\Theta)(\cos\Theta-\rho)\sin\Theta}{(1+\rho^2-2\rho\cos\Theta)^2} + \frac{(1+\rho\cos\Theta)(\cos\Theta+\rho)\sin\Theta}{(1+\rho^2+2\rho\cos\Theta)^2}\right] \quad (C5.5.6)$$

with $\rho=r/R$. T-stresses and geometric functions are given in Figs. C5.15-C5.17 and in Tables C5.2-C5.4.

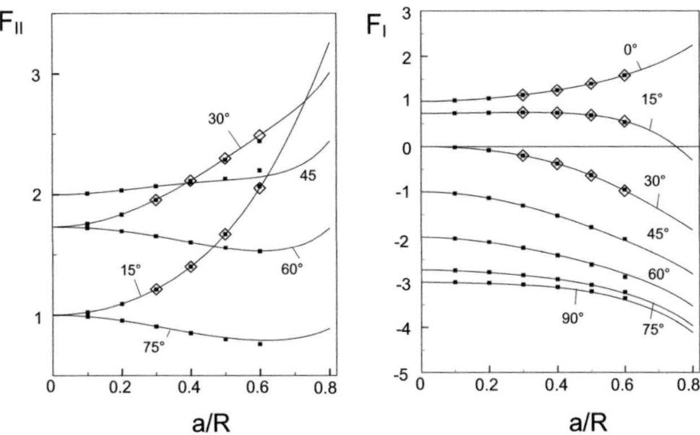

Fig. C5.15 Geometric functions for mode-II and mode-I stress intensity factors at several angles Θ. Curves: obtained with weight functions [C5.5]; solid squares: Atkinson et al. [C5.3]; open squares: Sato and Kawamata [C5.8].

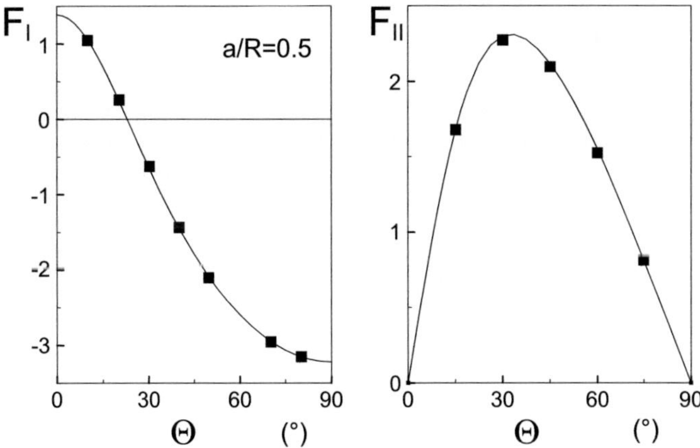

Fig. C5.16 Geometric functions for $a/R=0.5$ as a function of the angle Θ. Curves: obtained with the weight function procedure; squares: results from Atkinson et al. [C5.3] and Awaji and Sato [C5.4].

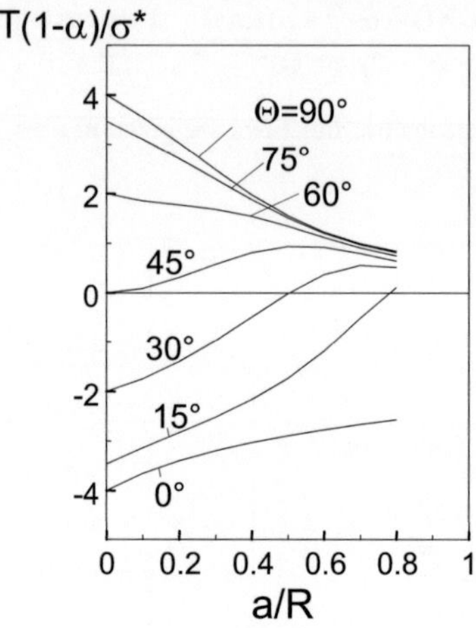

Fig. C5.17 T-stress for the Brazilian disk as a function of the angle Θ.

Table C5.2 T-stress $T(1-a/R)/\sigma^*$ for the Brazilian disk test.

a/R	Θ=0°	15°	30°	45°	60°	75°	90°
0	-4.000	-3.464	-2.000	0.000	2.000	3.464	4.000
0.1	-3.656	-3.136	-1.745	0.091	1.855	3.104	3.552
0.2	-3.398	-2.829	-1.396	0.312	1.773	2.711	3.029
0.3	-3.197	-2.515	-0.969	0.581	1.684	2.294	2.485
0.4	-3.033	-2.1623	-0.491	0.813	1.543	1.882	1.980
0.5	-2.895	-1.732	-0.013	0.936	1.344	1.508	1.552
0.6	-2.775	-1.182	0.370	0.919	1.115	1.200	1.223
0.7	-2.668	-0.505	0.557	0.795	0.903	0.971	0.970
0.8	-2.574	0.119	0.518	0.642	0.741	0.815	0.820

Table C5.3 Geometric function F_{II} for the Brazilian disk tests.

a/R	Θ=0°	15°	30°	45°	60°	75°	90°
0	0.	1.000	1.732	2.000	1.732	1.000	0.
0.1	0.	1.023	1.758	2.010	1.724	0.988	0.
0.2	0.	1.092	1.835	2.036	1.698	0.955	0.
0.3	0.	1.214	1.957	2.069	1.656	0.907	0.
0.4	0.	1.400	2.116	2.097	1.603	0.856	0.
0.5	0.	1.670	2.299	2.119	1.554	0.813	0.
0.6	0.	2.053	2.491	2.146	1.530	0.792	0.
0.7	0.	2.578	2.697	2.220	1.564	0.808	0.
0.8	0.	3.260	3.009	2.441	1.720	0.889	0.

Table C5.4 Geometric function F_I for the Brazilian disk tests.

a/R	$\Theta=0°$	15°	30°	45°	60°	75°	90°
0	1.000	0.732	0	-1.000	-2.000	-2.732	-3.000
0.1	1.017	0.737	-0.020	-1.037	-2.033	-2.750	-3.016
0.2	1.063	0.746	-0.084	-1.141	-2.120	-2.793	-3.031
0.3	1.137	0.752	-0.200	-1.308	-2.248	-2.854	-3.062
0.4	1.241	0.742	-0.379	-1.527	-2.406	-2.940	-3.118
0.5	1.384	0.693	-0.635	-1.789	-2.594	-3.065	-3.220
0.6	1.578	0.562	-0.973	-2.083	-2.819	-3.250	-3.393
0.7	1.846	0.263	-1.381	-2.413	-3.108	-3.525	-3.665
0.8	2.244	-0.302	-1.843	-2.824	-3.530	-3.965	-4.112

C5.6   Mixed boundary conditions

C5.6.1 Constant radial displacement and zero shear traction

The internally cracked circular disk under constant radial displacement and disappearing shear tractions along the circumference is illustrated in Fig. C5.18.

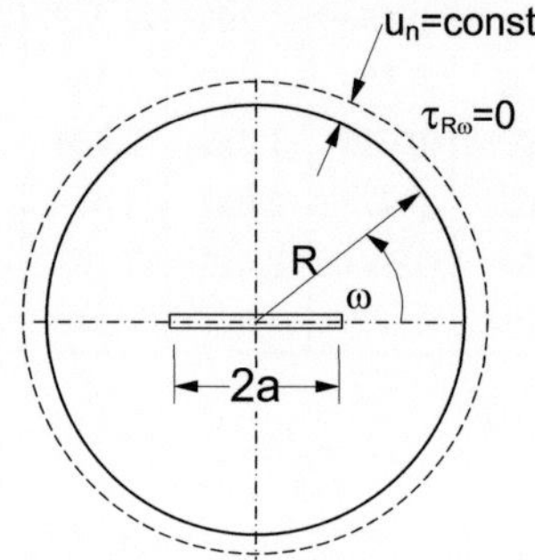

Fig. C5.18 Boundary conditions u_n = const., $\tau_{R\omega}$ = 0.

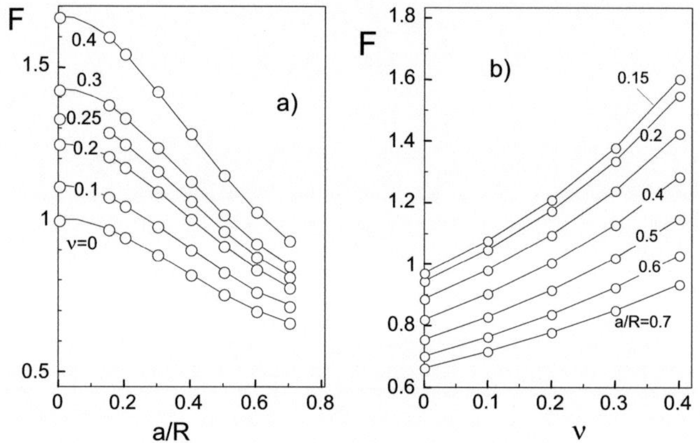

Fig. C5.19 Geometric function F according to eq.(C5.6.1)

The stress intensity factor for the loading case of u_n = const, $\tau_{R\omega}$ = 0 is defined by

$$K = \sigma^* \sqrt{\pi a}\, F(a/R, \nu)\,, \quad \sigma^* = \frac{u_n E}{R} \tag{C5.6.1}$$

where E is Young's modulus and ν Poisson's ratio.

The geometric function F is plotted versus a/R and ν in Fig. C5.19. For the special case of $\nu = 0.25$ and $\alpha \leq 0.7$ a fit relation reads

$$F \cong \tfrac{4}{3} - 2.154\,\alpha^2 + 3.200\,\alpha^4 - 1.987\,\alpha^6 \tag{C5.6.2}$$

The T-stress normalised to the stress σ^* is represented in Fig. C5.20. The higher-order coefficients A_1 and B_1, see eq.(A1.1.4), are compiled in Tables C5.5 and C5.6

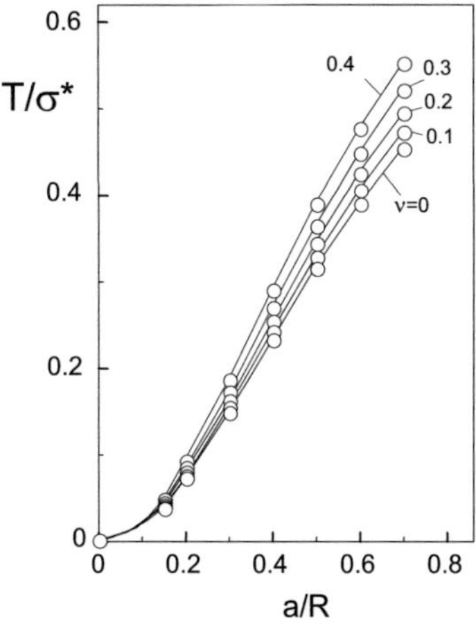

Fig. C5.20 T-stress as a function of crack size and Poisson's ratio.

For $\nu = 0.25$ and $\alpha = a/R \leq 0.7$ we find

$$T/\sigma^* \cong 2.597\,\alpha^2 - 2.685\,\alpha^3 + 0.6495\,\alpha^4 \tag{C5.6.3}$$

Table C5.5 Coefficient A_1 according to eq.(A1.1.4).

	$\nu=0$	0.1	0.2	0.3	0.4
$a/R=0.15$	-0.1255	-0.1393	-0.1565	-0.1784	-0.2073
0.2	-0.1060	-0.1175	-0.1317	-0.1497	-0.1734
0.3	-0.0826	-0.0911	-0.1016	-0.1147	-0.1316
0.4	-0.0692	-0.0757	-0.0836	-0.0933	-0.1056
0.5	-0.0624	-0.0674	-0.0734	-0.0807	-0.0897
0.6	-0.0617	-0.0656	-0.0702	-0.0758	-0.0825
0.7	-0.0689	-0.0722	-0.0760	-0.0805	-0.0858

Table C5.6 Coefficient B_1 according to eq.(A1.1.4)

	ν=0	0.1	0.2	0.3	0.4
a/R=0.15	0.0018	0.0019	0.0020	0.0020	0.0022
0.2	0.0036	0.0035	0.0034	0.0033	0.0031
0.3	0.0105	0.0101	0.0097	0.0093	0.0089
0.4	0.0211	0.0202	0.0193	0.0184	0.0174
0.5	0.0346	0.0330	0.0313	0.0296	0.0277
0.6	0.0506	0.0480	0.0453	0.0424	0.0392
0.7	0.0704	0.0665	0.0624	0.0579	0.0531

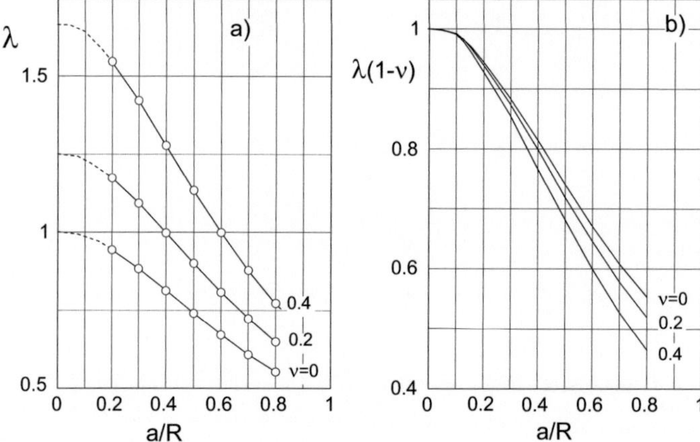

Fig. C5.21 Crack opening displacement δ at $x=0$ (for δ and x see Fig. C5.3) as a function of Poisson's ratio.

The crack opening displacements at $x=0$, represented as

$$\delta = \frac{2a\sigma^*}{E'}\lambda(\alpha) \ , \quad \alpha = a/R \tag{C5.6.4}$$

with σ* defined in (C5.6.1), are shown in Fig. C5.21.

C5.6.2 Constant radial traction and zero tangential displacements

The internally cracked circular disk under constant radial traction σ_n and zero tangential displacements along the circumference is illustrated in Fig. C5.22.

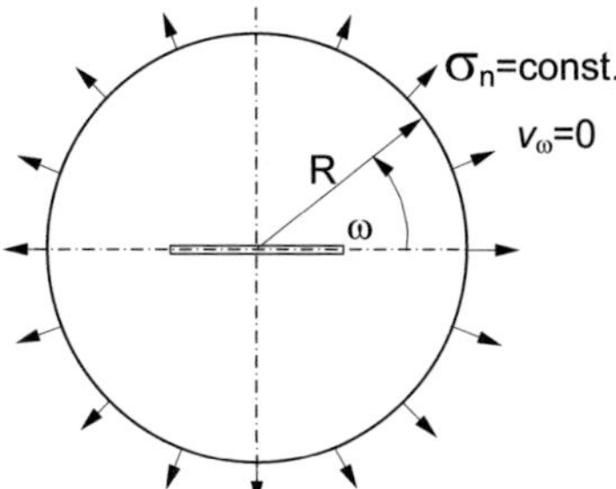

Fig. C5.22 Boundary conditions σ_n = const, v_ω = 0.

The stress intensity factor for the loading case of σ_n = constant, $v_\omega = 0$ is represented by eq.(C5.6.1) with now σ_n instead of σ^*. The related geometric function is shown in Fig. C5.23a. For $\nu = 0.25$ and $a/R \le 0.7$ an approximation is given by

$$F \cong 1 + 0.8162\alpha^2 + 3.8905\alpha^3 - 6.3161\alpha^4 + 2.0754\alpha^5, \alpha = a/R \qquad (C5.6.5)$$

The T-stress is represented in Fig. C5.23b. A fit relation is

$$T/\sigma_n \cong -0.7379\alpha^2 - 7.7055\alpha^3 + 16.00\alpha^4 - 7.9212\alpha^5 \qquad (C5.6.6)$$

Only a minor influence of ν on F and T/σ_n is visible in Fig. C5.23. From the additionally introduced results for the boundary conditions of $\tau_{R\omega} = 0$ instead of $v_\omega = 0$ (see dashed curves), we find an influence of the different tangential boundary conditions only, if $\alpha > 0.4$.

The higher-order coefficients A_1 and B_1 are compiled in Table C5.7 for $\nu = 0.25$.

Table C5.7 Coefficients A_1 and B_1 for $\nu = 0.25$ according to eq.(A1.1.4).

a/R	A_1	B_1
0.2	-0.1166	-0.0100
0.3	-0.0974	-0.0403
0.4	-0.0800	-0.0959
0.5	-0.0548	-0.1917
0.6	-0.0103	-0.3472
0.7	0.0706	-0.5967

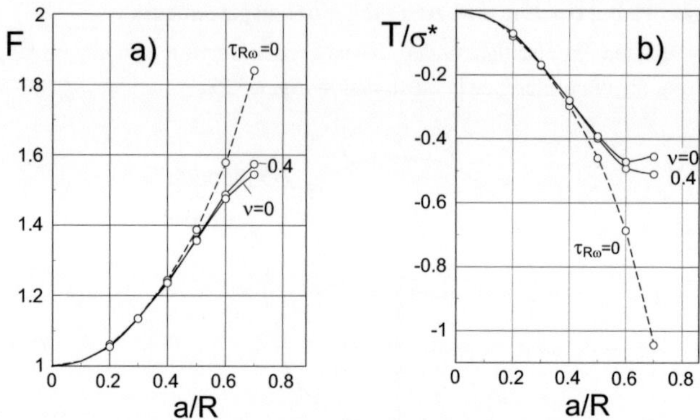

Fig. C5.23 Geometric function and T-stress under the boundary conditions of $\sigma_n=$ const., $v_\omega=0$ (dashed curves: stress boundary conditions $\sigma_n=$ const., $\tau_{R\omega}=0$).

C5.7 Full displacement boundary conditions

The internally cracked circular disk under constant radial displacement u_n and zero tangential displacement v_ω is shown in Fig. C5.24. The stress intensity factor solution expressed by the geometric function F (see eq.(C5.6.1)) is represented in Fig. C5.25a. The T-stress term is shown in Fig. C5.25b.

For $v=0.25$ the results are approximated by

$$F \cong \tfrac{4}{3} - 2.5727\,\alpha^2 + 2.0487\,\alpha^3 + 0.9988\,\alpha^4 - 1.4003\,\alpha^5 , \quad \alpha = a/R \qquad \text{(C5.7.1)}$$

$$T/\sigma^* \cong 3.271\,\alpha^2 - 5.628\,\alpha^3 + 3.826\,\alpha^4 \qquad \text{(C5.7.2)}$$

The higher order coefficients A_1 and B_1 are compiled in Tables C5.8 and C5.9.

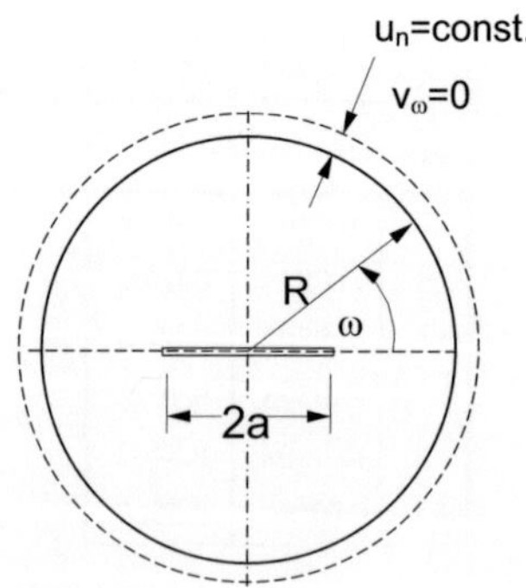

Fig. C5.24 Boundary conditions $u_n=$ const., $v_\omega=0$.

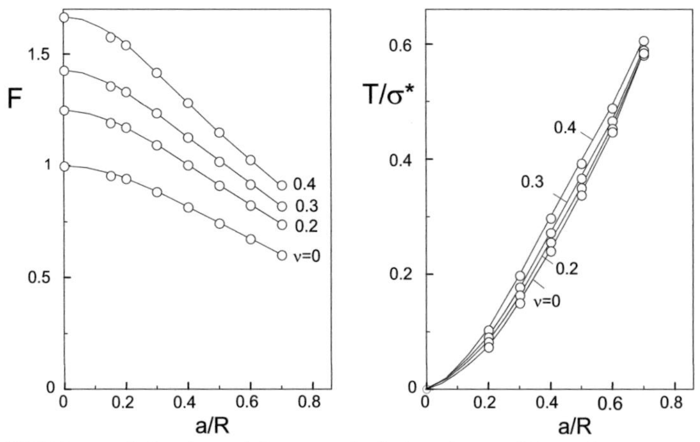

Fig. C5.25 Geometric function and T-stress under the boundary conditions of $u_n = $ const., $v_\omega = 0$.

Table C5.8 Coefficient A_1 according to eq.(A1.1.4).

	$v=0$	0.2	0.3	0.4
$a/R=0.2$	-0.106	-0.132	-0.150	-0.174
0.3	-0.082	-0.102	-0.116	-0.133
0.4	-0.067	-0.084	-0.096	-0.110
0.5	-0.057	-0.073	-0.083	-0.095
0.6	-0.049	-0.064	-0.074	-0.085
0.7	-0.041	-0.057	-0.067	-0.079

Table C5.9 Coefficient B_1 according to eq.(A1.1.4).

	$v=0$	0.2	0.3	0.4
$a/R=0.2$	0.003	0.006	0.007	0.010
0.3	0.008	0.014	0.018	0.023
0.4	0.013	0.023	0.030	0.040
0.5	0.012	0.028	0.039	0.052
0.6	0.000	0.021	0.036	0.053
0.7	-0.040	-0.009	0.010	0.033

C5.8 Partially loaded disks

C5.8.1 Stress boundary conditions

The case of different stress boundary conditions over parts of the circumference is dealt with in Section C5.2. Results for the stress intensity factor K are expressed by the geometric function F according to

$$K = \sigma_n \sqrt{\pi a}\, F(\gamma, a/R) \tag{C5.8.1}$$

and represented in Fig. C5.26.

The T-stresses are illustrated in Fig. C5.27 as a function of the loading angle γ and crack size a/R. In Tables C5.10 and C5.11 the next higher-order coefficients of the stress function, eq.(A1.1.4), are given.

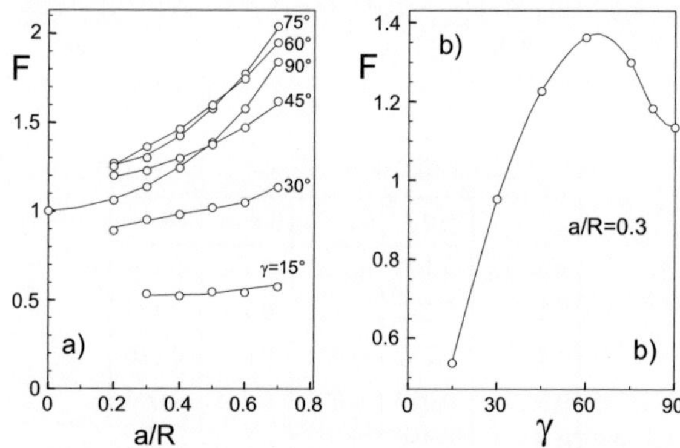

Fig. **C5.26** Geometric function F according to eq.(C5.6.1).

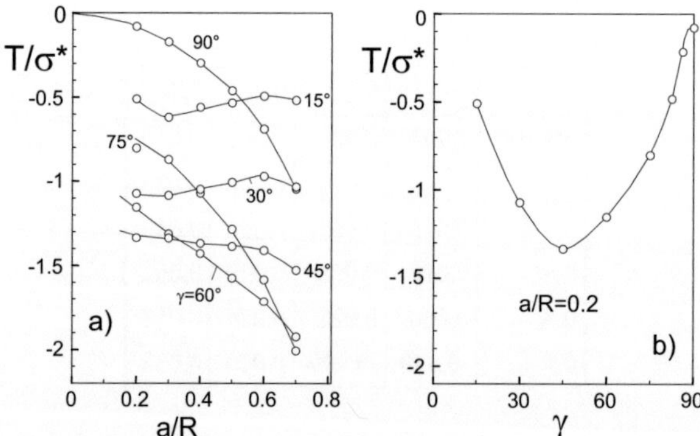

Fig. **C5.27** T-stress as a function of crack size and loading angle γ.

Table C5.10 Coefficients A_1 and B_1 according to eq.(A1.1.4) for $\gamma = 45°$.

	A_1	B_1
a/R=0.2	-0.115	-0.1659
0.3	-0.0766	-0.2554
0.4	-0.0393	-0.3977
0.5	0.0056	-0.5672
0.6	0.0598	-0.7338
0.7	0.1344	-0.9290

Table C5.11 Coefficients A_1 and B_1 according to eq.(A1.1.4) for $\gamma = 90°$.

	A_1	B_1
a/R = 0.2	-0.117	-0.0116
0.3	-0.0979	-0.0359
0.4	-0.0828	-0.0796
0.5	-0.0640	-0.1465
0.6	-0.0346	-0.2473
0.7	0.0179	-0.4107

C5.8.2   Mixed boundary conditions in the loading region

An internally cracked circular disk with constant radial displacements u_n over the angle 2γ and zero normal traction σ_n acting on the remaining part of the surface is shown in Fig. C5.28. In this loading case, the shear traction along the circumference is chosen to be $\tau_{R\omega} = 0$.

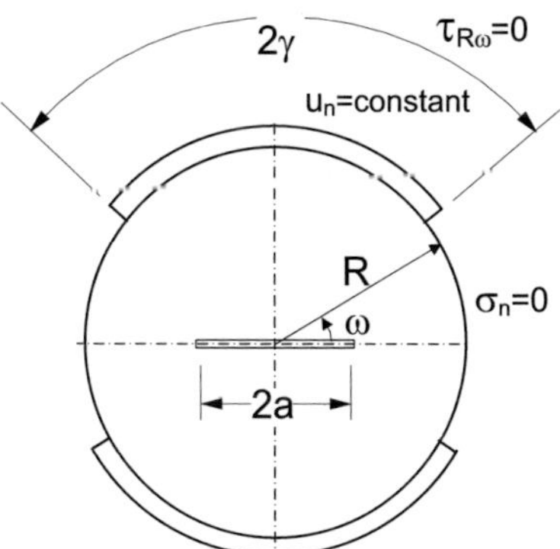

Fig. C5.28 Partially loaded, internally cracked disk under mixed boundary conditions: constant radial displacement over the angle 2γ, zero normal traction over the remaining part, zero shear traction along the whole circumference.

The stress intensity factor for the loading case of $u_n =$ constant, $\tau_{R\omega} = 0$ is defined by

$$K = \sigma^* \sqrt{\pi a}\, F(a/R, \nu, \gamma)\ , \quad \sigma^* = \frac{u_n E}{R} \tag{C5.8.2}$$

($E =$ Young's modulus, $F =$ geometric function). Results of boundary collocation computations are represented in Fig. C5.29 for a Poisson's ratio of $\nu = 0.25$ and several loading angles γ. The influence of the Poisson's ratio is shown in Fig. C5.30. The T-stress is represented in Fig. C5.31.

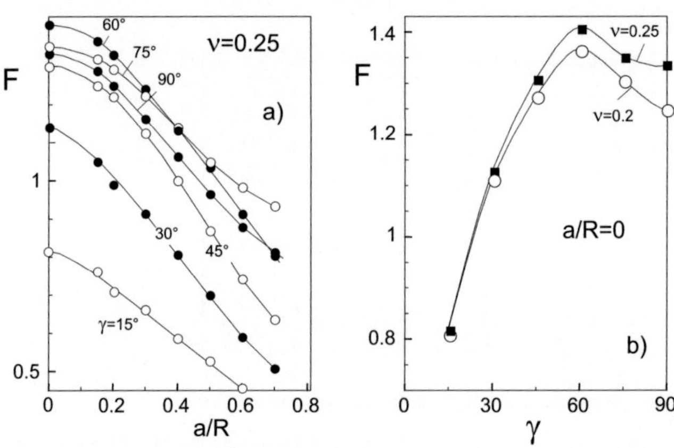

Fig. C5.29 Geometric function F, eq.(C5.8.2), as a function of crack size and loading angle.

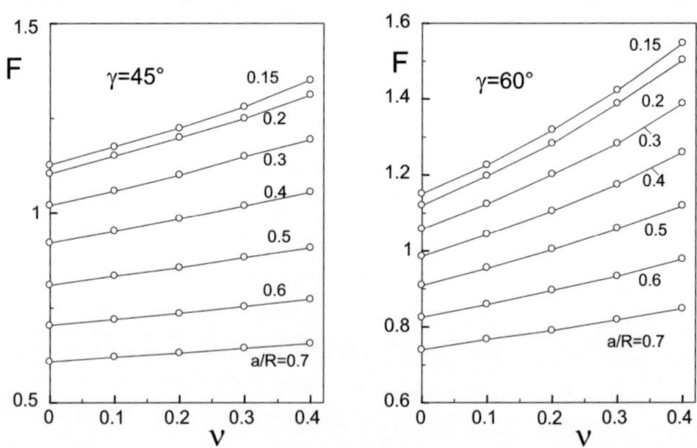

146

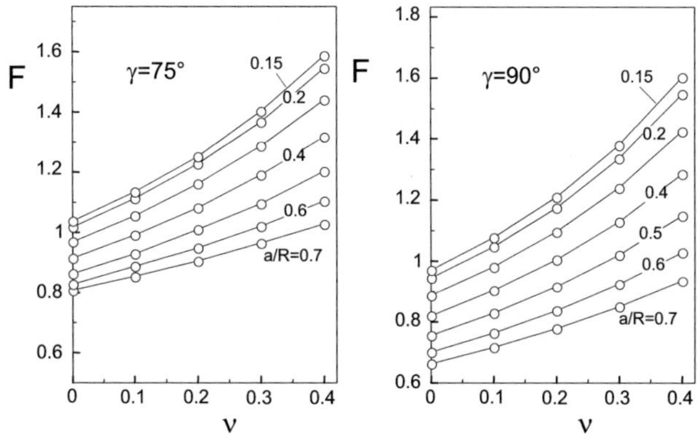

Fig. C5.30 Influence of Poisson's ratio ν on the geometric function F.

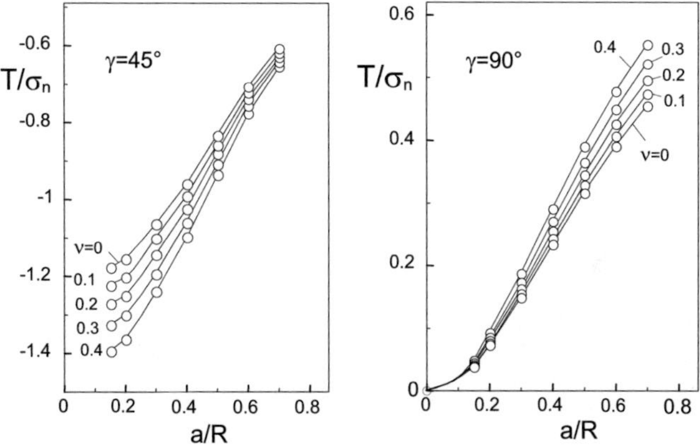

Fig. C5.31 T-stress as a function of crack size, Poisson's ratio, and loading angle γ.

C5.8.3 Displacement boundary conditions in the loading region

The internally cracked circular disk with constant radial displacements u_n, zero tangential displacements v_ω over the angle 2γ, and traction-free surfaces elsewhere is shown in Fig. C5.32. The geometric function according to eq.(C5.8.2) is plotted in Fig. C5.33 as a function of γ, a/R, and ν. The T-stress is shown in Fig. C5.34.

In Fig. C5.35, the geometric function and the T-stresses are plotted for the two boundary conditions in the loading region: u_n = constant, v_ω =0 (solid curves) and u_n = constant, $\tau_{R\omega} = 0$ (dashed curves). Only very small differences can be detected.

147

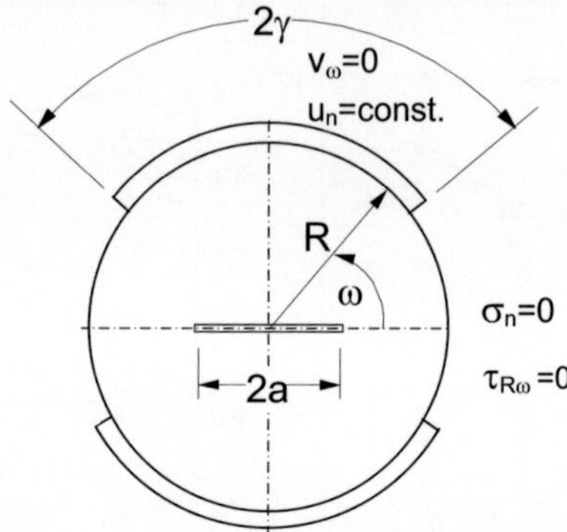

Fig. C5.32 Partially loaded, internally cracked disk under mixed boundary conditions: constant radial and zero tangential displacements over the angle 2γ, zero traction elsewhere.

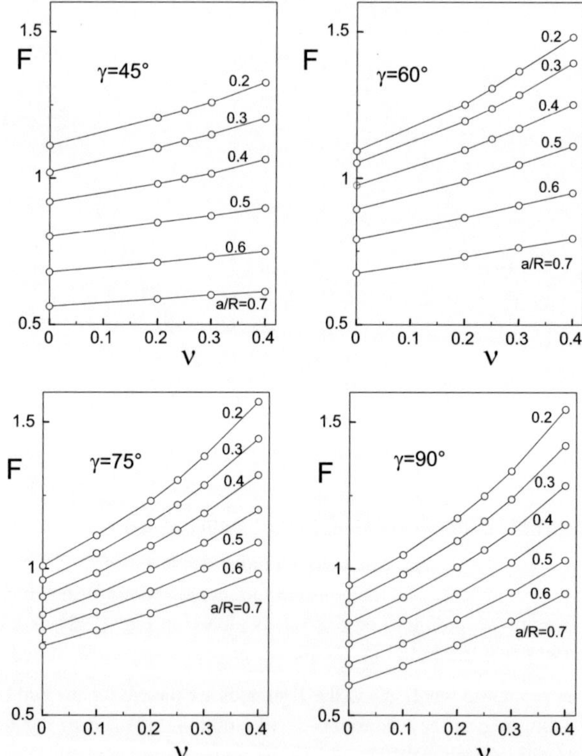

Fig. C5.33 Influence of Poisson's ratio ν on the geometric function *F*.

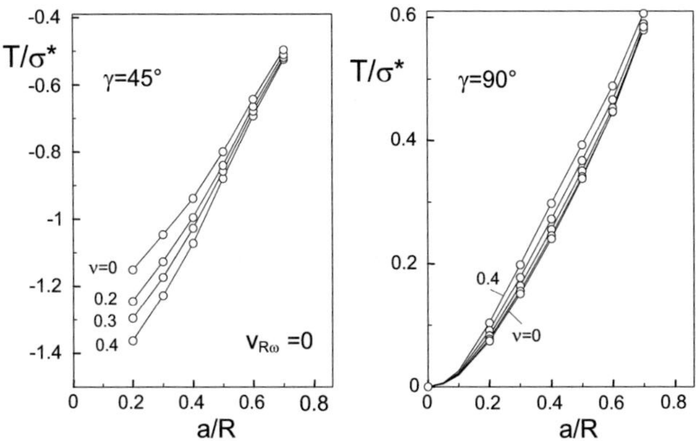

Fig. C5.34 T-stress as a function of crack size, Poisson's ratio, and loading angle γ.

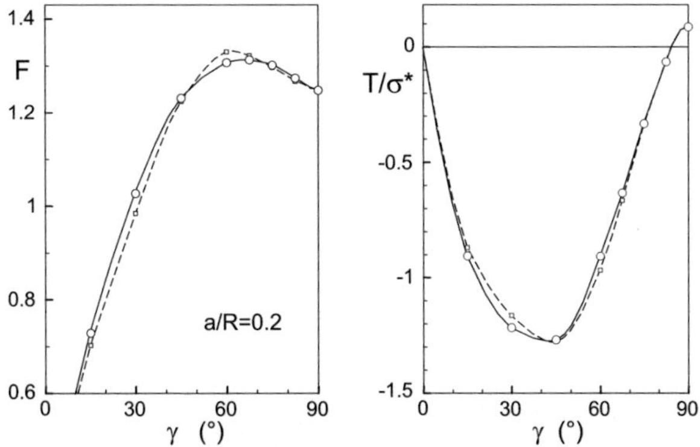

Fig. C5.35 Influence of a tangential boundary condition in the loading range on F and T-stress. Solid curve u_n=const., v_ω=0; dashed curve: u_n=const., $\tau_{R\omega}$=0.

References C5

[C5.1] Fett, T., Munz, D., Stress intensity factors and weight functions, Computational Mechanics Publications, Southampton, 1997.

[C5.2] Tada, H., Paris, P.C., Irwin, G.R., The stress analysis of cracks handbook, Del Research Corporation, 1986.

[C5.3] Atkinson, C., Smelser, R.E., Sanchez, J., Combined mode fracture via the cracked Brazilian disk test, Int. J. Fract. **18**(1982), 279-291.

[C5.4] Awaji, H., Sato, S., Combined mode fracture toughness measurement by the disk test, J. Engng. Mat. Tech. **100**(1978), 175-182.

[C5.5] Fett, T., Mode-II weight function for circular disks with internal radial crack and application to the Brazilian disk test, Int. J. Fract. **89**(1998), L9-L13.

[C5.6] Fett, T., Stress intensity factors and weight functions for special crack problems, Report FZKA 6025, Forschungszentrum Karlsruhe, 1998.

[C5.7] Fett, T., T-stresses for components with one-dimensional cracks, FZKA 6170, Forschungszentrum Karlsruhe, 1998.

[C5.8] Sato, S., Kawamata, K., Combined-mode fracture toughness of reactor-grade graphite at high temperature, High Temp.-High Press. **12**(1980), 23-32.

<u>C6</u>

Rectangular plate with an internal crack

C6.1   Stress conditions at the plate ends

The geometric data of the rectangular plate with an internal crack are illustrated in Fig. C6.1. At the ends of the plate constant tractions σ act. Data for the stress intensity factor

$$K_I = \sigma F \sqrt{\pi a}, \quad F = F' / \sqrt{1 - a/W} \qquad (C6.1.1)$$

from boundary collocation are represented by Fig. C6.2 and Table C6.1.

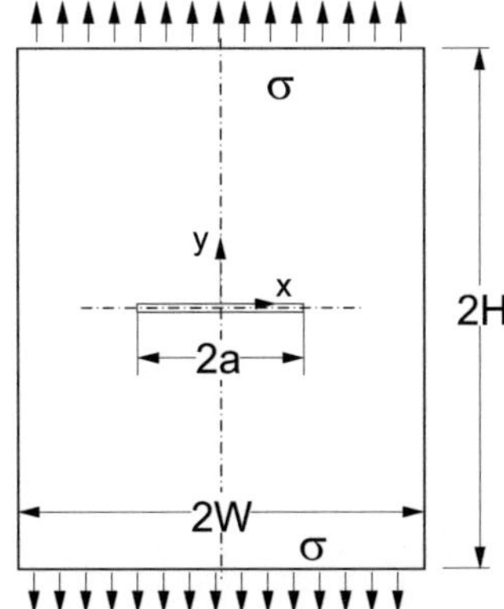

Fig. C6.1 Rectangular plate with a central internal crack (geometric data).

Table C6.1 Normalised geometric function F' for tension.

	H/W=1.5	1.25	1.00	0.75	0.5	0.35
α=0	1.00	1.00	1.00	1.00	1.00	1.00
0.2	0.916	0.924	0.940	0.977	1.051	1.182
0.3	0.888	0.905	0.940	1.008	1.147	1.373
0.4	0.869	0.890	0.942	1.053	1.262	1.562
0.5	0.851	0.877	0.943	1.099	1.391	1.742
0.6	0.827	0.856	0.937	1.130	1.533	1.938
0.7	0.816	0.826	0.914	1.125	1.668	2.197
0.8	0.814	0.818	0.840	1.088	1.689	2.41
1.0	0.826	0.826	0.826	0.826	0.826	0.826

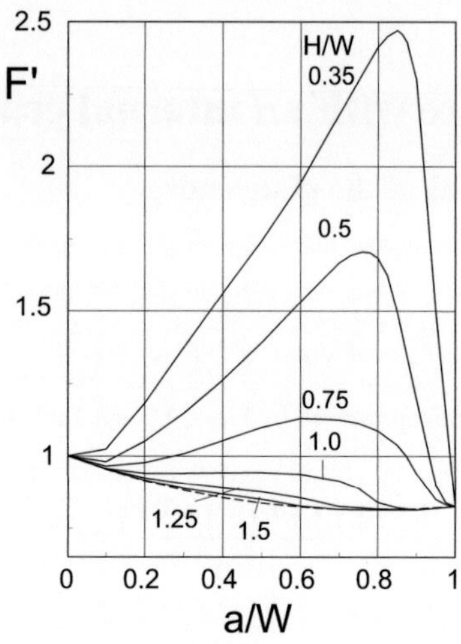

Fig. C6.2 Stress intensity factor for tensile loading.

T-stress results from BCM-computations are shown in Fig. C6.3a and Table C6.2 for different height-to-width ratios.

Table C6.2 T-stress term normalised according to $T(1-\alpha)/\sigma$ for different crack and plate geometries.

$\alpha = a/W$	H/W=0.35	0.50	0.75	1.00	1.25
0	-1.0	-1.0	-1.0	-1.0	-1.0
0.1	-0.97	-0.96	-0.92	-0.91	-0.9
0.2	-0.95	-0.92	-0.88	-0.85	-0.83
0.3	-0.766	-0.855	-0.85	-0.809	-0.777
0.4	-0.455	-0.745	-0.805	-0.756	-0.716
0.5	-0.110	-0.616	-0.738	-0.692	-0.656
0.6	0.145	-0.502	-0.647	-0.620	-0.596
0.7	0.215	-0.400	-0.543	-0.55	-0.53
0.8	0.13	-0.291	-0.45	-0.46	-0.47
0.9	-0.10	-0.25	-0.38	-0.41-	-0.43
1.0	-0.413	-0.413	-0.413	-0.413	-0.413

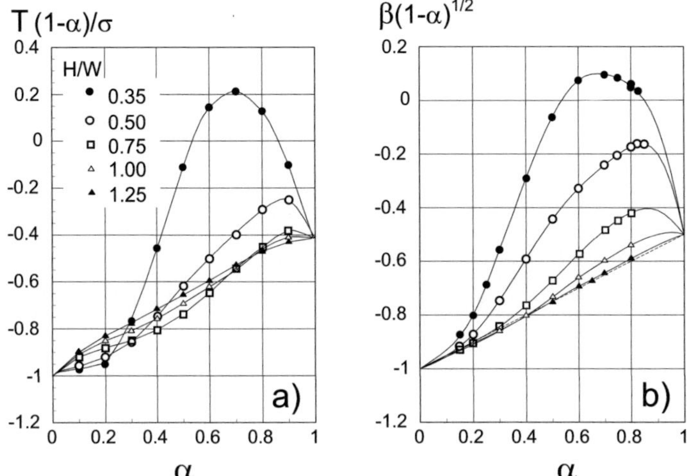

Fig. C6.3 Internal crack in a rectangular plate under tension, a) T-stress, b) biaxiality ratio.

The biaxiality ratio, as defined by eq.(A1.3.12), is plotted in Fig. C6.3b and additionally given in Table C6.3.

For a long plate ($H/W > 1.5$) and $\alpha = a/W < 0.8$, the biaxiality ratio β can be approximated by

$$\beta \cong -\frac{1-0.5\alpha}{\sqrt{1-\alpha}} \qquad (C6.1.2)$$

For the evaluation of arbitrarily distributed stresses in the uncracked plate (e.g. thermal stresses) application of the Green's function procedure is recommended. An approximate computation of T is possible by use of the T-solution for tension, exclusively. This approximation reads

$$T \cong \frac{3}{2a}(1+T_t/\sigma_0)\int_0^a (1-x^2/a^2)\sigma_y(x)dx - \sigma_y\big|_{x=a} + \sigma_x\big|_{x=a} \qquad (C6.1.3)$$

with T_t given by the data in Table C6.2. The related stress intensity factor (necessary for the computation of the biaxiality ratio β) can be calculated by using the weight function procedure. A rough approximation of the weight function reads

$$h \cong \sqrt{\frac{1+x/a}{\pi a}}\left[\frac{1}{\sqrt{1-x/a}} + 2(F-1)\sqrt{1-x/a}\right] \qquad (C6.1.4)$$

with the geometric function F for constant stress as given in Table C6.1. Coefficients for a weight function representation according to

$$h = \sqrt{\frac{2}{\pi a}} \left[\frac{1}{\sqrt{1-\rho}} + D_0\sqrt{1-\rho} + D_1(1-\rho)^{3/2} \right], \quad \rho = |x/a| \qquad (C6.1.5)$$

are compiled in Tables C6.4 and C6.5.

Table C6.3 Biaxiality ratio normalised by $\beta\,(1-\alpha)^{1/2}$ for different crack and plate geometries.

a/W	H/W=0.35	0.50	0.75	1.00	1.25
0	-1.0	-1.0	-1.0	-1.0	-1.0
0.1	-0.93	-0.95	-0.955	-0.955	-0.95
0.2	-0.801	-0.872	-0.90	-0.91	-0.905
0.3	-0.558	-0.746	-0.843	-0.860	-0.858
0.4	-0.291	-0.591	-0.764	-0.803	-0.805
0.5	-0.063	-0.443	-0.672	-0.734	-0.749
0.6	0.075	-0.328	-0.573	-0.661	-0.693
0.7	0.098	-0.241	-0.483	-0.598	-0.645
0.8	0.055	-0.173	-0.418	-0.54	-0.59
0.9	-0.1	-0.2	-0.41	0.5	-0.54
1.0	-0.5	-0.5	-0.5	-0.5	-0.5

Table C6.4 Coefficient D_0.

H/W	1.00	0.5	0.35
α=0	0.165	0.165	0.165
0.1	0.210	0.395	0.395
0.2	0.366	0.813	1.071
0.3	0.600	1.362	1.89
0.4	0.897	2.038	2.78
0.5	1.264	2.879	3.85
0.6	1.699	3.879	5.29
0.7	2.095	4.804	6.85
0.8	2.75	6.173	9.17

Table C6.5 Coefficient D_1.

H/W	1.00	0.5	0.35
α=0	0.278	0.278	0.278
0.1	0.263	0.260	0.195
0.2	0.227	0.171	0.555
0.3	0.239	0.343	0.963
0.4	0.259	0.651	1.57
0.5	0.300	1.127	2.27
0.6	0.397	1.996	3.20
0.7	0.775	3.906	5.85
0.8	0.848	5.686	9.65

The Williams coefficients A_1, B_1, A_2, and B_2, as defined by eq.(A1.1.4), are entered in Tables C6.6-C6.9.

Table C6.6 Coefficient A_1 for different crack and plate geometries.

a/W	H/W=0.35	0.50	0.75	1.00	1.25
0.2	-0.0651	-0.0817	-0.0837	-0.0824	-0.0817
0.3	0.0117	-0.0508	-0.0674	-0.0685	-0.0686
0.4	0.1223	-0.0074	-0.0493	-0.0575	-0.0603
0.5	0.2665	0.0557	-0.022	-0.0452	-0.0549
0.6	0.4560	0.1584	0.0216	-0.0300	-0.0485
0.7	0.7797	0.3607	0.0893	-0.0133	-0.1178
0.8	0.7242	0.7987	0.1645	-0.3734	-0.2886

Table C6.7 Coefficient B_1 for different crack and plate geometries.

$\alpha = a/W$	H/W=0.35	0.50	0.75	1.00	1.25
0.2	-0.2608	-0.0792	-0.0180	-0.0064	-0.0019
0.3	-0.5306	-0.1920	-0.0527	-0.0197	-0.0053
0.4	-0.7606	-0.3129	-0.1065	-0.0409	-0.0089
0.5	-0.9124	-0.4263	-0.1787	-0.0655	-0.0086
0.6	-0.9652	-0.5736	-0.2694	-0.0812	-0.0041
0.7	-1.096	-0.9091	-0.3629	-0.0555	0.333
0.8	-1.429	-1.709	-0.3075	1.154	0.8425

Table C6.8 Coefficient A_2.

a/W	H/W=0.50	0.75	1.25
0.2	0.1977	0.136	0.113
0.3	0.2126	0.118	0.070
0.4	0.2372	0.139	0.057
0.5	0.2797	0.188	0.057
0.6	0.4367	0.278	0.079
0.65	0.6322	0.352	0.119
0.7	0.9848	0.462	-0.079
0.8	2.748	0.911	-0.463

Table C6.9 Coefficient B_2.

a/W	H/W=0.50	0.75	1.25
0.2	-0.06174	-0.023	-0.003
0.3	0.0133	-0.032	-0.005
0.4	0.1697	-0.031	-0.003
0.5	0.3255	-0.032	0.000
0.6	0.3194	-0.063	-0.004
0.65	0.1475	-0.104	-0.022
0.7	-0.2523	-0.190	0.025
0.8	-2.747	-0.816	0.092

C6.2 Mixed boundary conditions at the ends

A rectangular plate with an internal crack is loaded by application of a constant displacement v but disappearing shear stress (Fig. C6.4). The geometric function for K, T-stress, and higher coefficients A_1 and B_1 are given in Tables C6.10-C6.16. The characteristic stress σ_0 is defined by the constant plate end displacements v as

$$\sigma_0 = \frac{v}{H} E \qquad\qquad (C6.2.1)$$

(E= Young's modulus).

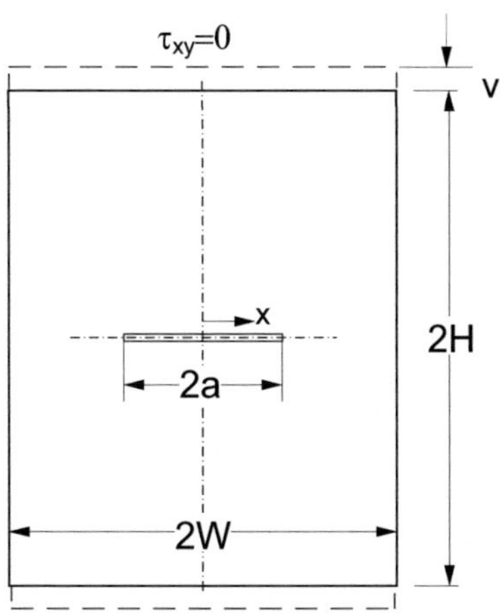

Fig. C6.4 Internally cracked plate with mixed boundary conditions at the ends.

Table C6.10 Geometric function F for stress intensity factor solution.

a/W	H/W=0.25	0.50	0.75	1.00	1.25
0.00	1.00	1.00	1.00	1.00	1.00
0.25	0.570	0.790	0.889	0.937	0.959
0.3	0.518	0.735	0.852	0.913	0.944
0.4	0.446	0.642	0.778	0.860	0.907
0.5	0.399	0.573	0.737	0.805	0.865
0.6	0.364	0.523	0.652	0.751	0.823
0.7	0.338	0.485	0.603	0.702	0.778
0.8	0.319	0.455	0.562	0.667	-

Table C6.11 T-stress data T/σ_0.

a/W	$H/W=0.25$	0.50	0.75	1.00	1.25
0.00	-1.00	-1.00	-1.00	-1.00	-1.00
0.25	-0.606	-0.756	-0.869	-0.932	-0.964
0.3	-0.596	-0.707	-0.832	-0.910	-0.952
0.4	-0.592	-0.646	-0.770	-0.869	-0.928
0.5	-0.592	-0.626	-0.737	-0.840	-0.912
0.6	-0.594	-0.637	-0.734	-0.833	-0.913
0.7	-0.600	-0.674	-0.760	-0.857	-0.965
0.8	-0.635	-0.740	-0.831	-0.98	-

Table C6.12 Biaxiality ratio β.

a/W	$H/W=0.25$	0.50	0.75	1.00	1.25
0.00	-1.00	-1.00	-1.00	-1.00	-1.00
0.25	-1.064	-0.957	-0.977	-0.995	-1.005
0.3	-1.151	-0.962	-0.976	-0.997	-1.008
0.4	-1.327	-1.007	-0.990	-1.010	-1.022
0.5	-1.485	-1.093	-1.037	-1.044	-1.054
0.6	-1.630	-1.219	-1.125	-1.110	-1.109
0.7	-1.777	-1.389	-1.260	-1.220	-1.240
0.8	-1.993	-1.627	-1.477	-1.474	-

Table C6.13 Coefficient A_1 for the internally cracked plate.

a/W	$H/W=0.25$	0.50	0.75	1.00	1.25
0.25	-0.0734	-0.0624	-0.0648	-0.0668	-0.0682
0.3	-0.0735	-0.0575	-0.0581	-0.0599	-0.0614
0.4	-0.0740	-0.0533	-0.0499	-0.0503	-0.0515
0.5	-0.0742	-0.0527	-0.0457	-0.0439	-0.0448
0.6	-0.0743	-0.0532	-0.0430	-0.0393	-0.0396
0.7	-0.0748	-0.0528	-0.0398	-0.0349	-0.0416
0.8	-0.0758	-0.0488	-0.0348	-0.0392	

Table C6.14 Coefficient B_1 for the internally cracked plate.

a/W	H/W=0.25	0.50	0.75	1.00	1.25
0.25	0.2239	0.0514	0.0167	0.0071	0.0038
0.3	0.2384	0.0699	0.0258	0.0116	0.0063
0.4	0.2454	0.1005	0.0466	0.0232	0.0140
0.5	0.2457	0.1220	0.0675	0.0374	0.0261
0.6	0.2468	0.1385	0.0853	0.0542	0.0428
0.7	0.2544	0.1524	0.1001	0.0721	0.0873
0.8	0.2822	0.1634	0.1222	0.1262	

Table C6.15 Coefficient A_2 for the internally cracked plate.

a/W	H/W=0.25	0.50	0.75	1.00	1.25
0.25	-0.087	0.025	0.057	0.068	0.072
0.3	-0.092	0.000	0.035	0.048	0.053
0.4	-0.090	-0.024	0.009	0.026	0.033
0.5	-0.089	-0.032	0.000	0.018	0.024
0.6	-0.089	-0.030	0.004	0.021	0.028
0.7	-0.092	-0.011	0.029	0.049	0.037
0.8	-0.079	0.059	0.109	0.125	

Table C6.16 Coefficient B_2 for the internally cracked plate.

a/W	H/W=0.25	0.50	0.75	1.00	1.25
0.25	-0.035	0.038	0.018	0.008	0.004
0.3	-0.077	0.032	0.020	0.010	0.006
0.4	-0.104	0.014	0.020	0.011	0.008
0.5	-0.106	0.000	0.011	0.007	0.007
0.6	-0.101	-0.011	-0.009	-0.006	-0.002
0.7	-0.084	-0.042	-0.052	-0.061	-0.033
0.8	-0.072	-0.159	-0.188	-0.276	

C6.3   Displacement boundary conditions at the ends

If free deformation in x-direction is suppressed, u=0, pure displacement boundary conditions are fulfilled (Fig. C6.5). The geometric function, T-stress, and higher coefficients A_1 and B_1 are given in Tables C6.17-C6.31. The characteristic stress σ_0 is again defined by eq.(C6.2.1).

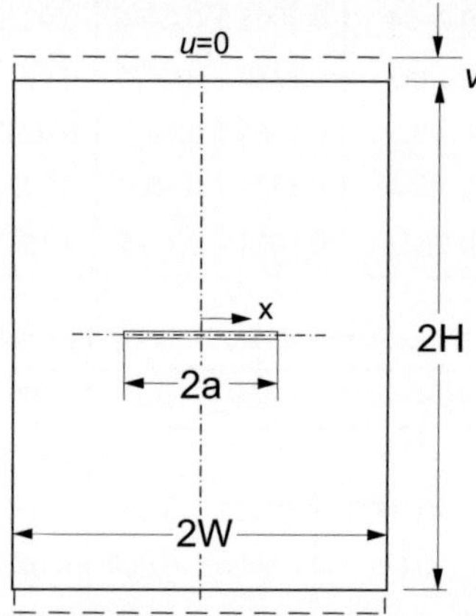

Fig. C6.5 Internally cracked plate with pure displacement conditions at the ends.

Table C6.17 T-stress T/σ_0 for $H/W = 0.25$.

a/W	$v=0$	0.1	0.2	0.3	0.4
0	-1.000				
0.3	-0.612	-0.567	-0.527	-0.492	-0.462
0.4	-0.600	-0.563	-0.533	-0.509	-0.491
0.5	-0.598	-0.568	-0.545	-0.529	-0.520
0.6	-0.602	-0.578	-0.561	-0.551	-0.549

Table C6.18 Geometric function F for $H/W = 0.25$.

a/W	$v = 0$	0.1	0.2	0.3	0.4
0	1.000				
0.3	0.518	0.519	0.527	0.541	0.561
0.4	0.447	0.449	0.456	0.467	0.483
0.5	0.399	0.402	0.408	0.417	0.430
0.6	0.365	0.367	0.372	0.380	0.391

Table C6.19 Biaxiality ratio β for $H/W = 0.25$.

a/W	$\nu = 0$	0.1	0.2	0.3	0.4
0	-1.000				
0.3	-1.182	-1.090	-1.000	-0.992	-0.825
0.4	-1.343	-1.253	-1.169	-1.090	-1.016
0.5	-1.499	-1.413	-1.336	-1.268	-1.210
0.6	-1.651	-1.575	-1.508	-1.451	-1.404

Table C6.20 Coefficient A_1 for $H/W = 0.25$.

a/W	$\nu=0$	0.1	0.2	0.3	0.4
0.3	-0.0757	-0.0777	-0.0813	-0.0865	-0.0932
0.4	-0.0759	-0.0780	-0.0816	-0.0866	-0.0931
0.5	-0.0761	-0.0783	-0.0817	-0.0864	-0.0924
0.6	-0.0767	-0.0787	-0.0817	-0.0857	-0.0908

Table C6.21 Coefficient B_1 for $H/W = 0.25$.

a/W	$\nu=0$	0.1	0.2	0.3	0.4
0.3	0.2287	0.2302	0.2380	0.2520	0.2723
0.4	0.2376	0.2386	0.2455	0.2582	0.2768
0.5	0.2411	0.2444	0.2530	0.2668	0.2858
0.6	0.2451	0.2575	0.2740	0.2945	0.319

Table C6.22 T-stress T/σ_0 for $H/W = 0.50$.

a/W	$\nu=0$	0.1	0.2	0.3	0.4
0	-1.000				
0.3	-0.729	-0.697	-0.673	-0.657	-0.648
0.4	-0.675	-0.656	-0.643	-0.636	-0.634
0.5	-0.660	-0.650	-0.645	-0.645	-0.651
0.6	-0.667	-0.665	-0.666	-0.671	-0.679
0.7	-0.697	-0.698	-0.701	-0.707	-0.715

Table C6.23 Geometric function F for $H/W = 0.50$.

a/W	$\nu = 0$	0.1	0.2	0.3	0.4
0	1.000				
0.3	0.731	0.735	0.745	0.762	0.786
0.4	0.640	0.642	0.649	0.661	0.677
0.5	0.572	0.574	0.579	0.587	0.599
0.6	0.522	0.523	0.527	0.533	0.541
0.7	0.484	0.485	0.487	0.490	0.495

Table C6.24 Biaxiality ratio β for $H/W = 0.50$.

a/W	$\nu = 0$	0.1	0.2	0.3	0.4
0	-1.000				
0.3	-0.998	-0.949	-0.904	-0.863	-0.825
0.4	-1.056	-1.022	-0.991	-0.963	-0.937
0.5	-1.152	-1.132	-1.114	-1.099	-1.087
0.6	-1.278	-1.269	-1.263	-1.259	-1.257
0.7	-1.440	-1.439	-1.439	-1.440	-1.443

Table C6.25 Coefficient A_1 for $H/W = 0.50$.

a/W	$\nu = 0$	0.1	0.2	0.3	0.4
0.3	-0.0586	-0.0597	-0.0614	-0.0637	-0.0665
0.4	-0.0548	-0.0561	-0.0579	-0.0601	-0.0628
0.5	-0.0541	-0.0554	-0.0571	-0.0591	-0.0614
0.6	-0.0542	-0.0552	-0.0565	-0.0580	-0.0597
0.7	-0.0540	-0.0543	-0.0549	-0.0557	-0.0567

Table C6.26 Coefficient B_1 for $H/W = 0.50$.

a/W	$\nu = 0$	0.1	0.2	0.3	0.4
0.3	0.0717	0.0806	0.0904	0.1012	0.1129
0.4	0.1000	0.1089	0.1190	0.1303	0.1429
0.5	0.1172	0.1257	0.1348	0.1446	0.1550
0.6	0.1309	0.1370	0.1433	0.1499	0.1569
0.7	0.1499	0.1489	0.1492	0.1509	0.1540

Table C6.27 T-stress T/σ_0 for $H/W = 1.0$.

a/W	$\nu = 0$	0.1	0.2	0.3	0.4
0	-1.000				
0.3	-0.910	-0.911	-0.918	-0.930	-0.947
0.4	-0.871	-0.870	-0.873	-0.880	-0.892
0.5	-0.845	-0.842	-0.843	-0.847	-0.855
0.6	-0.842	-0.838	-0.837	-0.838	-0.842
0.7	-0.872	-0.867	-0.864	-0.863	-0.865
0.8	-0.958	-0.960	-0.963	-0.967	-0.973

Table C6.28 Geometric function F for $H/W = 1.0$.

a/W	$\nu = 0$	0.1	0.2	0.3	0.4
0	1.000				
0.3	0.905	0.915	0.929	0.948	0.971
0.4	0.851	0.857	0.866	0.879	0.895
0.5	0.795	0.798	0.803	0.811	0.822
0.6	0.744	0.744	0.746	0.750	0.757
0.7	0.699	0.698	0.698	0.700	0.703
0.8	0.666	0.665	0.665	0.667	0.669

Table C6.29 Biaxiality ratio β for $H/W = 1.0$.

a/W	$\nu = 0$	0.1	0.2	0.3	0.4
0	-1.000				
0.3	-1.006	-0.996	-0.988	-0.981	-0.975
0.4	-1.024	-1.015	-1.008	-1.002	-0.997
0.5	-1.063	-1.056	-1.050	-1.045	-1.040
0.6	-1.132	-1.127	-1.122	-1.117	-1.113
0.7	-1.247	-1.242	-1.238	-1.234	-1.231
0.8	-1.440	-1.444	-1.448	-1.451	-1.454

Table C6.30 Coefficient A_1 for $H/W = 1.0$.

a/W	$\nu=0$	0.1	0.2	0.3	0.4
0.3	-0.0599	-0.0602	-0.0608	-0.0616	-0.0626
0.4	-0.0507	-0.0506	-0.0507	-0.0510	-0.0514
0.5	-0.0451	-0.0447	-0.0445	-0.0444	-0.0444
0.6	-0.0416	-0.0410	-0.0405	-0.0401	-0.0398
0.7	-0.0388	-0.0380	-0.0374	-0.0369	-0.0365
0.8	-0.0329	-0.0338	-0.0346	-0.0353	-0.0359

Table C6.31 Coefficient B_1 for $H/W = 1.0$.

a/W	$\nu=0$	0.1	0.2	0.3	0.4
0.3	0.0123	0.0127	0.0127	0.0124	0.0118
0.4	0.0245	0.0248	0.0248	0.0245	0.0238
0.5	0.0402	0.0399	0.0395	0.0389	0.0381
0.6	0.0594	0.0583	0.0572	0.0561	0.0549
0.7	0.0842	0.0817	0.0797	0.0781	0.0770
0.8	0.1202	0.1227	0.1252	0.1278	0.1304

C6.4 Sub-surface cracks

A sub-surface crack is shown in Fig. C6.6. A stress intensity factor solution for constant crack-face loading σ_0 was proposed by Isida [C1] for the region of $a/(d+a)=\alpha<0.9$ as

$$F_A = 1+\sum_{n=2}^{19} C_n\alpha^n \, , \quad F_B = 1+\sum_{n=2}^{19}(-1)^n C_n\alpha^n \qquad (C6.4.1)$$

with

$$
\begin{array}{llllll}
C_2 = 0.25 & C_3 = 0.125 & C_4 = 0.1328 & C_5 = 0.0781 & C_6 = 0.0967 & C_7 = 0.0671 \\
C_8 = 0.0836 & C_9 = 0.0618 & C_{10} = 0.0766 & C_{11} = 0.0585 & C_{12} = 0.0724 & C_{13} = 0.0562 \\
C_{14} = 0.0697 & C_{15} = 0.0544 & C_{16} = 0.0678 & C_{17} = 0.0529 & C_{18} = 0.0662 & C_{19} = 0.0517 .
\end{array}
$$

Values of the weight function for the sub-surface cracks are shown in Fig. C6.6. The integration has to be performed according to

$$K_A = \int_{-a}^{a} h_A(a,\xi)\,\sigma(\xi)\,d\xi \, , \quad K_B = \int_{-a}^{a} h_B(a,\xi)\,\sigma(\xi)\,d\xi . \qquad (C6.4.2)$$

A simple solution (for not too high ratios a/d) was proposed in [C6.2]

$$h_A(x,a) \cong \sqrt{\frac{1-\rho}{\pi a}}\left[\frac{1}{\sqrt{1-\rho}}+2(F_A-1)\sqrt{1+\rho}\right] \qquad (C6.4.3)$$

$$h_B(x,a) \cong \sqrt{\frac{1+\rho}{\pi a}}\left[\frac{1}{\sqrt{1+\rho}}+2(F_B-1)\sqrt{1-\rho}\right] \qquad (C6.4.4)$$

($\rho=\xi/a$) where F_A and F_B are the geometric functions for the sub-surface crack having a constant crack surface loading $\sigma=\sigma_0=$ constant, see (C6.4.1).

This approximate solution is indicated by the curves in Fig. C6.6. Results of Aliabadi et al. [C6.3] are entered as squares. The data from the weight function procedure are shown by the circles. Deviations between the numerical results and the approximation (C6.4.3, C6.4.4) are less than 1% in this case.

In Fig. C6.7 the deviations between the weight function for the sub-surface crack and the weight function for the crack in an infinite body

$$h'_\infty = \sqrt{a\pi}\,h_\infty = \sqrt{\frac{1\mp\rho}{1\pm\rho}} \qquad (C6.4.5)$$

are plotted. In (C6.4.5), the upper signs are related to location A, the lower ones to location B. The differences between the numerical results (circles) and the approximate solution, eq.(C6.4.4) (curves), clearly indicate the occurrence of an antisymmetric term.

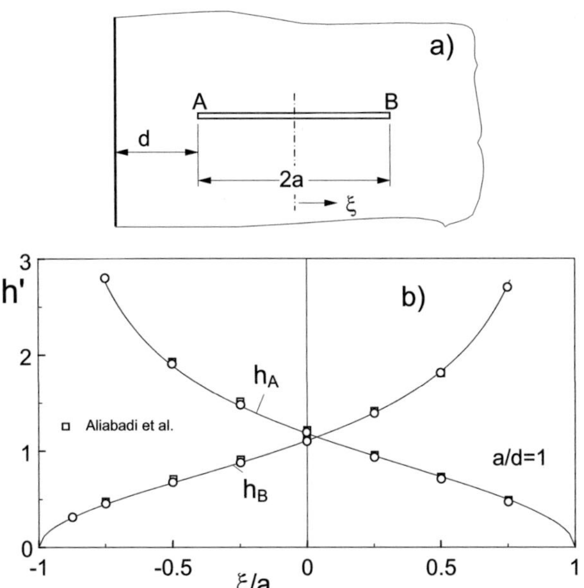

Fig. C6.6 a) Subsurface crack, b) curves: Fett and Munz [C6.2] and eqs.(C6.4.3, C6.4.4)), circles [C6.4], squares: Aliabadi et al.[C6.3]. Normalisation: $h' = h\sqrt{\pi a}$.

In order to improve the approximate weight function, the next term of the weight function is considered by

$$h_A(\rho,a) = \frac{1}{\sqrt{\pi a}}\sqrt{\frac{1-\rho}{1+\rho}}[1+2(F_A-1)(1+\rho)-C_A\rho(1+\rho)] \qquad (C6.4.6)$$

$$h_B(\rho,a) = \frac{1}{\sqrt{\pi a}}\sqrt{\frac{1+\rho}{1-\rho}}[1+2(F_B-1)(1-\rho)-C_B\rho(1-\rho)] \qquad (C6.4.7)$$

As can be seen from these relations, the next correction term is antisymmetric with respect to $\rho=0$ (see Fig. C6.7). The numerical results can be described well by eqs.(C6.4.6,C6.4.7) up to $\alpha = 0.7$ ($a/d = 2.3$) with maximum deviations of less than 3%. The coefficients C_A and C_B obtained by curve fitting are

$$C_A \cong 1.492\alpha^{4.676} \quad , \quad C_B \cong 0.221\alpha^{3.433}, \quad \alpha = a/(a+d). \qquad (C6.4.8)$$

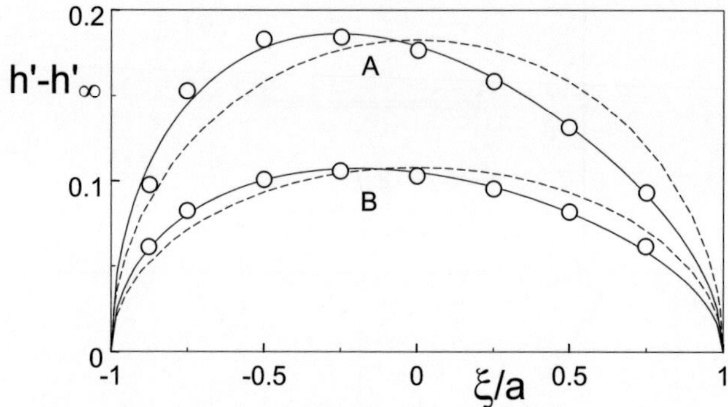

Fig. C6.7 Deviation of the (normalized) weight function for the sub-surface crack, h', from the weight function solution for a crack in an infinite body h'_∞ (circles: numerical results, dashed curves: eqs.(C6.4.3, C6.4.4), continuous curves: eqs.(C6.4.6, C6.4.7)).

C6.5 Transverse loading

An edge-cracked plate under transverse traction σ_x is illustrated in Fig. C6.8. Under this loading, the stress intensity factor may be defined by

$$K = \sigma_x F\sqrt{\pi a} \tag{C6.5.1}$$

The geometric function F is plotted in Fig. C6.9 for several values of a/W, H/W, and d/W. Figure C6.10 represents the T-stresses.

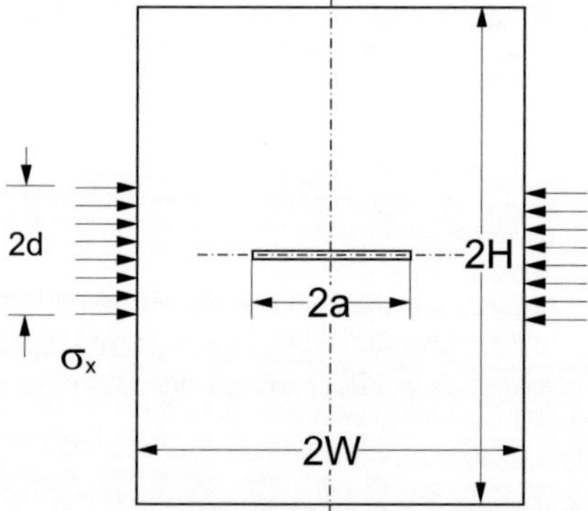

Fig. C6.8 Internally cracked plate with transverse loading.

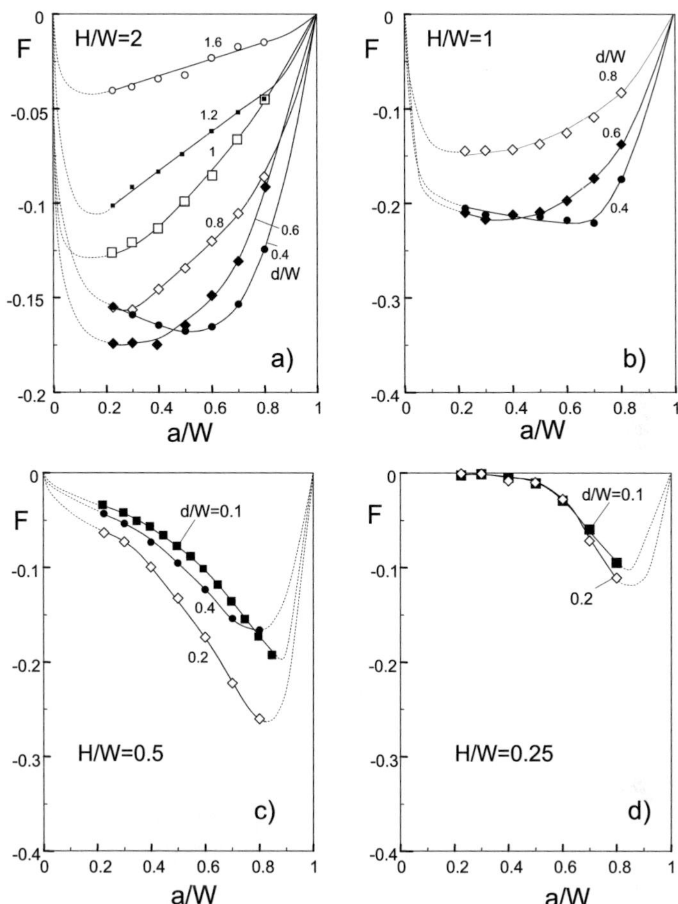

Fig. C6.9 Geometric function F according to eq.(C6.5.1).

167

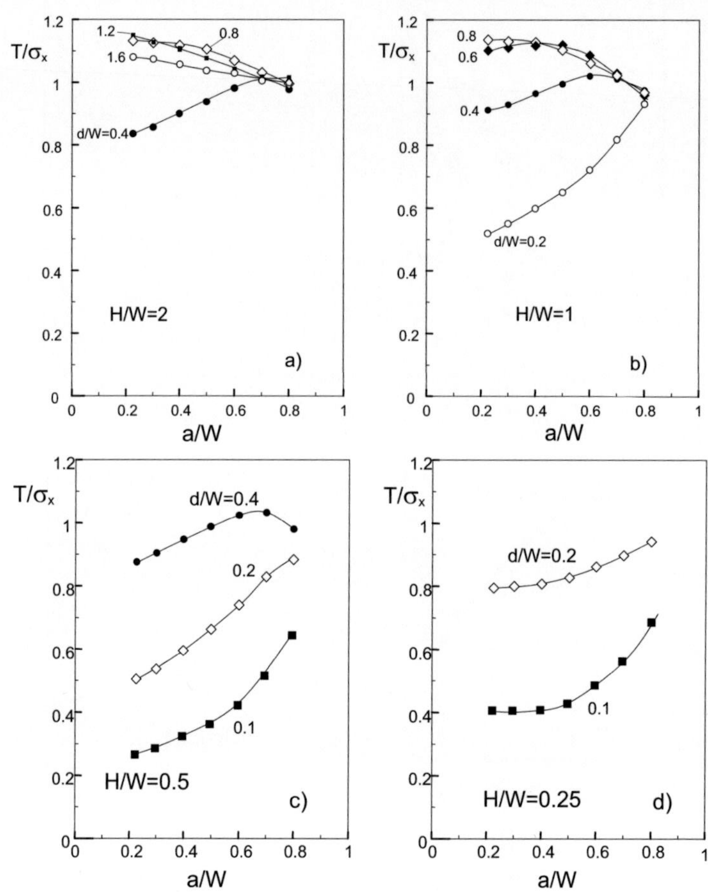

Fig. C6.10 T-stresses represented as T/σ_x.

References C6

[C6.1] Isida, M., Stress intensity factors for the tension of an eccentrically cracked strip, J. Appl. Mech. **33**(1965), 674.

[C6.2] Fett, T., Munz, D., Stress intensity factors and weight functions, Computational Mechanics Publications, Southampton, 1997.

[C6.3] Aliabadi, M.H., Cartwright, D.J., Doney, W., Bueckner weight functions for cracks near a half-plane, Engng. Fract. Mech. **37**(1990), 437-446.

[C6.4] Fett, T., T-stress and stress intensity factor solutions for 2-dimensional cracks, VDI-Verlag, 2002, Düsseldorf.

C7

Edge-cracked circular disk

Edge-cracked circular disks are often used as fracture mechanics test specimens. Examples are the RCT specimen and a modification of the Brazilian disk test. Figure C7.1 shows the geometric data.

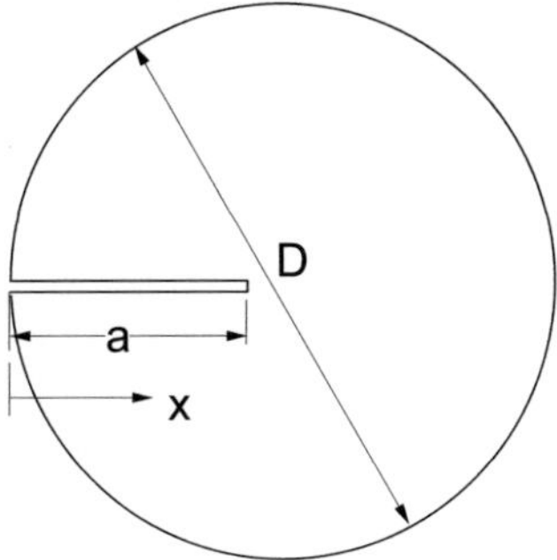

Fig. C7.1 Geometric data of an edge-cracked circular disk.

C7.1 Circumferentially loaded disk (traction boundary conditions)

A circular disk is loaded by constant normal tractions σ_n along the circumference (for loading see Fig. C7.2)

$$\sigma_n = \text{const} , \quad \tau_{R\omega} = 0 \tag{C7.1.1}$$

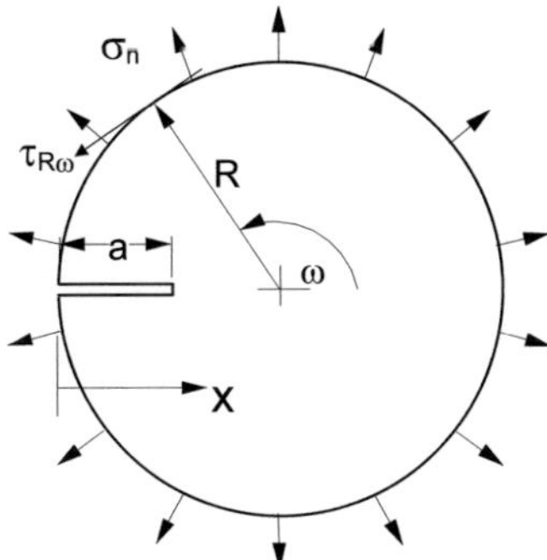

Fig. C7.2 Edge-cracked circular disk under pure stress boundary conditions.

The stress intensity factor solution for this loading case is

$$K_1 = \sigma_n F \sqrt{\pi a}, \quad F = \frac{1.1215}{(1-\alpha)^{3/2}}, \alpha = a/D \tag{C7.1.2}$$

For a single-edge-cracked disk a weight function is given in [C7.1] as

$$h(x,a) = \sqrt{\frac{2}{\pi a}} \left[\frac{\rho}{\sqrt{1-\rho}} + D_0\sqrt{1-\rho} + D_1(1-\rho)^{3/2} + D_2(1-\rho)^{5/2} \right] \tag{C7.1.3}$$

with $\rho = x/a$ and the coefficients of

$$\begin{aligned}
D_0 &= (1.5721 + 2.4109\alpha - 0.8968\alpha^2 - 1.4311\alpha^3)/(1-\alpha)^{3/2} \\
D_1 &= (0.4612 + 0.5972\alpha + 0.7466\alpha^2 + 2.2131\alpha^3)/(1-\alpha)^{3/2} \\
D_2 &= (-0.2537 + 0.4353\alpha - 0.2851\alpha^2 - 0.5853\alpha^3)/(1-\alpha)^{3/2}
\end{aligned} \tag{C7.1.4}$$

In this case, it holds [C7.1]

$$B_0(1-\alpha)^2 = -0.11851 \tag{C7.1.5}$$

and

$$\frac{T}{\sigma_n} = -4B_0 = \frac{0.474}{(1-\alpha)^2} \tag{C7.1.6}$$

The biaxiality ratio is given by

$$\beta = \frac{0.4227}{\sqrt{1-\alpha}} \tag{C7.1.7}$$

Figure C7.3a shows the Green's function for several crack depths. For $a/D \leq 0.6$, it can be expressed by

$$t = \frac{1}{a(1-a/D)^2}(C_0\sqrt{1-x/a} + C_1(1-x/a)^{3/2} + C_2(1-x/a)^{5/2}) \tag{C7.1.8}$$

with the coefficients

$$C_0 = 0.328 - 0.2927\alpha^2$$

$$C_1 = 0.179 + 0.1134\alpha + 1.746\alpha^2 \tag{C7.1.9}$$

$$C_2 = 0.663 - 0.4171\alpha - 1.308\alpha^2$$

In Fig. C7.3b a common curve unique curve approximates all data. As a consequence of Fig. C3.7b, an approximation can be given by

$$t \cong \frac{1}{a(1-a/D)^2}(0.3\sqrt{1-x/a}+0.3945(1-x/a)^{3/2}+0.4025(1-x/a)^{5/2}) \quad \text{(C7.1.10)}$$

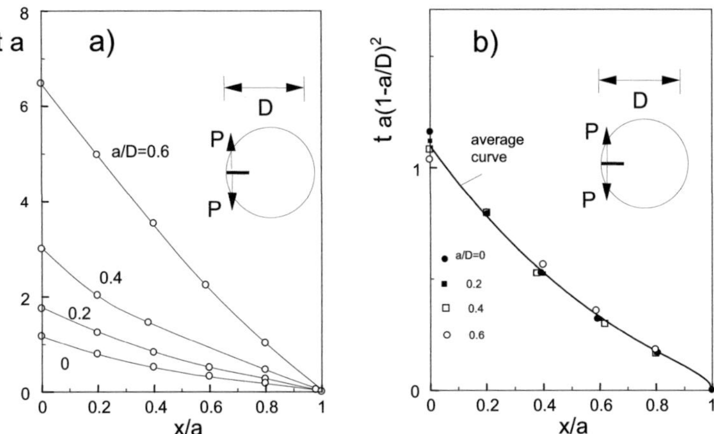

Fig. C7.3 Green's function for the edge-cracked disk: a) Results for several relative crack depths, b) results of a) in a normalized representation.

A rough estimation of the Green's function can be derived from the reference solution eq.(C7.1.6) as

$$t \approx \tfrac{1}{a}E_0(1-x/a) \quad \text{(C7.1.11)}$$

with

$$E_0 = \frac{0.9481}{(1-\alpha)^2} , \quad \alpha = a/D \quad \text{(C7.1.12)}$$

This relation is introduced in Fig. C7.4a as the dashed straight line. In this rough approximation, the T-stress reads

$$T \approx \frac{0.9481}{(1-\alpha)^2}\int_0^1 (1-\rho)\sigma_y(\rho)d\rho -\sigma_y\big|_{x=a}+\sigma_x\big|_{x=a} , \quad \rho = x/a \quad \text{(C7.1.13)}$$

The Green's function at the crack mouth, $x=0$, is plotted in Fig. C7.4b. The straight-line fit of the data points is

$$t_{(x=0)} \cong \frac{1.164 - 0.2147\,\dfrac{a}{D}}{a(1-a/D)^2}$$ (C7.1.14)

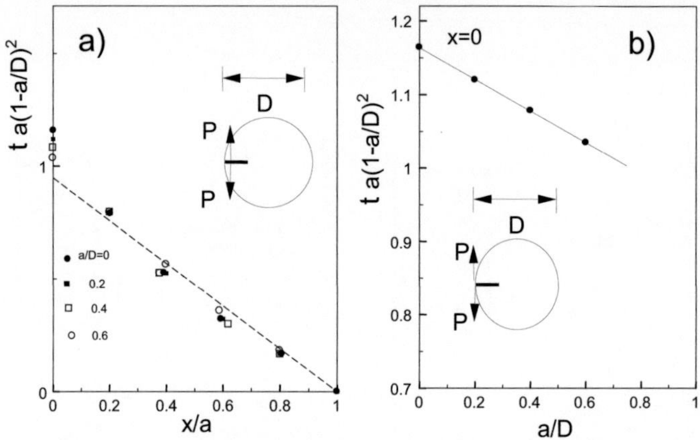

Fig. C7.4 Green's function for the edge-cracked disk: a) Results of Fig. C7.3b fitted by a straight line relation according to eq.(C7.1.11), b) Green's function at the crack mouth (x/a=0).

Further coefficients of the Williams stress function [C7.2] are

$$A_1 = \frac{-0.02279 + 0.1322\,\alpha}{(1-\alpha)^{5/2}\sqrt{\alpha}}$$ (C7.1.15)

$$B_1 = \frac{0.04812 - 0.1185\alpha}{(1-\alpha)^3\,\alpha}$$ (C7.1.16)

$$A_2 = \frac{-0.00680 - 0.03416\alpha + 0.0991\alpha^2}{(1-\alpha)^{7/2}\,\alpha^{3/2}}$$ (C7.1.17)

$$B_2 = \frac{-0.01787 + 0.09627\alpha - 0.11851\alpha^2}{(1-\alpha)^4\,\alpha^2}$$ (C7.1.18)

For special applications also crack opening displacements δ at the crack mouth $x = 0$ are of interest. Figure C7.5 represents the displacements under constant normal traction σ_n in the form of

$$\delta = \frac{2a\sigma_n}{E'}(1-\alpha)^2\lambda(\alpha)$$ (C7.1.19)

The results of boundary collocation computations are represented by the circles. From a least-squares fit, one obtains

$$\lambda \cong 1.454 + 0.526\,\alpha \qquad\qquad (C7.1.20)$$

The dashed curve in Fig. C7.5 is the solution for the single edge-cracked endless parallel strip as reported by Tada [C7.3].

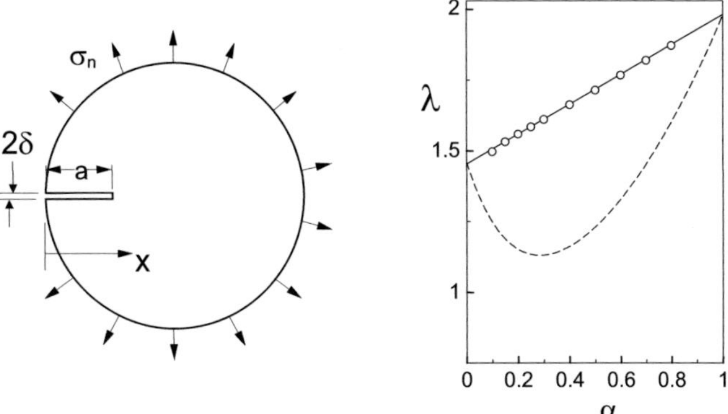

Fig. C7.5 Crack-mouth displacements $(x=0)$ according to eq.(C7.1.19); circles: edge-cracked disk, dashed curve: results for the single edge-cracked endless parallel strip, reported by Tada [C7.3].

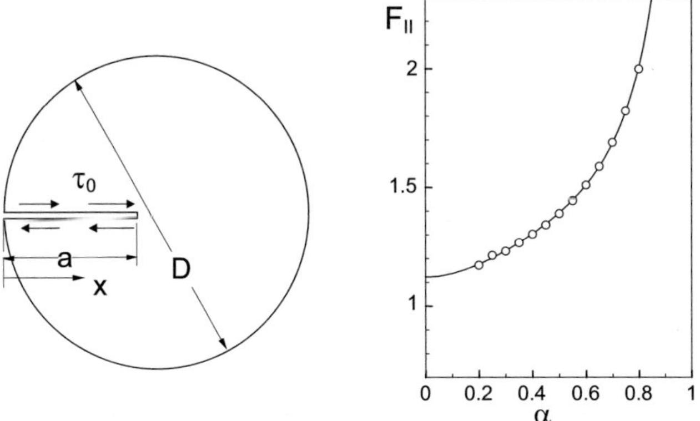

Fig. C7.6 Geometric function for loading of the crack faces by constant shear traction.

The stress intensity factor for mode-II loading by constant shear traction τ_0 is

$$K_{II} = \tau_0 F_{II} \sqrt{\pi a} \qquad \text{(C7.1.21)}$$

The related geometric function F_{II} is plotted in Fig. C7.6. A fit relation for $\alpha = a/D \leq 0.8$ is given by

$$F_{II} \cong \frac{1.1216 - 0.5608\alpha + 1.3433\alpha^2 - 1.9734\alpha^3 + 0.8954\alpha^4}{\sqrt{1-\alpha}} \qquad \text{(C7.1.22)}$$

A mode-II weight function is

$$h_{II}(x,a) = \sqrt{\frac{2}{\pi a}}\left[\frac{1}{\sqrt{1-\rho}} + D_0\sqrt{1-\rho} + D_1(1-\rho)^{3/2} + D_2(1-\rho)^{5/2}\right] \qquad \text{(C7.1.23)}$$

with the coefficients compiled in Table C7.1, which can be interpolated by cubic splines. For $a/D \leq 0.8$ the coefficients are approximated as

$$D_0 = \frac{0.407 + 0.2393\alpha + 0.04666\alpha^2 - 0.547\alpha^3}{(1-\alpha)^{3/2}} \qquad \text{(C7.1.24a)}$$

$$D_1 = \frac{0.6863 - 1.1193\alpha + 0.9002\alpha^2 - 0.8361\alpha^3}{(1-\alpha)^{3/2}} \qquad \text{(C7.1.24b)}$$

$$D_2 = \frac{-0.3117 + 0.552\alpha - 0.2758\alpha^2 + 0.1259\alpha^3}{(1-\alpha)^{3/2}} \qquad \text{(C7.1.24c)}$$

Higher-order coefficients for the antisymmetric stress function, eq.(A1.1.12), are compiled in Table C7.2.

Table C7.1 Coefficients for the mode-II weight function eq.(C7.1.23).

a/D	D_0	D_1	D_2
0.1	0.4981	0.6931	-0.305
0.2	0.6228	0.6853	-0.296
0.25	0.7069	0.6856	-0.290
0.3	0.8032	0.6896	-0.284
0.35	0.9118	0.6965	-0.278
0.4	1.034	0.7044	-0.272
0.45	1.172	0.7096	-0.264
0.5	1.332	0.7064	-0.253
0.55	1.523	0.6861	-0.236

0.6	1.759	0.6352	-0.210
0.65	2.062	0.5321	-0.169
0.7	2.470	0.3394	-0.103
0.75	3.052	-0.013	0.006
0.8	3.944	-0.674	0.198

Table C7.2 Higher-order coefficients for mode-II loading according to eq.(A1.1.12).

a/D	$\hat{B}_0$	$\hat{A}_1$	$\hat{B}_1$
0.2	0.175	0.217	-0.832
0.3	0.164	0.144	-0.441
0.4	0.178	0.089	-0.276
0.5	0.215	0.039	-0.159
0.6	0.294	-0.018	0.000
0.7	0.471	-0.094	0.412
0.8	0.981	-0.229	2.231

C7.2   Diametrically loaded disk

A disk of unit thickness is considered, which is diametrically loaded by a pair of tensile forces P (Fig. C7.7). The forces may act perpendicularly to the crack plane. In this case, the stresses are given by

$$\frac{\sigma_y}{\sigma^*} = \frac{4}{[1+(1-\xi)^2]^2} - 1 \tag{C7.2.1}$$

$$\frac{\sigma_x}{\sigma^*} = \frac{4(1-\xi)^2}{[1+(1-\xi)^2]^2} - 1 \ , \quad \sigma^* = \frac{P}{\pi RB} \tag{C7.2.2}$$

with
$$\xi = x/R \ , \quad R = D/2$$

as illustrated in Fig. C7.7.

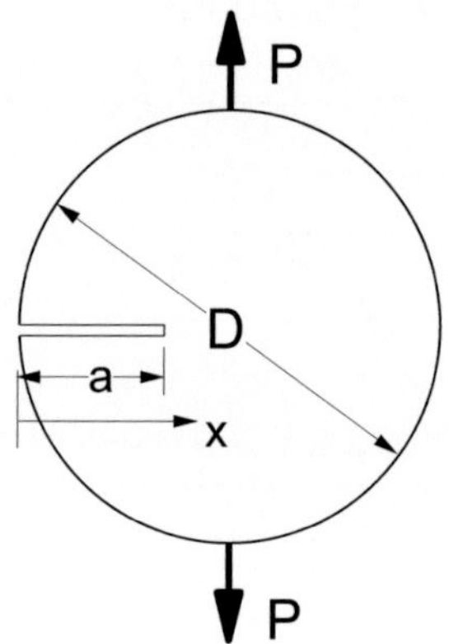

Fig. C7.7 Diametrically loaded circular disk.

The stress intensity factor results from application of eq.(A3.1.10a) with the weight function of eq.(C7.1.3) and the T-term from eq.(C7.1.8)

$$T \cong \frac{0.9481\sigma^*}{2(1-\alpha)^2(a/R)^2}\left[4\left(1-\frac{a}{R}\right)\arctan\left(1-\frac{a}{R}\right)+2\frac{a}{R}-\frac{a^2}{R^2}-\pi\left(1-\frac{a}{R}\right)\right]- \tag{C7.2.3}$$

$$-\sigma_y\Big|_{x=a}+\sigma_x\Big|_{x=a}$$

The geometric function of the stress intensity factor, defined as

$$K_I = \sigma * F\sqrt{\pi a}, \quad \sigma* = \frac{P}{\pi BR} \tag{C7.2.4}$$

is plotted in Fig. C7.9a as the curve. This curve may be approximated by

$$F \cong \frac{-24.095\alpha + 9.41\alpha^2 + 18.516\tanh(1.4131\alpha)}{(1-\alpha)^{3/2}} \tag{C7.2.5}$$

with $\alpha = a/D$.

A simplified approximation represented by the straight dashed lines is

$$F(1-\alpha)^{3/2} \approx \begin{cases} 3.18\alpha & for\ \alpha \leq 0.55 \\ 1.75 & else \end{cases} \tag{C7.2.6}$$

The biaxiality ratio is plotted as the solid curve in Fig. C7.9b and compiled in Table C7.3. Application of the simple set-up eq.(C7.1.13) results in the dashed curve. A very good agreement of the two solutions is visible. This is an indication of an adequate description of the Green's function by the set-up of eq.(C7.1.11) using one regular term only.

In addition to the Green's function computations, the biaxiality ratios were determined directly using the boundary collocation method (BCM). The results are entered as circles. An excellent agreement is found between the BCM results and those obtained from the Green's function representation.

The result of Fig. C7.9b can be described by

$$\beta \cong \frac{-1.236 + 0.6925\alpha - 8.8735\alpha^2 + 52.901\alpha^3 - 85.839\alpha^4 + 54.96\alpha^5 - 12.181\alpha^6}{\sqrt{1-\alpha}} \tag{C7.2.7}$$

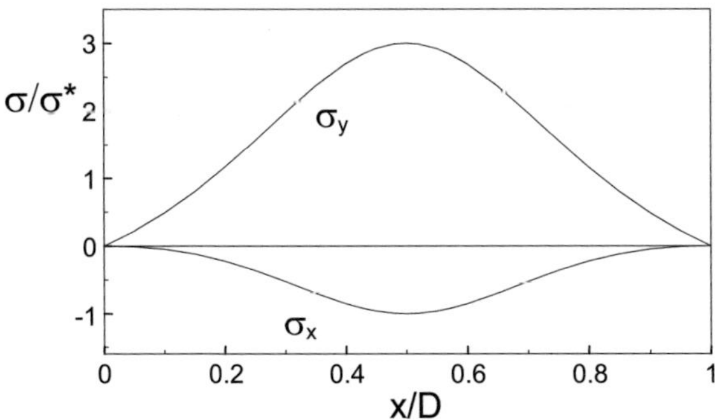

Fig. C7.8 Stresses along the x-axis in a diametrically loaded disk.

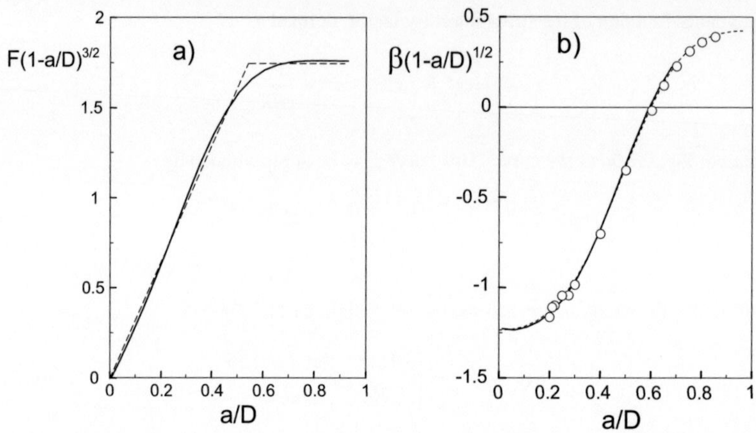

Fig. C7.9 Edge-cracked circular disk diametrically loaded by a pair of forces; a) geometric function for the stress intensity factor, b) Biaxiality ratio, solid line: eqs.(C7.1.8) and (C7.1.9), dashed line: single-term approximation (C7.1.13), circles: BCM results.

Table C7.3 T-stress and biaxiality ratio for diametrical point forces.

a/D	$T(1-a/D)^2$	$\beta(1-a/D)^{1/2}$
0	0	-1.236
0.1	-0.365	-1.220
0.2	-0.735	-1.139
0.3	-0.975	-0.965
0.4	-0.921	-0.688
0.5	-0.532	-0.336
0.6	0.003	0.002
0.7	0.431	0.246
0.8	0.655	0.373

C7.3   Radially loaded disk under mixed boundary conditions

Edge-cracked circular disks under mixed boundary conditions are shown in Fig. C7.10. In Fig. C7.10a the load is given by constant radial tractions σ_n at the circumference with disappearing tangential displacements. In Fig. C7.10b a constant radial displacement u and disappearing shear along the circumference are prescribed.

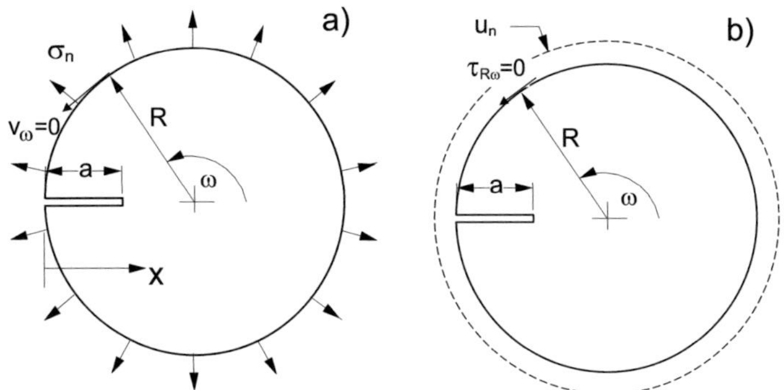

Fig. C7.10 Edge-cracked disk under mixed boundary conditions; a) constant normal traction, disappearing circumferential displacements, b) constant radial displacement, disappearing shear traction at the surface.

C7.3.1 Case u_n = constant, $\tau_{R\omega}$=0

With the characteristic stress

$$\sigma^* = \frac{u_n E}{R} \tag{C7.3.1}$$

the stress intensity factor is

$$K_I = \sigma^* \sqrt{\pi a}\, F(a/D, v) \tag{C7.3.2}$$

The geometric function F is plotted in Fig. C7.11a for several Poisson's ratios. For the special value of $v = 0.25$ the results are fitted as

$$F \cong \tfrac{4}{3} + 0.953\alpha + 20.157\alpha^2 - 107.35\alpha^3 + 156.09\alpha^4 - 72.69\alpha^5 \tag{C7.3.3}$$

(with $\alpha = a/D$). Using a modified geometric function of the form

$$F^* = (1 - v)F \tag{C7.3.4}$$

a coincidence of the curves becomes visible at $a/D\rightarrow 0$ (see Fig. C7.11b).

The T-stresses are shown in Fig. C7.12. An approximation for $\nu = 0.25$ and $\alpha = a/D \leq 0.75$ is

$$T / \sigma^* \cong -3.137\alpha^2 + 40.744\alpha^3 - 76.904\alpha^4 + 41.623\alpha^5 \qquad (C7.3.5)$$

For $\nu = 0.25$ a 3-terms weight function is given by

$$h_{\mathrm{I}} = \sqrt{\frac{2}{\pi a}} \left[\frac{1}{\sqrt{1-\rho}} + D_0 \sqrt{1-\rho} + D_1 (1-\rho)^{3/2} + D_2 (1-\rho)^{5/2} \right], \quad \rho = x/a, \qquad (C7.3.6)$$

with the coefficients compiled in Table C7.4.

The higher-order coefficients of eq.(A1.1.4) are compiled in Tables C7.5 and C7.6.

Table C7.4 Coefficients for the mode-I weight function eq.(C7.3.6).

a/D	D_0	D_1	D_2
0	-0.1380	1.0932	-0.4278
0.1	0.3626	1.0805	-0.3919
0.2	0.6191	1.3460	-0.4279
0.3	0.3251	1.5681	-0.4920
0.4	-0.4503	1.6524	-0.5605
0.5	-1.3284	1.4336	-0.5753
0.6	-1.1857	-0.6181	-0.1554
0.7	1.1930	-6.5147	1.1825

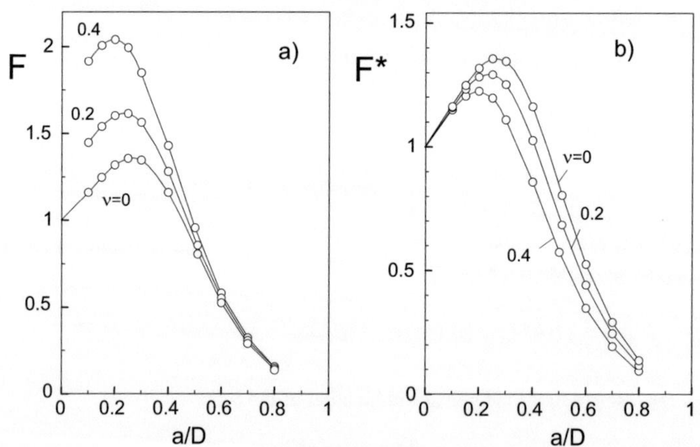

Fig. C7.11 Stress intensity factor for the boundary conditions of $u_{\mathrm{n}} = $ constant, $\tau_{R\omega} = 0$. For F see eq.(C7.3.2) and for F^* eq.(C7.3.4).

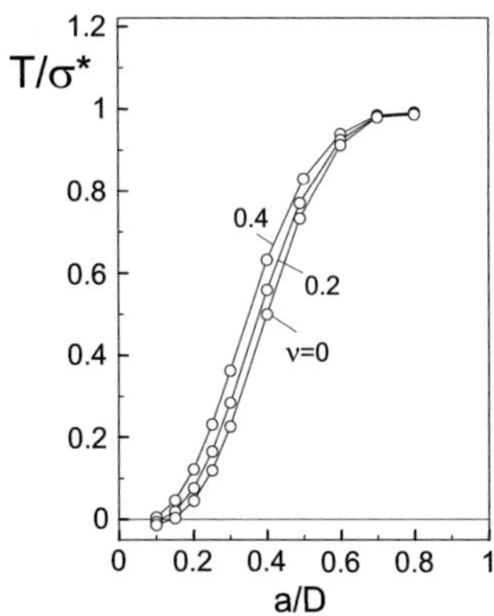

Fig. C7.12 T-stress for the boundary conditions of $u_n = $ const., $\tau_{R\omega} = 0$.

Table C7.5 Coefficient A_1 according to eq.(A1.1.4).

a/D	v=0	0.2	0.4
0.1	-0.0983	-0.1209	-0.158
0.2	-0.0389	-0.0450	-0.0546
0.3	0.0189	0.0239	0.0305
0.4	0.0721	0.0802	0.0901
0.5	0.1035	0.110	0.1166
0.6	0.1113	0.1153	0.1196
0.7	0.1036	0.1072	0.1108
0.8	0.0879	0.0920	0.0960

Table C7.6 Coefficient B_1 according to eq.(A1.1.4).

a/D	v=0	0.2	0.4
0.1	0		
0.2	0	0	
0.3	0	0	
0.4	0.001	0	0
0.5	0.007	0.006	0.005
0.6	0.017	0.013	0.010
0.7	0.030	0.024	0.017
0.8	0.047	0.037	0.027

Figure C7.13 represents the crack opening displacements δ (for δ see Fig. C7.5) under constant radial displacements and disappearing shear traction at the circumference as

$$\delta = \frac{2a\sigma^*}{E'}\lambda(a/D) \qquad (C7.3.7)$$

with σ* given by eq.(C7.3.1).

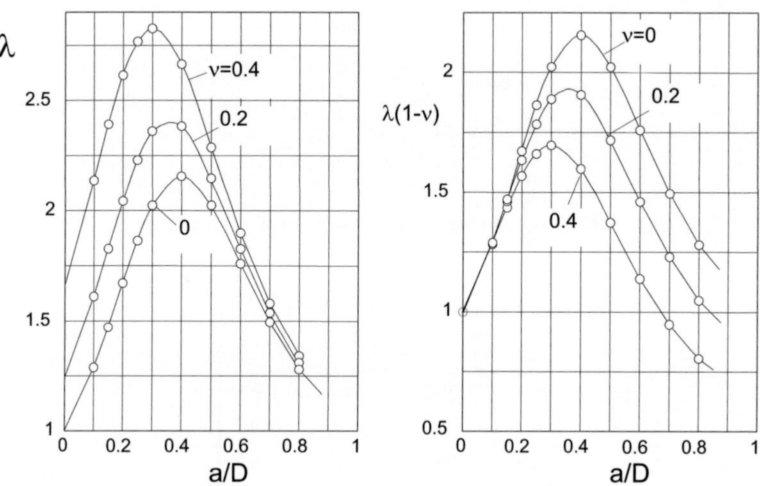

Fig. C7.13 Crack-mouth displacement represented by eq.(C7.3.7). Boundary conditions: $u_n = $ const., $\tau_{R\omega} = 0$.

C7.3.2 Case $\sigma_n = $ constant, $v_\omega = 0$

In this case, the stress intensity factor is

$$K_I = \sigma_n\sqrt{\pi a}\,F(a/D,v) \qquad (C7.3.8)$$

The geometric function is plotted in Fig. C7.14a for several values of v. Figure C7.14b represents the T-stress.

For the special value of $v = 0.25$ the geometric function is fitted for $\alpha \leq 0.75$ as

$$F \cong 0.6163 + 0.2603\alpha + 0.7739\alpha^2 - 0.4497\alpha^3 \qquad (C7.3.9)$$

For $v = 0.25$ a 3-terms weight function is given by eq.(C7.3.6) with the coefficients compiled in Table C7.7.

The higher-order coefficients A_1 and B_1 of eq.(A1.1.4) are listed in Tables C7.8 and C7.9.

Table C7.7 Coefficients for the mode-I weight function eq.(C7.3.6).

a/D	D_0	D_1	D_2
0	-2.004	2.3359	-.801
0.1	-1.8374	2.2169	-.766
0.2	-1.6113	2.0637	-.720
0.3	-1.3594	1.9233	-.675
0.4	-1.1044	1.8252	-.639
0.5	-0.8624	1.7833	-.614
0.6	-0.6434	1.8003	-.603
0.7	-0.454	1.8723	-.605
0.8	-0.300	1.9927	-.619

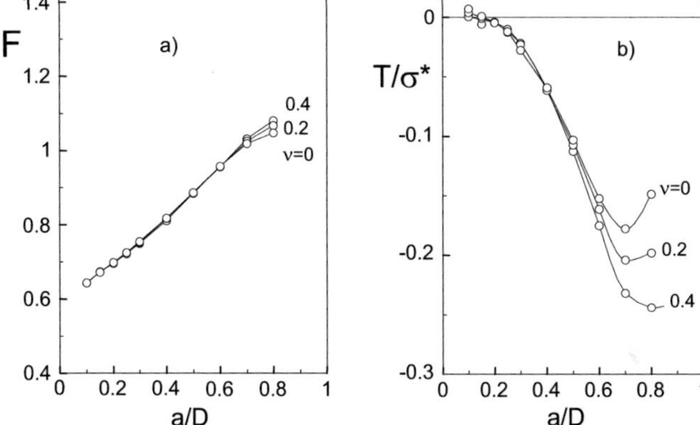

Fig. C7.14 Geometric function for the stress intensity factor K and T-stress under the boundary conditions of σ_n = const., v_ω = 0.

Table C7.8 Coefficient A_1 according to eq.(A1.1.4).

a/D	$v=0$	0.2	0.4
0.1	-0.176	-0.176	-0.176
0.2	-0.124	-0.123	-0.122
0.3	-0.100	-0.099	-0.098
0.4	-0.080	-0.080	-0.080
0.5	-0.058	-0.059	-0.061
0.6	-0.029	-0.031	-0.034
0.7	0.018	0.013	0.009
0.8	0.100	0.096	0.088

Table C7.9 Coefficient B_1 according to eq.(A1.1.4).

a/D	$\nu=0$	0.2	0.4
0.1	0.047	0.040	0.024
0.2	-0.059	-0.067	-0.078
0.3	-0.087	-0.092	-0.095
0.4	-0.130	-0.130	-0.129
0.5	-0.191	-0.186	-0.181
0.6	-0.281	-0.270	-0.260
0.7	-0.420	-0.405	-0.392
0.8	-0.664	-0.657	-0.626

C7.4 Disk under displacement boundary conditions

An edge-cracked circular disk under pure displacement boundary conditions is shown in Fig. C7.15. In this case a constant radial displacement and disappearing tangential displacements are prescribed.

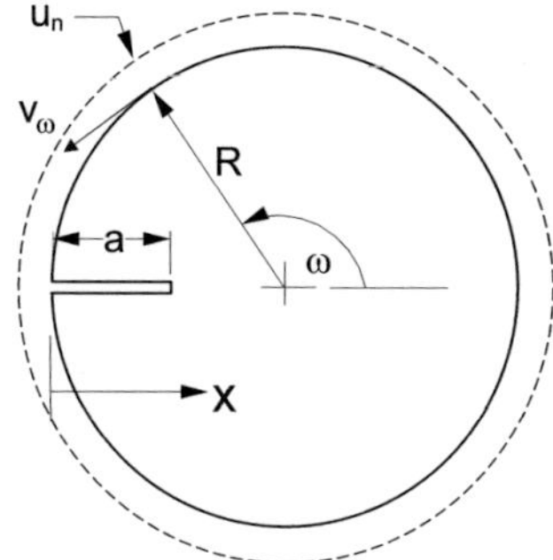

Fig. C7.15 Edge-cracked disk under pure displacement boundary conditions: constant radial displacement, disappearing tangential displacements.

Case: u_n = constant, v_ω = 0

With the characteristic stress value given by eq.(C7.3.1), the stress intensity factor is

$$K_I = \sigma^* \sqrt{\pi a}\, F(a/D, v) \tag{C7.4.1}$$

The geometric functions F and F^* according to eq.(C7.3.4) are plotted in Fig. C7.16 for several Poisson's ratios. The T-stresses are shown in Fig. C7.17.

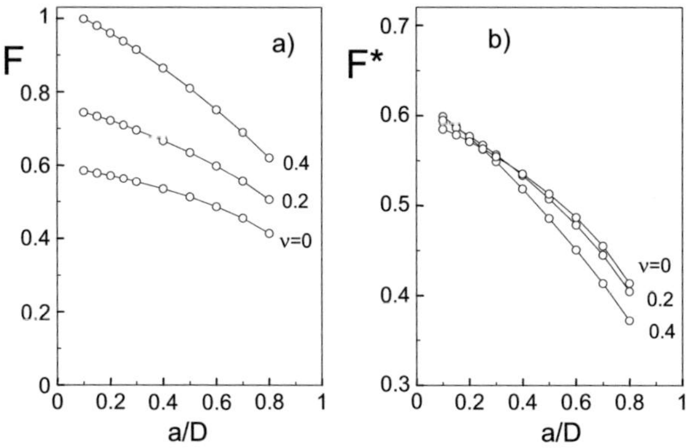

Fig. C7.16 Geometric functions for the boundary conditions of u_n = const., v_ω = 0. For F see eq.(C7.4.1), for F^* eq.(C7.3.4).

For the special value of $\nu = 0.25$ the results are fitted as

$$F \cong 0.814 - 0.175\alpha - 0.2192\alpha^2, \alpha = a/D \qquad (C7.4.2)$$

and the related T-stress (for $0.1 \leq a/D \leq 0.8$) by

$$T/\sigma^* \cong 0.0388 + 0.5568\,\alpha - 1.1934\,\alpha^2 + 1.448\,\alpha^3 \qquad (C7.4.3)$$

For $\nu = 0.25$ a 3-terms weight function is given by eq.(C7.3.6) with the coefficients compiled in Table C7.10.

Table C7.10 Coefficients for the mode-I weight function eq.(C7.3.6).

a/D	D_0	D_1	D_2
0	-2.0340	2.3560	-0.8068
0.1	-2.1094	2.4119	-0.8230
0.2	-2.2105	2.4972	-0.8468
0.3	-2.3398	2.6175	-0.8795
0.4	-2.5015	2.7814	-0.9230
0.5	-2.7020	3.0014	-0.9804
0.6	-2.9501	3.2955	-1.0558
0.7	-3.2593	3.6902	-1.1553
0.8	-3.6498	4.2264	-1.2886

Higher-order coefficients of eq.(A1.1.4) are compiled in Tables C7.11 and C7.12.

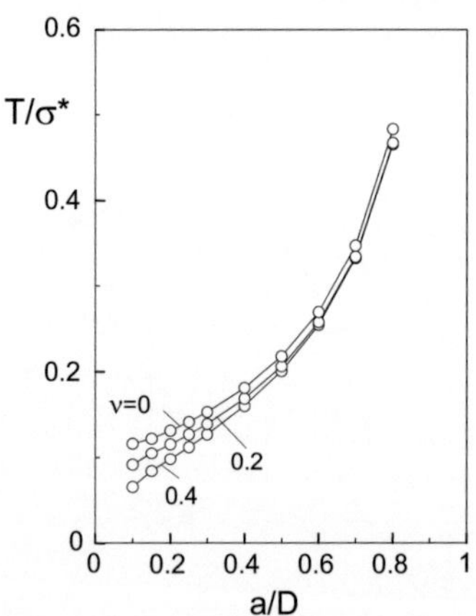

Fig. C7.17 T-stress for the boundary conditions of $u_n = $ constant, $v_\omega = 0$.

Table C7.11 Coefficient A_1 according to eq.(A1.1.4).

a/D	$\nu=0$	0.2	0.4
0.1	-0.171	-0.213	-0.283
0.2	-0.120	-0.149	-0.197
0.3	-0.097	-0.120	-0.157
0.4	-0.083	-0.102	-0.133
0.5	-0.073	-0.090	-0.115
0.6	-0.064	-0.080	-0.102
0.7	-0.057	-0.071	-0.091
0.8	-0.047	-0.062	-0.082

Table C7.12 Coefficient B_1 according to eq.(A1.1.4).

a/D	$\nu=0$	0.2	0.4
0.1	0.257	0.248	0.260
0.2	0.136	0.142	0.159
0.3	0.099	0.107	0.125
0.4	0.082	0.090	0.108
0.5	0.071	0.080	0.098
0.6	0.061	0.073	0.092
0.7	0.048	0.063	0.085
0.8	0.014	0.037	0.067

C7.5   Brazilian disk (edge-cracked)

The diametric compression test carried out on an edge-cracked circular plate of diameter $D=2R$ is illustrated in Fig. C7.18. This arrangement is called Brazilian disk test.

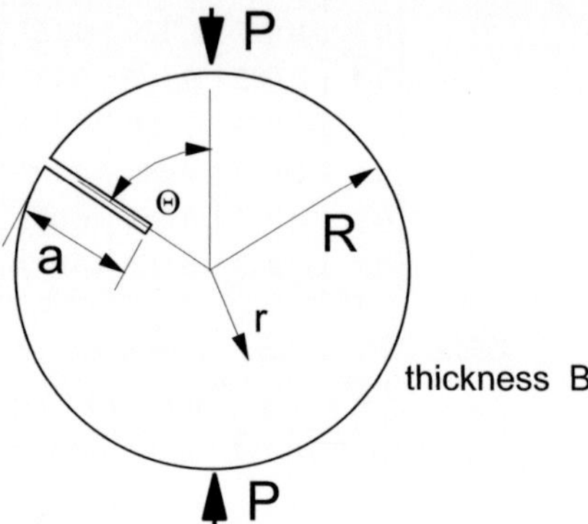

Fig. C7.18 Brazilian disk test with edge-cracked disk.

The tangential, radial, and shear stress components (σ_φ, σ_r, and $\tau_{r\varphi}$) in an uncracked Brazilian disk were given by Erdlac (quoted in [C7.4]) as

$$\sigma_\varphi = \sigma_n = \frac{2P}{\pi BR}\left[\frac{1}{2} - \frac{(1-\rho\cos\Theta)\sin^2\Theta}{(1+\rho^2-2\rho\cos\Theta)^2} - \frac{(1+\rho\cos\Theta)\sin^2\Theta}{(1+\rho^2+2\rho\cos\Theta)^2}\right], \quad \rho = r/R \quad \text{(C7.5.1)}$$

$$\sigma_r = \frac{2P}{\pi BR}\left[\frac{1}{2} - \frac{(1-\rho\cos\Theta)(\cos\Theta-\rho)^2}{(1+\rho^2-2\rho\cos\Theta)^2} - \frac{(1+\rho\cos\Theta)(\cos\Theta+\rho)^2}{(1+\rho^2+2\rho\cos\Theta)^2}\right] \quad \text{(C7.5.2)}$$

$$\sigma_{r\varphi} = \frac{2P}{\pi BR}\left[\frac{(1-\rho\cos\Theta)(\cos\Theta-\rho)\sin\Theta}{(1+\rho^2-2\rho\cos\Theta)^2} + \frac{(1+\rho\cos\Theta)(\cos\Theta+\rho)\sin\Theta}{(1+\rho^2+2\rho\cos\Theta)^2}\right] \quad \text{(C7.5.3)}$$

Using eq.(C7.1.8), the T-stress has been determined. The T-stress term, evaluated for several relative crack depths a/W and several angles Θ, is compiled in Table C7.13.

The mode-I stress intensity factor K_I (necessary for the computation of the biaxiality ratio) is compiled in Table C7.14 and plotted in Fig. C7.19a in form of the geometric function F which in this case is defined as

$$K_\mathrm{I} = \sigma^* F\sqrt{\pi a}, \quad \sigma^* = \frac{P}{\pi BR}$$

The biaxiality ratio β is compiled in Table C7.15 and plotted in Fig. C7.19b. In both representations the values of β are given in the normalized form $\beta(1-a/D)^{1/2}$.

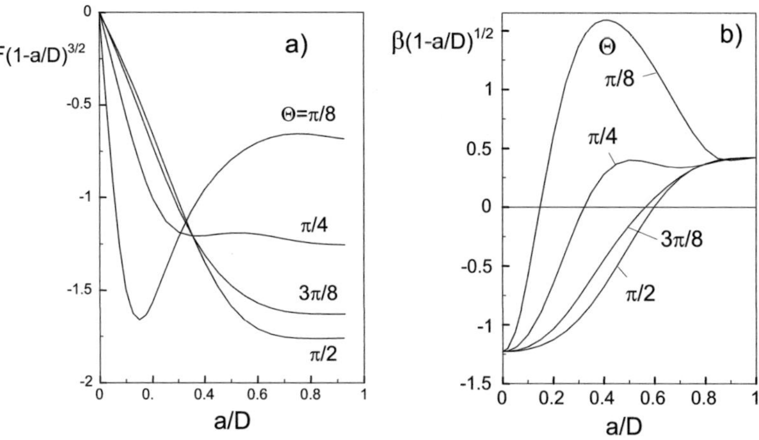

Fig. C7.19 Brazilian disk test with an edge-cracked disk: a) Geometric function for the mode-I stress intensity factor, b) biaxiality ratio $\beta(1-a/D)^{1/2}$.

Table C7.13 T-stress T/σ^* for the Brazilian disk test ($\sigma^*=P/(\pi BR)$).

a/D	$\Theta = \pi/16$	$\pi/8$	$\pi/4$	$3\pi/8$	$7\pi/16$	$\pi/2$
0	0.000	0.000	0.000	0.000	0.000	0.000
0.05	1.89	1.079	0.379	0.229	0.204	0.197
0.1	-1.93	1.12	0.769	0.516	0.468	0.454
0.15	-4.53	0.00	1.032	0.846	0.792	0.775
0.2	-5.45	-1.42	1.045	1.185	1.165	1.157
0.25	-5.66	-2.56	0.777	1.488	1.561	1.581
0.3	-5.65	-3.34	0.296	1.699	1.936	2.009
0.35	-5.59	-3.85	-0.285	1.773	2.226	2.375
0.4	-5.57	-4.22	-0.864	1.689	2.360	2.591
0.45	-5.59	-4.49	-1.388	1.435	2.263	2.557
0.5	-5.70	-4.75	-1.846	0.993	1.862	2.174
0.55	-5.90	-5.00	-2.272	0.315	1.09	1.358
0.6	-6.20	-5.33	-2.751	-0.70	-0.16	0.028
0.65	-6.55	-5.60	-3.450	-2.24	-2.03	-1.941
0.7	-7.1	-6.04	-4.600	-4.58	-4.72	-4.825
0.75	-7.9	-6.8	-6.946	-87.37	-8.9	-9.290

Table C7.14 Geometric function $F\times(1-a/D)^{3/2}$ for the Brazilian disk test.

a/D	$\Theta=\pi/8$	$\pi/4$	$3\pi/8$	$\pi/2$
0	0	0	0	0
0.05	-0.905	-0.283	-0.168	-0.144
0.1	-1.495	-0.565	-0.348	-0.300
0.2	-1.568	-1.003	-0.723	-0.650
0.3	-1.221	-1.182	-1.063	-1.015
0.4	-0.958	-1.200	-1.316	-1.346
0.5	-0.795	-1.186	-1.477	-1.584
0.6	-0.702	-1.192	-1.570	-1.709
0.65	-0.675	-1.204	-1.597	-1.738
0.7	-0.660	-1.217	-1.613	-1.753
0.75	-0.655	-1.230	-1.622	-1.759

Table C7.15 Biaxiality ratio $\beta(1-a/D)^{1/2}$ for the Brazilian disk test.

a/D	$\Theta = \pi/16$	$\pi/8$	$\pi/4$	$3\pi/8$	$7\pi/16$	$\pi/2$
0	-1.228	-1.228	-1.228	-1.228	-1.228	-1.228
0.05	-0.62	-1.072	-1.206	-1.227	-1.224	-1.225
0.1	0.535	-0.60	-1.097	-1.198	-1.214	-1.212
0.15	1.43	0.006	-0.910	-1.137	-1.176	-1.187
0.2	1.98	0.590	-0.662	-1.046	-1.116	-1.138
0.25	2.30	1.040	-0.384	-0.924	-1.031	-1.064
0.3	2.46	1.345	-0.117	-0.779	-0.920	-0.965
0.35	2.53	1.519	0.106	-0.620	-0.785	-0.838
0.4	2.52	1.586	0.265	-0.459	-0.632	-0.688
0.45	2.47	1.570	0.357	-0.305	-0.465	-0.517
0.5	2.38	1.492	0.393	-0.165	-0.293	-0.337
0.55	2.25	1.367	0.390	-0.039	-0.132	-0.160
0.6	2.09	1.208	0.369	0.073	0.017	0.00
0.65	1.87	1.016	0.347	0.172	0.146	0.138
0.7	1.63	0.827	0.338	0.256	0.248	0.246
0.75	1.36	0.65	0.348	0.324	0.323	0.323
1	0.423	0.423	0.423	0.423	0.423	0.423

C7.6 Edge-cracked disk with thermal stresses

In a thermally loaded circular disk the stresses in the absence of a crack consist of the circumferential stress component σ_φ and of the radial stress distribution σ_r. The two stress components can be computed from the temperature distribution $\Theta(r)$ with $r = R\text{-}x$ (see e.g. [C7.5])

$$\sigma_r = \alpha_T E\left(\frac{1}{R^2}\int_0^R \Theta r\,dr - \frac{1}{r^2}\int_0^r \Theta r\,dr\right) \tag{C7.6.1}$$

$$\sigma_\varphi = \alpha_T E\left(\frac{1}{R^2}\int_0^R \Theta r\,dr + \frac{1}{r^2}\int_0^r \Theta r\,dr - \Theta\right) \tag{C7.6.2}$$

with the thermal expansion coefficient α_T. The temperature distributions may often be represented by

$$\Theta(r) = \Theta_0\left[1 + B_2\left(\frac{r}{R}\right)^2 + B_4\left(\frac{r}{R}\right)^4\right] \tag{C7.6.3}$$

with the maximum temperature occurring in the centre of the disk ($r = 0$). In this case, the related stresses are given by

$$\sigma_\varphi = \alpha_T E\Theta_0\left[\frac{1}{4}B_2 + \frac{1}{6}B_4 - \frac{3}{4}B_2\left(\frac{r}{R}\right)^2 - \frac{5}{6}B_4\left(\frac{r}{R}\right)^4\right] \tag{C7.6.4}$$

$$\sigma_r = \alpha_T E\Theta_0\left[\frac{1}{4}B_2\left(1 - \frac{r^2}{R^2}\right) + \frac{1}{6}B_4\left(1 - \frac{r^4}{R^4}\right)\right] \tag{C7.6.5}$$

For the special case of $\Theta = 0$ at the circumference it must hold

$$B_4 = -(B_2 + 1) \tag{C7.6.6}$$

with the stresses

$$\sigma_r = \alpha_T E\Theta_0\left[-\tfrac{1}{6} + \tfrac{1}{12}B_2 - \tfrac{1}{4}B_2\left(\frac{r}{R}\right)^2 + \tfrac{1}{6}(1 + B_2)\left(\frac{r}{R}\right)^4\right] \tag{C7.6.7}$$

$$\sigma_\varphi = \alpha_T E\Theta_0\left[-\tfrac{1}{6} + \tfrac{1}{12}B_2 - \tfrac{3}{4}B_2\left(\frac{r}{R}\right)^2 + \tfrac{5}{6}(1 + B_2)\left(\frac{r}{R}\right)^4\right] \tag{C7.6.8}$$

As an example of application, the coefficient B_2 may be chosen to be $B_2=-2$. Figure C7.20a shows the temperature distribution. The related stresses are given in Fig. C7.20b.

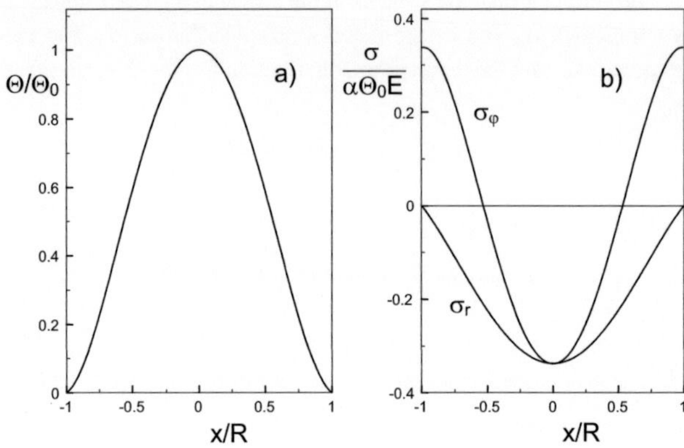

Fig. C7.20 a) Temperature distribution and b) stresses in a thermally heated disk (for $B_2=-2$).

The stress intensity factor can be computed with the weight function for the edge-cracked disk. The resulting K is given in Fig. C7.21a in normalized form. The T-stress computed by use of the 3-terms Green's function (C7.1.8, C7.1.9) is represented in Fig. C7.21b as the solid curve. The 3-terms Green's function with averaged coefficients, eq.(C7.1.10) is shown by the dash-dotted line.

The approximate T-stress solution obtained by using the single-term Green's function eq.(C7.1.13) reads

$$T \cong \alpha_T E\Theta_0 \left\{ 0.9481\left[\tfrac{1}{4} + (1+B_2)\left(\tfrac{1}{12} - \tfrac{2}{9}\frac{a}{R} + \tfrac{1}{9}\left(\frac{a}{R}\right)^2 \right) \right] \right.$$
$$\left. -\left(1-\frac{a}{R}\right)^2\left[\tfrac{1}{2} + (1+B_2)\left(1 - \tfrac{4}{3}\frac{a}{R} + \tfrac{2}{3}\left(\frac{a}{R}\right)^2 \right) \right] \right\} \tag{C7.6.9}$$

and for the specially chosen value of $B_2=-2$

$$T \cong \alpha_T E\Theta_0 \left\{ 0.9481\left[\tfrac{1}{6} + \left(\tfrac{2}{9}\frac{a}{R} - \tfrac{1}{9}\left(\frac{a}{R}\right)^2 \right) \right] + \left(1-\frac{a}{R}\right)^2\left[\tfrac{1}{2} - \left(\tfrac{4}{3}\frac{a}{R} - \tfrac{2}{3}\left(\frac{a}{R}\right)^2 \right) \right] \right\} \tag{C7.6.10}$$

This solution is introduced in Fig. C7.21b as the dashed curve.

The biaxiality ratio β, as defined by eq.(A1.3.12), is plotted in Fig. C7.21c. Very high β-values occur for $a/D > 0.5$. The main reason is the very small stress intensity factor, which disappears at approximately $a/D = 0.75$.

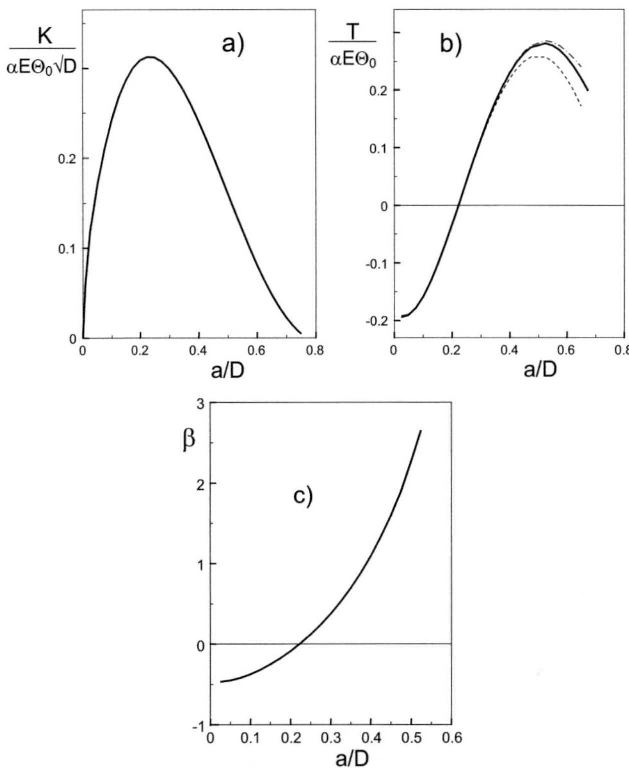

Fig. C7.21 a) Stress intensity factor and b) T-stress for a disk under thermal loading (solid curve: eqs.(C7.1.8, C7.1.9), dashed curve: single-term approximation (C7.1.11), dash dotted: approximation (C7.1.10)), c) biaxiality ratio.

References C7:

[C7.1] Fett, T., Munz, D., Stress intensity factors and weight functions, Computational Mechanics Publications, Southampton, 1997.

[C7.2] Williams, M.L., On the stress distribution at the base of a stationary crack, J. Appl. Mech. **24**(1957), 109-114.

[C7.3] Tada, H., Paris, P.C., Irwin, G.R., The stress analysis of cracks handbook, Del Research Corporation, 1986.

[C7.4] Atkinson, C., Smelser, R.E., Sanchez, J., Combined mode fracture via the cracked Brazilian disk test, Int. J. Fract. **18**(1982), 279-291

[C7.5] Timoshenko, S.P., Goodier, J.N., Theory of Elasticity, McGraw-Hill Kogagusha, Ltd., Tokyo.

C8

Single-edge-cracked rectangular plates

C8.1   Rectangular plate under pure tension

The edge-cracked rectangular plate under constant *tensile loading* is shown in Fig. C8.1.

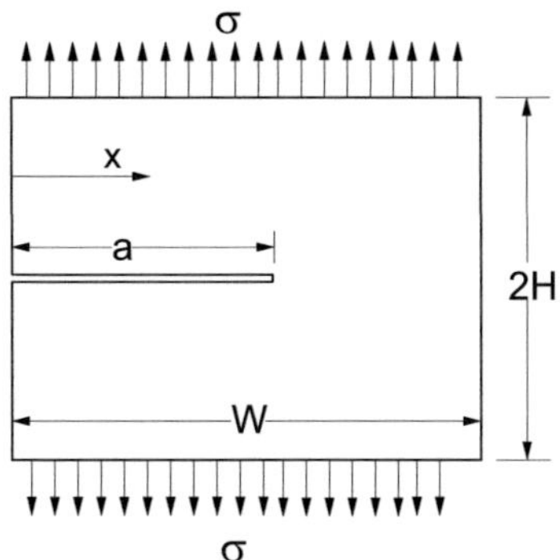

Fig. C8.1 Edge-cracked plate under tensile load.

The stress intensity factor solution for this case is given by

$$K = \sigma\sqrt{\pi a}\,F_t \qquad (C8.1.1)$$

with the geometric function compiled in Table C8.1.

A 2-terms weight function for the edge-cracked plate reads [C8.1]

$$h = \sqrt{\frac{2}{\pi a}}\left[\frac{1}{\sqrt{1-\rho}} + D_0\sqrt{1-\rho} + D_1(1-\rho)^{3/2}\right], \quad \rho = x/a \qquad (C8.1.2)$$

with the coefficients of Tables C8.2 and C8.3. The T-stress is plotted in Fig. C8.2 and compiled in Table C8.4. In Fig. C8.2 the biaxiality ratios for $H/W=0.5$ and 1.0 are compared with a solution by Leevers and Radon [C8.2] available for these geometries. The agreement is very good.

Figure C8.3 shows the biaxiality ratio β (see also Table C8.5).

For a long plate ($H/W=1.5$) the stress intensity factor is given by [C8.1]

$$F_t = \frac{1.1215}{(1-\alpha)^{3/2}}\left[1 - 0.23566(1-\alpha) + \tfrac{1}{150}(1-\alpha)^2 + 3\alpha^2(1-\alpha)^7 + 0.229\exp(-7.52\tfrac{\alpha}{1-\alpha})\right] \qquad (C8.1.3)$$

with $\alpha=a/W$. The T-stress is described by

$$\frac{T}{\sigma} = \frac{-0.526 + 0.641\,\alpha + 0.2049\,\alpha^2 + 0.755\,\alpha^3 - 0.7974\,\alpha^4 + 0.1966\alpha^5}{(1-\alpha)^2} \quad \text{(C8.1.4)}$$

In this case, the biaxiality ratio reads

$$\beta = \frac{-0.469 + 0.1456\,\alpha + 1.3394\,\alpha^2 + 0.4369\,\alpha^3 - 2.1025\,\alpha^4 + 1.0726\,\alpha^5}{\sqrt{1-\alpha}} \quad \text{(C8.1.5)}$$

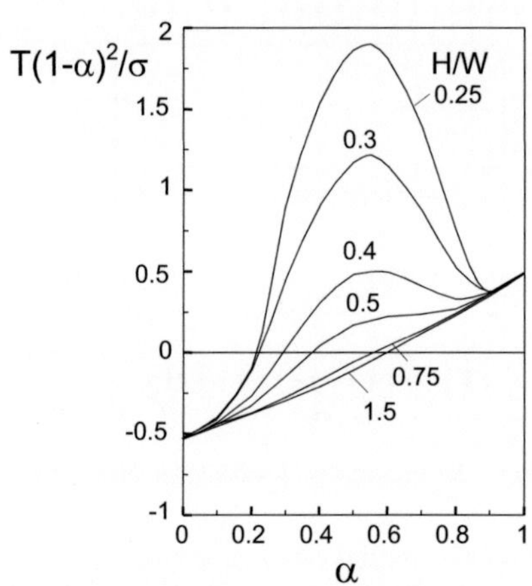

Fig. C8.2 T-stress under tensile loading.

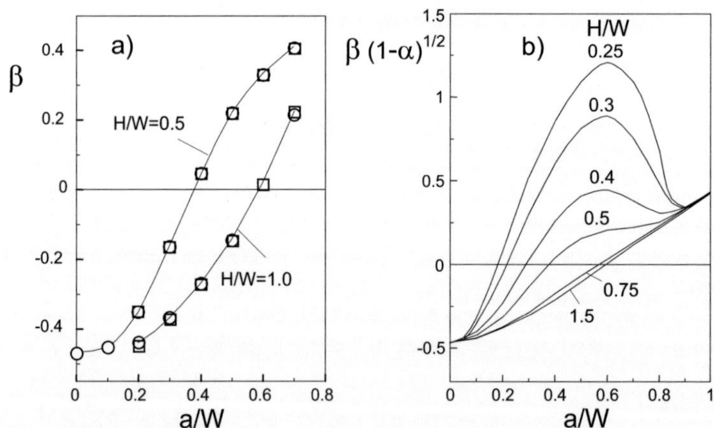

Fig. C8.3 Biaxiality ratios β: a) data from Table C8.5 (circles) compared with data reported by Leevers and Radon [C8.2] (squares), b) influence of the specimen height.

Table C8.1 Geometric function for tension $F_t \cdot (1-a/W)^{3/2}$.

	$H/W=1.5$	1.25	1.00	0.75	0.5	0.4	0.3	0.25
$\alpha=0$	1.1215	1.1215	1.1215	1.1215	1.1215	1.1215	1.1215	1.1215
0.1	1.0170	1.0172	1.0174	1.0182	1.0352	1.0649	1.1455	1.2431
0.2	0.9800	0.9799	0.9798	0.9877	1.0649	1.1625	1.3619	1.5358
0.3	0.9722	0.9723	0.9729	0.9840	1.0821	1.2134	1.4892	1.7225
0.4	0.9813	0.9813	0.9819	0.9915	1.0819	1.2106	1.5061	1.7819
0.5	0.9985	0.9986	0.9989	1.0055	1.0649	1.1667	1.4298	1.7013
0.6	1.0203	1.0203	1.0204	1.0221	1.0496	1.1073	1.2898	1.5061
0.7	1.0440	1.0441	1.0441	1.0442	1.0522	1.0691	1.1498	1.2685
0.8	1.0683	1.0683	1.0683	1.0690	1.0691	1.0734	1.0861	1.1201
1.0	1.1215	1.1215	1.1215	1.1215	1.1215	1.1215	1.1215	1.1215

Table C8.2 Coefficient $D_0(1-a/W)^{3/2}$ for the weight function (C8.1.2).

a/W	$H/W=1.5$	1.25	1.00	0.75	0.5	0.4
0.2	1.001	1.001	1.003	1.010	1.249	1.347
0.3	1.298	1.302	1.326	1.317	1.539	1.816
0.4	1.581	1.581	1.598	1.616	1.836	2.036
0.5	1.827	1.829	1.835	1.859	1.973	2.122
0.6	1.996	1.996	1.998	2.001	2.060	2.110
0.7	2.070	2.071	2.071	2.079	2.104	2.094
0.8	2.015	2.015	2.017	2.054	2.064	2.094

Table C8.3 Coefficient $D_1(1-a/W)^{3/2}$ for the weight function (C8.1.2).

a/W	$H/W=1.5$	1.25	1.00	0.75	0.5	0.4
0.2	0.1963	0.200	0.2100	0.2245	0.300	0.634
0.3	0.3072	0.301	0.2641	0.3422	0.460	0.784
0.4	0.4909	0.4909	0.4661	0.4887	0.624	0.970
0.5	0.7329	0.7300	0.7213	0.7183	0.857	1.170
0.6	1.074	1.074	1.072	1.077	1.186	1.368
0.7	1.526	1.525	1.525	1.513	1.516	1.629
0.8	2.128	2.128	2.128	2.066	2.050	2.018

Table C8.4 T-stress for a plate under tension $T/\sigma\cdot(1-a/W)^2$.

a/W	H/W=1.5	1.00	0.75	0.50	0.40	0.30	0.25
0	-0.526	-0.526	-0.526	-0.526	-0.526	-0.526	-0.526
0.1	-0.452	-0.452	-0.452	-0.444	-0.432	-0.416	-0.400
0.2	-0.374	-0.376	-0.373	-0.334	-0.270	-0.084	0.143
0.3	-0.299	-0.298	-0.282	-0.148	0.030	0.449	0.890
0.4	-0.208	-0.205	-0.175	0.040	0.310	0.912	1.526
0.5	-0.106	-0.102	-0.070	0.167	0.473	1.165	1.858
0.6	0.006	0.008	0.032	0.220	0.490	1.142	1.812
0.7	0.122	0.123	0.134	0.234	0.404	0.869	1.387
0.8	0.232	0.234	0.240	0.268	0.324	0.524	0.760
0.9	0.352	0.353	0.356	0.364	0.372	0.376	0.380
1.0	0.474	0.474	0.474	0.474	0.474	0.474	0.474

Table C8.5 Biaxiality ratio $\beta(1-a/W)^{1/2}$.

a/W	H/W=1.5	1.00	0.75	0.50	0.40	0.30	0.25
0	-0.469	-0.469	-0.469	-0.469	-0.469	-0.469	-0.469
0.1	-0.444	-0.444	-0.444	-0.429	-0.406	-0.363	-0.322
0.2	-0.382	-0.384	-0.377	-0.314	-0.232	-0.062	0.093
0.3	-0.308	-0.306	-0.287	-0.137	0.025	0.302	0.517
0.4	-0.212	-0.209	-0.176	0.037	0.256	0.606	0.856
0.5	-0.106	-0.102	-0.070	0.157	0.405	0.815	1.092
0.6	0.006	0.008	0.031	0.210	0.443	0.885	1.203
0.7	0.117	0.118	0.128	0.222	0.378	0.756	1.093
0.8	0.217	0.219	0.225	0.251	0.302	0.482	0.679
1.0	0.423	0.423	0.423	0.423	0.423	0.423	0.423

Tables C8.6 and C8.7 represent some values for the coefficients A_1 and B_1 of the Williams series expansion. The next higher-order terms are compiled in Tables C8.8 and C8.9. For long plates ($H/W \geq 1.5$) the coefficients A_1 and B_1 can be approximated by

$$A_1 \cong \frac{-0.02279 + 0.04107\alpha + 0.03231\alpha^2 + 0.2470\alpha^3 - 0.3241\alpha^4 + 0.1358\alpha^5}{(1-\alpha)^{5/2}\sqrt{\alpha}} \qquad (C8.1.6)$$

$$A^*_1 \cong \frac{0.04813 - 0.1062\alpha - 0.08187\alpha^2 + 0.3276\alpha^3 - 0.4092\alpha^4 + 0.1511\alpha^5}{(1-\alpha)^3\,\alpha} \qquad (C8.1.7)$$

Table C8.6 Coefficients A_1 for tension.

	H/W=1.00	0.75	0.5	0.4	0.3	0.25
α=0.2	-0.0459	-0.0440	-0.0251	0.0061	0.0907	
0.3	-0.0140	-0.0084	0.0436	0.1219	0.3205	0.5414
0.4	0.0438	0.0537	0.1431	0.2782	0.6248	1.011
0.5	0.1655	0.1770	0.2933	0.4836	1.0043	1.595
0.6	0.4513	0.4606	0.5774	0.8001	1.477	2.294
0.7	1.254	1.257	1.335	1.5314	2.240	3.195
0.8	3.768	4.284	4.346	4.440	4.81	

Table C8.7 Coefficients B_1 for tension.

	H/W=1	0.75	0.5	0.4	0.3	0.25
α=0.2	0.2473	0.2379	0.1574	0.0561	-0.1510	
0.3	0.1453	0.1223	-0.0188	-0.1640	-0.4022	-0.5714
0.4	0.0551	0.0328	-0.1050	-0.2557	-0.4886	-0.5957
0.5	-0.0807	-0.0815	-0.1247	-0.2257	-0.4073	-0.4062
0.6	-0.3932	-0.3563	-0.1838	-0.0893	-0.0277	0.1377
0.7	-1.383	-1.313	-0.821	-0.2534	0.7099	1.446
0.8	-5.22	-5.90	-5.26	-4.04	0.866	

Table C8.8 Coefficients A_2 for tension.

	H/W=1	0.5	0.25
α=0.3	0.0111	0.0328	-0.7476
0.4	0.0888	-0.0130	-1.8675
0.5	0.2546	-0.0451	-3.4075
0.6	0.7246	0.1850	-5.415
0.7	2.4535	1.7412	-7.471
0.8	10.61	11.55	

Table C8.9 Coefficients B_2 for tension.

	H/W=1	0.5	0.25
α=0.3	-0.2882	-0.0631	3.368
0.4	-0.2302	0.2938	5.898
0.5	-0.3278	0.5297	8.845
0.6	-0.8237	0.3264	12.513
0.7	-3.088	-1.981	16.688
0.8	-16.39	-18.47	

C8.2   Rectangular plate under bending load

The plate under pure bending stresses

$$\sigma(x) = \sigma_b (1 - 2x/W) \quad , \tag{C8.2.1}$$

is shown in Fig. C8.4. The stress intensity factor is given by

$$K = \sigma_b \sqrt{\pi a}\, F_b \tag{C8.2.2}$$

with the geometric function F compiled in Table C8.10.

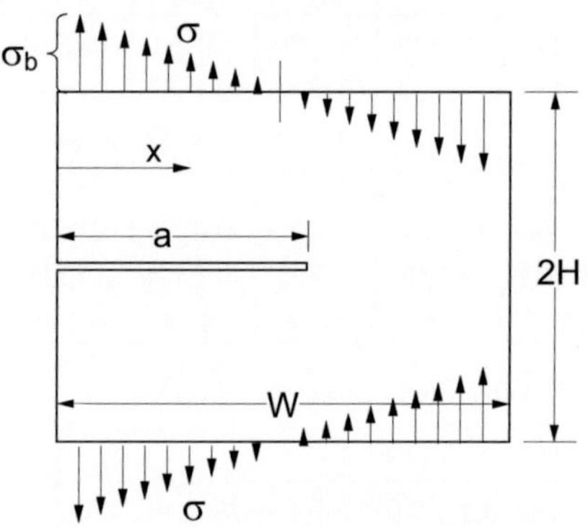

Fig.C8.4 Edge-cracked plate under bending load.

Table C8.10 Geometric function for bending $F_b \cdot (1-a/W)^{3/2}$.

a/W	$H/W=1.5$	1.25	1.00	0.75	0.5	0.4
0	1.1215	1.1215	1.1215	1.1215	1.1215	1.1215
0.2	0.7561	0.7561	0.7562	0.7628	0.8279	0.9130
0.3	0.6583	0.6583	0.6589	0.6677	0.7444	0.8475
0.4	0.5861	0.5861	0.5865	0.5930	0.6567	0.7505
0.5	0.5293	0.5293	0.5296	0.5332	0.5717	0.6388
0.6	0.4842	0.4842	0.4842	0.4852	0.5022	0.5367
0.7	0.4481	0.4479	0.4478	0.4478	0.4514	0.4621
0.8	0.4203	0.4188	0.4191	0.4185	0.4180	0.4185
1.0	0.374	0.374	0.374	0.374	0.374	0.374

Table C8.11 T-stress for a plate under bending $T/\sigma_b\cdot(1-a/W)^2$.

a/W	$H/W=1.5$	0.75	0.50	0.40	0.30	0.25
0	-0.526	-0.526	-0.526	-0.526	-0.526	-0.526
0.2	-0.150	-0.148	-0.114	-0.061	0.099	0.292
0.3	-0.039	-0.024	0.080	0.222	0.559	0.920
0.4	0.044	0.067	0.224	0.424	0.873	1.333
0.5	0.099	0.124	0.283	0.493	0.964	1.439
0.6	0.133	0.150	0.269	0.438	0.840	1.251
0.7	0.151	0.158	0.217	0.314	0.574	0.857
0.8	0.158	0.158	0.174	0.204	0.302	0.426
0.9	0.140	0.142	0.150	0.162	0.169	0.186
1.0	0.113	0.113	0.113	0.113	0.113	0.113

For a long plate ($H/W \geq 1.5$) the stress intensity factor is [C8.1]

$$F_b = \frac{1.1215}{(1-\alpha)^{3/2}}\left[\frac{5}{8} - \frac{5}{12}\alpha + \frac{1}{8}\alpha^2 + 5\alpha^2(1-\alpha)^6 + \frac{3}{8}\exp(-6.1342\alpha/(1-\alpha))\right] \quad (C8.2.3)$$

The T-stress can be expressed by

$$\frac{T}{\sigma_b} = \frac{-0.526 + 2.481\alpha - 3.553\alpha^2 + 2.6384\alpha^3 - 0.9276\alpha^4}{(1-\alpha)^2} \quad (C8.2.4)$$

with the bending stress σ_b defined in eq.(C8.2.1). For other H/W see Table C8.11.

The biaxiality ratio for a long plate ($H/W=1.5$) is approximated by

$$\beta = \frac{-0.469 + 1.2825\alpha + 0.6543\alpha^2 - 1.2415\alpha^3 + 0.07568\alpha^4}{\sqrt{1-\alpha}} \quad (C8.2.5)$$

(for other H/W see Table C8.12). Higher-order coefficients of the Williams stress function for bending are compiled in Tables C8.13 and C8.14.

For long plates ($H/W=1.5$) the coefficients A_1 and B_1 can be approximated by

$$A_1 \cong \frac{-0.02279 + 0.19661\alpha - 0.30552\alpha^2 + 0.247618\alpha^3 - 0.08037\alpha^4}{(1-\alpha)^{5/2}\sqrt{\alpha}} \quad (C8.2.6)$$

$$B_1 \cong \frac{0.04813 - 0.4224\alpha + 1.0005\alpha^2 - 1.0269\alpha^3 + 0.3799\alpha^4}{(1-\alpha)^3\alpha} \quad (C8.2.7)$$

Figure C8.5 shows the T-stresses. In Fig. C8.6a, the biaxiality ratios for bending are compared with those for tension.

In Fig. C8.6b the biaxiality ratios for $H/W=1.5$ are compared with a solution from literature [C8.3]. It should be noted that the results given by Sham [C8.3] were determined for a very long plate of $H/W=6$. Nevertheless, this solution (squares) is very close to the BCM results of Table C8.12 (curve: interpolated by application of cubic splines). This excellent agreement indicates that the plates are represented by the limit case of an "infinitely long plate" in both cases.

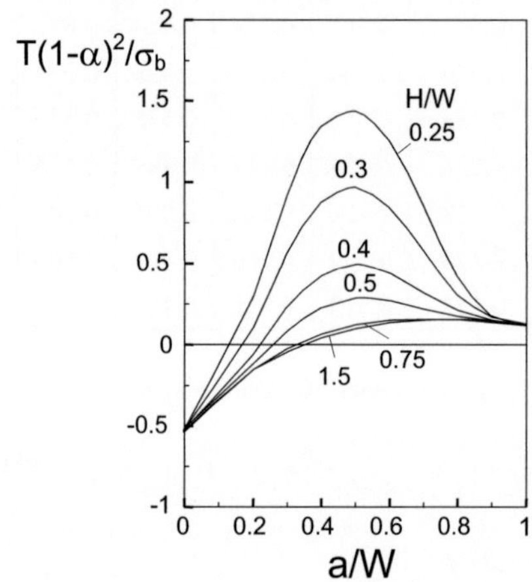

Fig. C8.5 T-stress under bending loading.

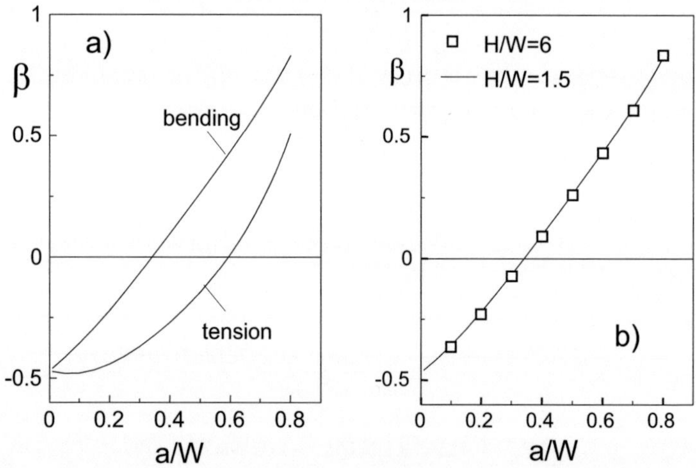

Fig. C8.6 a) Biaxiality ratio of an edge-cracked plate or bar under tension and bending, b) biaxiality ratio β (Table C8.12, curve) compared with data reported by Sham [C8.3] (squares) for bending.

Table C8.12 Biaxiality ratio for bending $\beta \cdot (1-a/W)^{1/2}$.

a/W	$H/W=1.5$	0.75	0.5	0.4
0	-0.469	-0.469	-0.469	-0.469
0.2	-0.198	-0.194	-0.138	-0.067
0.3	-0.059	-0.036	0.107	0.262
0.4	0.075	0.113	0.341	0.565
0.5	0.187	0.233	0.495	0.772
0.6	0.275	0.309	0.536	0.816
0.7	0.337	0.353	0.481	0.679
0.8	0.376	0.378	0.416	0.487
1.0	0.423	0.423	0.423	0.423

Table C8.13 Coefficient A_1 for bending.

a/W	$H/W=1.5$	1.25	1.00	0.75	0.5	0.4
0.2	0.024	0.024	0.024	0.025	0.0419	0.067
0.3	0.065	0.065	0.065	0.0696	0.1104	0.1722
0.4	0.116	0.118	0.1185	0.1257	0.1906	0.2887
0.5	0.201	0.201	0.2023	0.2104	0.2885	0.4148
0.6	0.362	0.362	0.3623	0.3684	0.4409	0.5751
0.7	0.746	0.744	0.744	0.747	0.792	0.900
0.8	2.059	2.051	2.045	2.043	2.049	2.09

Table C8.14 Coefficient B_1 for bending.

a/W	$H/W=1.5$	1.25	1.00	0.75	0.5	0.4
0.2			-0.035	-0.04	-0.103	-0.188
0.3	-0.1216	-0.127	-0.123	-0.141	-0.251	-0.363
0.4	-0.1944	-0.1958	-0.197	-0.213	-0.310	-0.408
0.5	-0.2884	-0.2872	-0.289	-0.289	-0.308	-0.348
0.6	-0.4666	-0.467	-0.464	-0.440	-0.315	-0.213
0.7	-0.96	-0.951	-0.950	-0.907	-0.598	-0.230
0.8	-3.07	-3.03	-3.00	-2.94	-2.52	-1.84

C8.3 Weight and Green's functions for plates of arbitrary height

The two loading cases of tension and bending allow for deriving a weight function for stress intensity factor computations according to eq.(A3.1.10a) as outlined in Section A3.2.2 (see also [C8.1], Chapter 3). For a 3 terms set-up

$$h = \sqrt{\frac{2}{\pi a}}\left[\frac{1}{\sqrt{1-x/a}} + D_0\sqrt{1-x/a} + D_1(1-x/a)^{3/2} + D_2(1-x/a)^{5/2}\right] \qquad (C8.3.1)$$

the additional crack-mouth condition $d^2h/dx^2=0$ at $x=0$ (see [C8.1]) can be used. Then the coefficients read

$$D_0 = -11 - \frac{35\pi}{128\sqrt{2}\alpha}(27F_b - (27-16\alpha)F_t) \qquad (C8.3.2)$$

$$D_1 = \frac{49}{3} + \frac{35\pi}{192\sqrt{2}\alpha}(81F_b - (81-64\alpha)F_t) \qquad (C8.3.3)$$

$$D_2 = -\frac{21}{5} - \frac{21\pi}{128\sqrt{2}\alpha}(21F_b - (21-16\alpha)F_t) \qquad (C8.3.4)$$

where the geometric functions from Tables C8.1 and C8.10 can be used. Since the coefficients D_0-D_1 contain differences between tensile and bending solutions (which are identical for $a/W{\rightarrow}0$), the application of (C8.3.2-C8.3.4) is recommended for $a/W{>}0.2$. Figure C8.7a shows the weight function h for a relative crack depth of $a/W{=}0.5$ for a long and a short plate. The Green's functions for T-stress are given in Fig. C8.7b. It is clearly visible that the effect of the specimen height is more pronounced for the Green's function t.

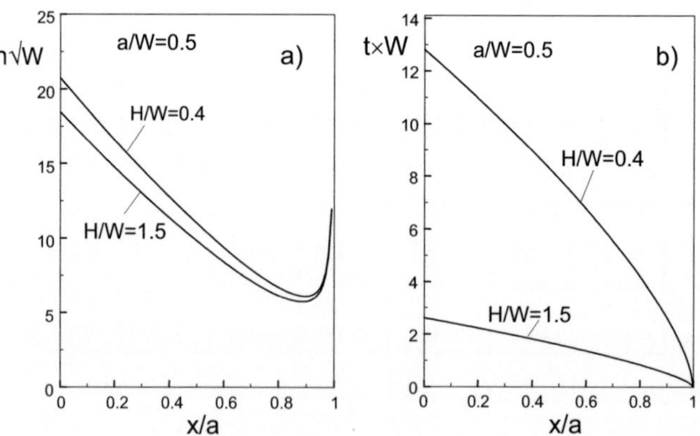

Fig. C8.7 Influence of the plate height H on a) the weight function h, b) the Green's function t.

The tension and bending solutions also allow for deriving a Green's function for T-stress

$$t = \tfrac{1}{a}[C_0\sqrt{1-x/a} + C_1(1-x/a)^{3/2}]$$ (C8.3.5)

According to eqs.(A4.3.6) and (A4.3.7), the coefficients follow as

$$C_0 = \frac{15}{16\alpha}\left((7-4\alpha)\frac{T_t}{\sigma_t} - 7\frac{T_b}{\sigma_b} + 10\alpha\right)$$ (C8.3.6)

$$C_1 = -\frac{35}{16\alpha}\left((5-4\alpha)\frac{T_t}{\sigma_t} - 5\frac{T_b}{\sigma_b} + 6\alpha\right)$$ (C8.3.7)

with the T-stresses T_t for tension and T_b for bending as obtained from Tables C8.4 and C8.11. In the special case of the long plate with $H/W \geq 1.5$, it holds

$$C_0 = \frac{15(-0.3889 + 1.8706\,\alpha - 2.0012\,\alpha^2 - 1.0544\,\alpha^3 + 2.283\,\alpha^4 - 0.3932\,\alpha^5)}{8(1-\alpha)^2}$$ (C8.3.8)

$$C_1 = \frac{35(0.5487 - 2.1127\,\alpha + 2.1180\,\alpha^2 + 1.1845\,\alpha^3 - 2.0864\,\alpha^4 + 0.3932\,\alpha^5)}{8(1-\alpha)^2}$$ (C8.3.9)

A Green's function approximation with integer exponents can be given by

$$t(x) = \tfrac{1}{a}[E_0(1-x/a) + E_1(1-x/a)^2]$$ (C8.3.10)

with the coefficients E_0 and E_1 compiled in Tables C8.15 and C8.16.

Table C8.15 Coefficient E_0/a for the Green's function, eq.(C8.3.10).

a/W	$H/W=1.5$	0.75	0.50	0.40	0.30
0.2	2.531	2.02	2.53	4.78	8.16
0.3	1.456	1.31	4.00	6.53	11.74
0.4	1.290	1.79	4.93	8.33	15.13
0.5	1.728	2.25	5.71	9.46	18.67
0.6	3.167	3.42	6.04	10.21	21.60
0.7	6.204	6.42	8.05	11.73	23.31

Table C8.16 Coefficient E_1/a for the Green's function, eq.(C8.3.10).

a/W	$H/W=1.5$	0.75	0.50	0.40	0.30
0.2	2.438	3.234	3.37	1.50	0.80
0.3	1.714	2.286	0.980	0.82	1.55
0.4	1.417	1.167	0.925	1.46	3.81
0.5	0.864	1.152	1.44	3.17	5.95
0.6	0.437	0.875	2.81	5.00	8.28
0.7	0.789	1.034	3.35	5.93	10.71

C8.4 Transverse loading

An edge-cracked plate under transverse tractions σ_x is illustrated in Fig. C8.8. Under this loading, the stress intensity factor is defined by

$$K = \sigma_x F\sqrt{\pi a} \qquad\qquad (C8.4.1)$$

The geometric function F is plotted in Fig. C8.9 for several values of a/W, H/W, and d/W. Figure C8.10 represents the T-stresses. For the long plate ($H/W \geq 1.3$) the data are compiled in Tables C8.17 and C8.18. The limit cases for F are: $F = 0$ for $d/H = 0$ and $d/H = 1$. For T it holds: $T = 0$ for $d/H = 0$ and $T/\sigma_x = 1$ for $d/H = 1$.

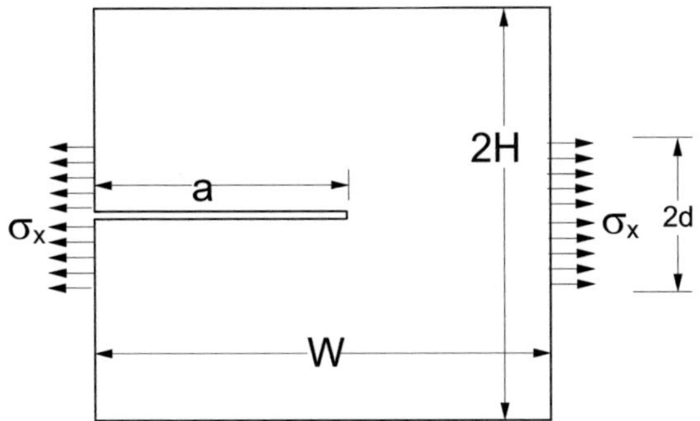

Fig. C8.8 Edge-cracked plate under transverse traction.

Table C8.17 Geometric function F for stress intensity factor representation eq.(C8.4.1).

a/W	$d/W=0.05$	0.1	0.2	0.4	0.6	0.8
0.2	0.0852	0.1488	0.2162	0.1911	0.1048	0.0461
0.3	0.0549	0.0950	0.1394	0.1300	0.0750	0.0341
0.4	0.0441	0.0734	0.1005	0.0870	0.0493	0.0221
0.5	0.0399	0.0631	0.0775	0.0560	0.0286	0.0118
0.6	0.0385	0.0568	0.0600	0.0307	0.0138	0.0043
0.7	0.0381	0.0504	0.0422	0.0147	0.0034	0.0000
0.8	0.0381	0.0410	0.0206	0.0033	-0.0009	-0.0013

Table C8.18 T-stress represented as T/σ_x.

a/W	d/W=0.05	0.1	0.2	0.4	0.6	0.8
0.2	0.4561	0.7713	1.0695	1.0888	1.0549	1.0242
0.3	0.3425	0.6158	0.9670	1.1540	1.1173	1.0630
0.4	0.2871	0.5332	0.8960	1.1750	1.1574	1.0907
0.5	0.2693	0.5072	0.8725	1.1797	1.1717	1.1011
0.6	0.2828	0.5317	0.9003	1.1866	1.1568	1.0920
0.7	0.3345	0.6161	0.9786	1.1627	1.1233	1.0661
0.8	0.4498	0.7805	1.0903	1.1157	1.0663	1.0326

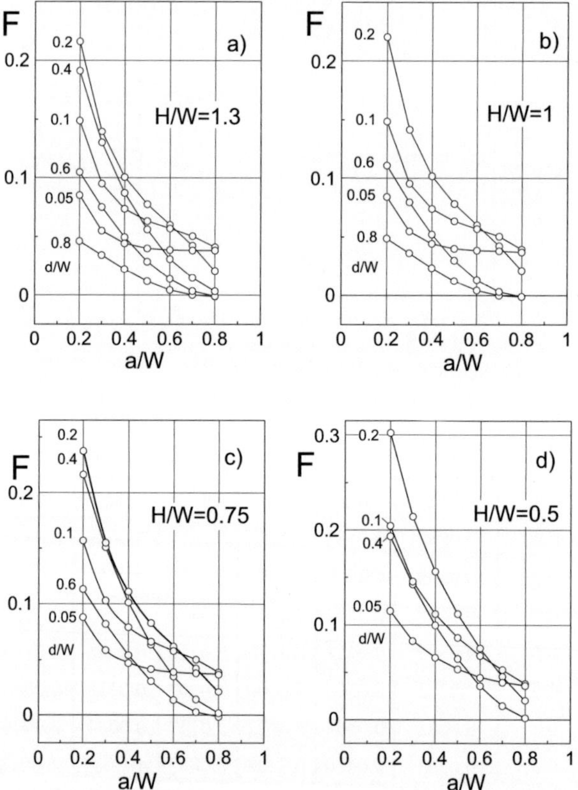

Fig. C8.9 Geometric function F according to eq.(C8.4.1).

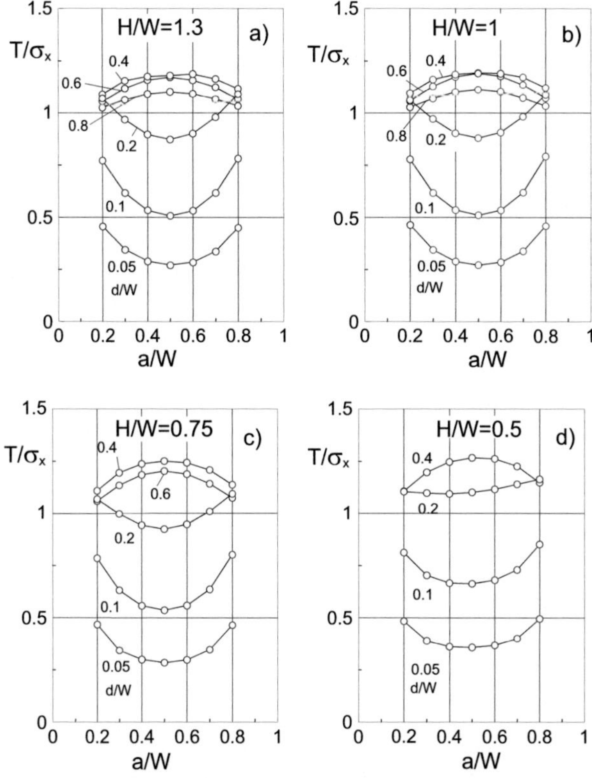

Fig. C8.10 T-stresses represented as T/σ_x.

C8.5   Shear loading on crack surfaces

An edge-cracked plate with crack faces loaded by constant shear stresses is illustrated by Fig. C8.11. The stress intensity factors K_{II} are compiled in Fig. C8.12 and Table C8.19 in the form of the geometric function F_{II} according to

$$K_{II} = \tau\, F_{II} \sqrt{\pi a} \qquad\qquad (C8.5.1)$$

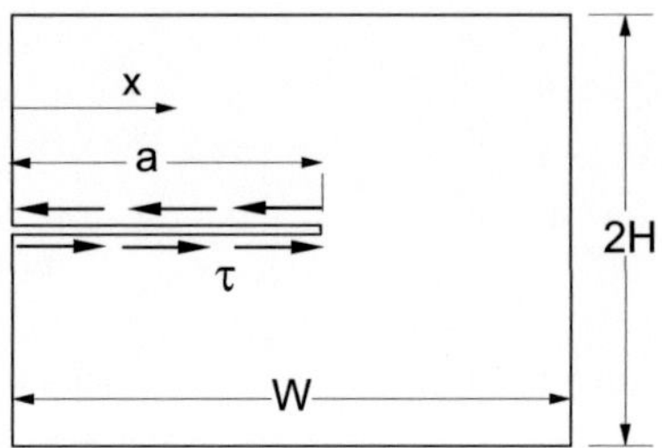

Fig. C8.11 Edge-cracked plate under shear loading.

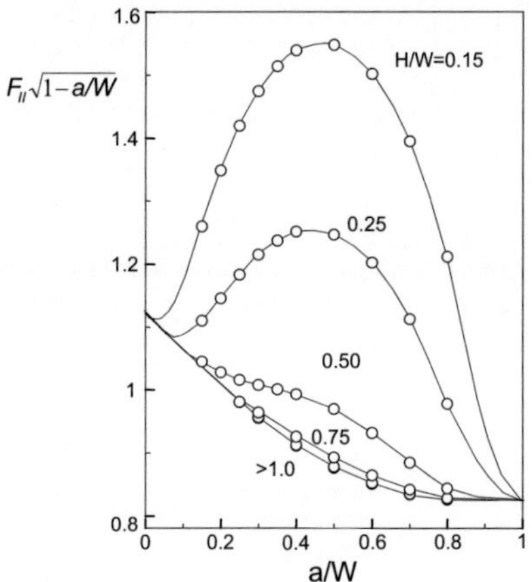

Fig. C8.12 Geometric function F_{II} for the edge-cracked plate under shear loading.

A representation of F_{II} is given for $H/W \geq 1.25$ by

$$F_{II} = \frac{1.1215 - 0.5608\alpha - 0.20\alpha^2 + 0.891\alpha^3 - 0.426\alpha^4}{\sqrt{1-\alpha}} \qquad\qquad (C8.5.2)$$

Approximate relations for any H/W are reported in [C8.1].

Table C8.19 Geometric function $F_{II}\sqrt{(1-\alpha)}$ according to eq.(C8.5.1).

a/W	H/W=1.25	1.00	0.75	0.5	0.25	0.15
0	1.1216	1.1216	1.1216	1.1216	1.1216	1.1216
0.2	1.007	1.008	1.011	1.027	1.146	1.348
0.3	0.955	0.958	0.964	0.983	1.215	1.475
0.4	0.912	0.913	0.926	0.993	1.252	1.539
0.5	0.876	0.878	0.893	0.969	1.246	1.549
0.6	0.851	0.852	0.864	0.932	1.202	1.502
0.7	0.834	0.835	0.842	0.885	1.112	1.395
0.8	0.826	0.826	0.828	0.843	0.978	1.212

A weight function h_{II} is given by

$$h_{II} = \sqrt{\frac{2}{\pi a}}\left[\frac{1}{\sqrt{1-x/a}} + D_0\sqrt{1-x/a} + D_1(1-x/a)^{3/2}\right] \qquad (C8.5.3)$$

with the coefficients D_0 and D_1 compiled in Tables C8.20 and C8.21.

Table C8.20 Coefficients for the weight function eq.(C8.5.3) at $H/W \geq 1.25$.

α	D_0	D_1
0	0.672	0.109
0.1	0.674	0.108
0.2	0.699	0.095
0.3	0.774	0.058
0.4	0.932	-0.018
0.5	1.219	-0.151
0.6	1.703	-0.372
0.7	2.535	-0.768

Table C8.21 Coefficients for the weight function eq.(C8.5.3) at H/W=0.5.

α	D_0	D_1
0	0.672	0.110
0.1	0.683	0.102
0.2	0.783	0.074
0.3	0.966	0.077
0.4	1.165	0.178
0.5	1.389	0.299
0.6	1.741	0.277
0.7	2.469	-0.149

References C8:

[C8.1] Fett, T., Munz, D., Stress intensity factors and weight functions, Computational Mechanics Publications, Southampton, 1997.
[C8.2] Leevers, P.S., Radon, J.C., Inherent stress biaxiality in various fracture specimen geometries, Int. J. Fract. **19**(1982), 311-325.
[C8.3] Sham, T.L., The determination of the elastic T-term using higher order weight functions, Int. J. Fract. **48**(1991), 81-102.

C9

Partially loaded rectangular plate

C9.1   Stress intensity factor solution

An edge-cracked rectangular plate with constant stresses acting on a part of the plate ends is illustrated in Fig. C9.1.

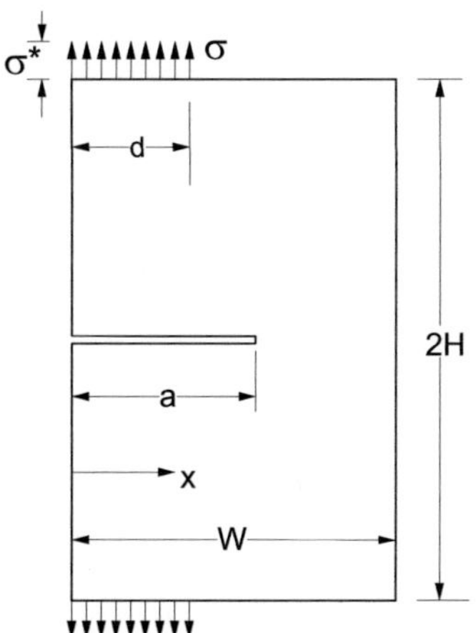

Fig. C9.1 Partially loaded edge-cracked rectangular plate.

The geometric function of the stress intensity factor defined by

$$K_I = \sigma * F \sqrt{\pi a}$$ (C9.1.1)

is compiled in Tables C9.1-C9.4 for several geometries [C9.1].

Table C9.1 Geometric function F for H/W=1.25.

a/W	d/W=0	0.25	0.5	0.75	1.0
0.3	0	1.049	1.643	1.859	1.637
0.4	0	1.245	1.990	2.318	2.103
0.5	0	1.546	2.538	2.968	2.825
0.6	0	2.054	3.472	4.080	4.034
0.7	0	3.138	5.274	6.191	6.327

Table C9.2 Geometric function F for H/W=1.00.

a/W	d/W=0	0.25	0.5	0.75	1.0
0.3	0	1.056	1.668	1.871	1.656
0.4	0	1.280	2.009	2.296	2.112
0.5	0	1.568	2.599	2.982	2.824
0.6	0	2.139	3.483	4.101	4.035
0.7	0	3.207	5.229	6.280	6.353

Table C9.3 Geometric function F for H/W=0.75.

a/W	d/W=0	0.25	0.5	0.75	1.0
0.3	0	1.100	1.697	1.864	1.681
0.4	0	1.302	2.038	2.295	2.135
0.5	0	1.614	2.612	3.012	2.842
0.6	0	2.129	3.435	4.099	4.043
0.7	0	3.174	5.209	6.284	6.357

Table C9.4 Geometric function F for H/W=0.50.

a/W	d/W=0	0.25	0.5	0.75	1.0
0.3	0	1.296	1.862	1.961	1.847
0.4	0	1.479	2.242	2.422	2.323
0.5	0	1.676	2.752	3.126	3.007
0.6	0	2.193	3.575	4.249	4.146
0.7	0	3.190	5.240	6.307	6.386

An example of application of this loading case may be demonstrated for a plate with $H/W =$ 1.25 loaded by a couple of point forces P at several locations d/W, as illustrated in Fig. C9.2. First, we determine the stress intensity factors for two values d_1 and d_2 with $d_1 = d-\varepsilon$ and $d_2 = d+\varepsilon$ ($\varepsilon \ll d$) by interpolation of the tabulated results applying cubic splines. The normal force P is given by

$$P = \sigma * (d_2 - d_1)B \tag{C9.1.2}$$

(B = thickness). The stress intensity factor for this case is

$$K_P = \left(\frac{K_2}{\sigma *} - \frac{K_1}{\sigma *} \right) \sigma * = \left(\frac{K_2}{\sigma *} - \frac{K_1}{\sigma *} \right) \frac{P}{B(d_2 - d_1)} \tag{C9.1.3}$$

(K_1=$K(x$=$d_1)$, K_2=$K(x$=$d_2)$) and in case of d_1, $d_2 \to d$ ($\varepsilon \to 0$)

$$K_P = \frac{\partial (K/\sigma *)}{\partial (d/W)} \frac{P}{WB}. \tag{C9.1.4}$$

Results for K_P are given in Fig. C9.3.

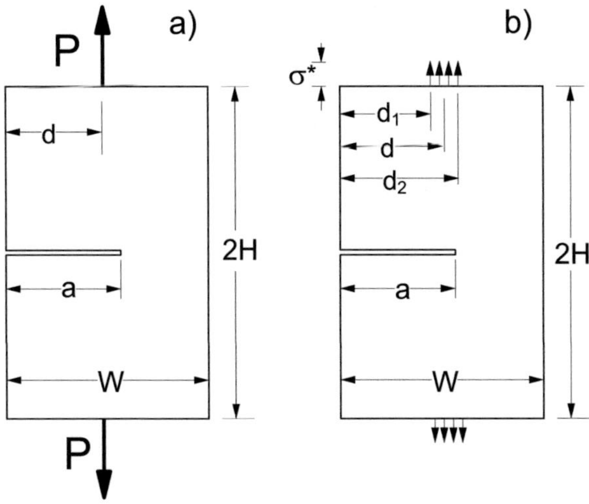

Fig. C9.2 Computation of stress intensity factors in plates loaded by a couple of point forces.

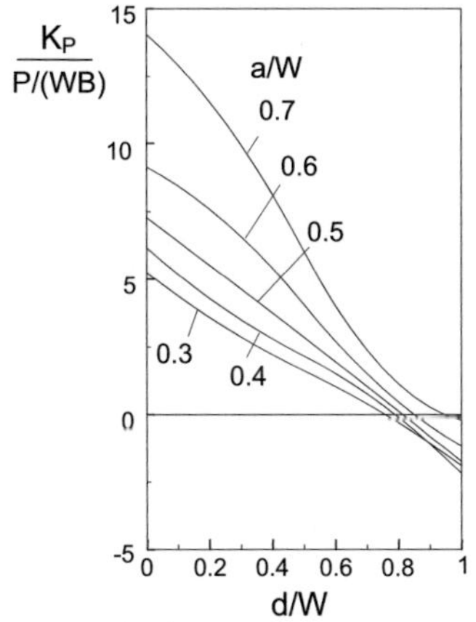

Fig. C9.3 Stress intensity factor caused by a couple of forces acting at location d ($H/W = 1.25$).

If an arbitrary smooth distribution of normal tractions acts on the ends of the plate, it is of advantage to evaluate

$$K = \frac{K_d}{\sigma^*} \sigma_n \Big|_{x=d=W} - \int_0^W \frac{K_d}{\sigma^*} \frac{d\sigma}{dx} dx . \qquad\qquad (C9.1.5)$$

C9.2 T-stress solution

The T-stress terms T_d and the biaxiality ratios for a constant stress over a distance d (Fig. C9.1) are entered into Tables C9.5-C9.12.

Due to the non-homogeneous traction at the plate ends, a stress component σ_x will be generated along the crack line in the uncracked component already.

Table C9.5 T-stress T_d/σ^* for H/W=1.25.

a/W	d/W=0	0.25	0.5	0.75	1.0
0.3	0	-0.196	-0.362	-0.501	-0.608
0.4	0	-0.072	-0.197	-0.372	-0.577
0.5	0	0.123	0.092	-0.102	-0.419
0.6	0	0.461	0.660	0.468	0.040
0.7	0	1.199	1.90	1.806	1.337

Table C9.6 T-stress T_d/σ^* for H/W=1.00.

a/W	d/W=0	0.25	0.5	0.75	1.0
0.3	0	-0.174	-0.360	-0.515	-0.606
0.4	0	-0.042	-0.193	-0.383	-0.570
0.5	0	0.157	0.117	-0.118	-0.409
0.6	0	0.522	0.680	0.474	0.051
0.7	0	1.329	1.959	1.917	1.366

Table C9.7 T-stress T_d/σ^* for H/W=0.75.

a/W	d/W=0	0.25	0.5	0.75	1.0
0.3	0	-0.094	-0.333	-0.524	-0.571
0.4	0	0.098	-0.115	-0.369	-0.485
0.5	0	0.348	0.251	-0.039	-0.277
0.6	0	0.703	0.808	0.560	0.199
0.7	0	1.456	2.052	2.011	1.485

Table C9.8 T-stress T_d/σ^* for H/W=0.50.

a/W	d/W=0	0.25	0.5	0.75	1.0
0.3	0	0.257	-0.119	-0.317	-0.299
0.4	0	0.722	0.457	0.136	0.110
0.5	0	1.157	1.195	0.783	0.666
0.6	0	1.614	2.007	1.668	1.372
0.7	0	2.250	3.174	3.007	2.593

Table C9.9 Biaxiality ratio $\beta(1-a/W)^{1/2}$ for H/W=1.25.

a/W	d/W=0.25	0.5	0.75	1.0
0.3	-0.156	-0.184	-0.225	-0.311
0.4	-0.045	-0.077	-0.124	-0.213
0.5	0.056	0.026	-0.024	-0.105
0.6	0.142	0.122	0.073	0.006
0.7	0.209	0.213	0.160	0.116

Table C9.10 Biaxiality ratio $\beta(1-a/W)^{1/2}$ for H/W=1.00.

a/W	d/W=0.25	0.5	0.75	1.0
0.3	-0.138	-0.181	-0.230	-0.306
0.4	-0.026	-0.074	-0.129	-0.209
0.5	0.071	0.032	-0.028	-0.102
0.6	0.154	0.124	0.073	0.008
0.7	0.227	0.205	0.167	0.118

Table C9.11 Biaxiality ratio $\beta(1-a/W)^{1/2}$ for H/W=0.75.

a/W	d/W=0.25	0.5	0.75	1.0
0.3	-0.071	-0.164	-0.235	-0.284
0.4	0.059	-0.044	-0.125	-0.176
0.5	0.153	0.068	-0.009	-0.069
0.6	0.209	0.149	0.086	0.031
0.7	0.251	0.216	0.175	0.128

a/W	$d/W=0.25$	0.5	0.75	1.0
0.3	0.166	-0.054	-0.135	-0.136
0.4	0.378	0.158	0.043	0.037
0.5	0.488	0.329	0.177	0.157
0.6	0.466	0.355	0.248	0.209
0.7	0.386	0.332	0.261	0.222

An example of application of this loading case may be demonstrated for a plate with $H/W=$ 1.25 loaded by a couple of point forces P at several locations d/W, as illustrated in Fig. C9.2a. The evaluation of the related T-stress term is explained in Fig. C9.2b.

The T-stress for this case is [C9.2]

$$T_P = \left(\frac{T_{d2}}{\sigma^*} - \frac{T_{d1}}{\sigma^*}\right)\sigma^* = \left(\frac{T_{d2}}{\sigma^*} - \frac{T_{d1}}{\sigma^*}\right)\frac{P}{B(d_2 - d_1)} \tag{C9.2.1}$$

and for the case of $d_1, d_2 \to d \, (\varepsilon \to 0)$

$$T_P = \frac{\partial(T_d/\sigma^*)}{\partial(d/W)}\frac{P}{WB} \tag{C9.2.2}$$

In Fig. C9.4 the T-stresses are plotted as a function of the relative crack length a/W.

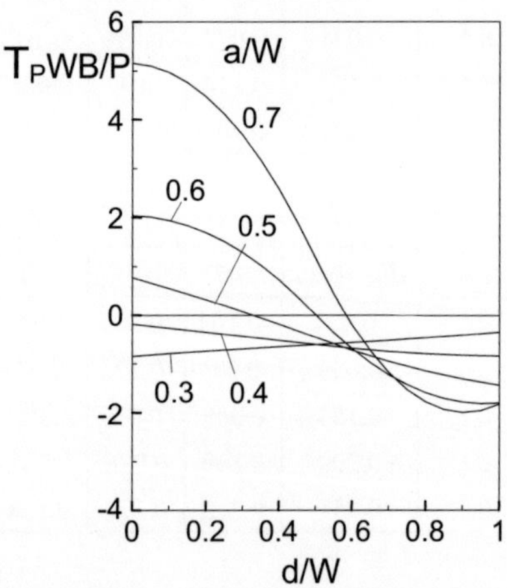

Fig. C9.4 T-stress caused by a couple of forces acting at location d ($H/W=1.25$).

The results compiled in Tables C9.5-C9.8 can be used to compute the T-stress for any given distribution of normal traction σ_n at the ends of the plate

$$T = \frac{1}{W} \int_0^W \frac{T_P}{\sigma^*} \sigma_n(x)dx \,, \quad \sigma^* = \frac{P}{WB} \,.$$ (C9.2.3)

If a smooth distribution of normal traction acts at the ends of the plate it is of advantage to rewrite also eq.(C9.2.3) and apply integration by parts. This leads to

$$T = \frac{T_d}{\sigma^*} \sigma_n \Big|_{x=d=W} - \int_0^W \frac{T_d}{\sigma^*} \frac{d\sigma}{dx} dx \,.$$ (C9.2.4)

As an example, the T-stress for bending was computed from eq.(C9.2.4). The results for two values of H/W are shown in Fig. C9.5 (circles) together with the data of Table C8.11 (curves), which were obtained directly from BCM computations. The agreement is good.

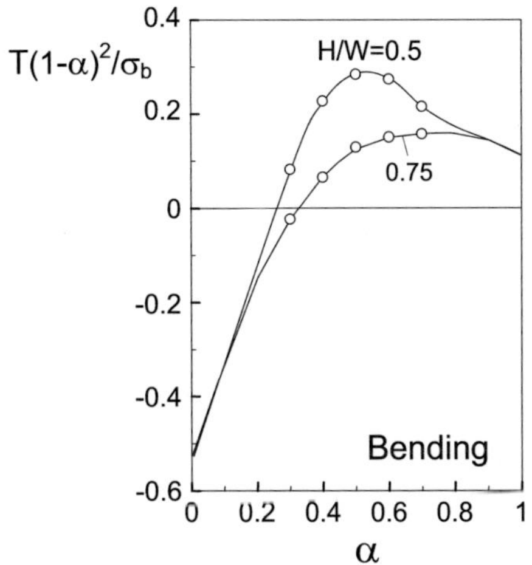

Fig. C9.5 Comparison of bending results obtained with eq.(C9.2.4) (circles) and BCM results (curves).

References C9

[C9.1] Fett, T., T-stress and stress intensity factor solutions for 2-dimensional cracks, VDI-Verlag, 2002, Düsseldorf.

[C9.2] Fett, T., T-stresses for components with one-dimensional cracks, FZKA 6170, Forschungs-zentrum Karlsruhe, 1998.

C10

Edge-cracked plate under mixed boundary conditions

C10.1   Mixed boundary conditions at the ends

The single-edge-cracked plate under displacement-controlled loading is shown in Fig. C10.1. In Fig. C10.1a the plate is extended in y-direction by a constant displacement v. Under plane stress conditions, the related stress in the uncracked plate is

$$\sigma_0 = \frac{v}{H} E \qquad\qquad (C10.1.1)$$

(E = Young's modulus). As second condition, disappearing shear traction at the ends of the plate may be prescribed, leading to a mixed boundary problem. The equivalent description of the crack problem is shown in Fig. C10.1b, where the crack faces are loaded by σ_0 and y-displacements at the ends of the plate are suppressed ($v = 0$). The rollers ensure free deformation in x direction.

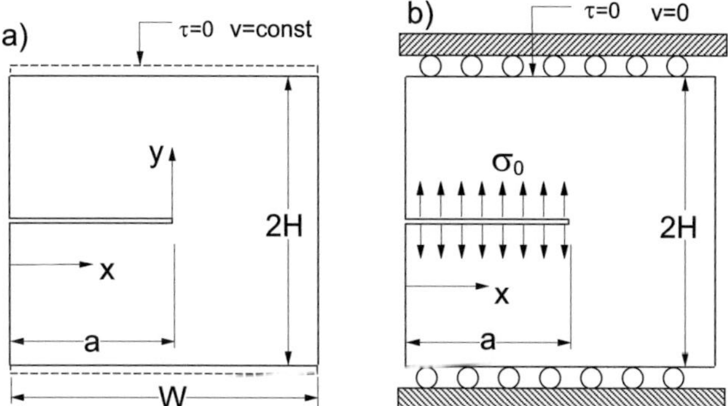

Fig. C10.1 Edge-cracked plate under displacement boundary conditions, a) loading by constant displacements v at the plate ends, b) equivalent crack face loading resulting from the superposition principle.

Results for the stress intensity factors are illustrated in Fig. C10.2a in the form of the geometric function F with $\sigma^* = \sigma_0$. Boundary collocation results are entered as circles. For $H/W \leq 0.5$ a simple representation of the results is given by [C10.1, C10.2]

$$F = \sqrt{\frac{H}{\pi a}} \tanh^{1/\gamma}\left(1.1215\sqrt{\frac{\pi a}{H}}\right)^{\gamma} \quad , \quad \gamma = 2.2 \qquad (C10.1.2)$$

This solution is indicated by the curves in Fig. C10.2a. Figure C10.2b illustrates the T-stress normalised to σ_0. In the case of $H/W=0.25$, the T-stress is nearly constant within the range of $0.4 \leq a/W \leq 0.7$. In order to allow interpolations, Tables C10.1 and C10.2 provide single values for F and T. The biaxiality ratio is compiled in Table C10.3 and higher stress function coefficients are given in Tables C10.4-C10.7.

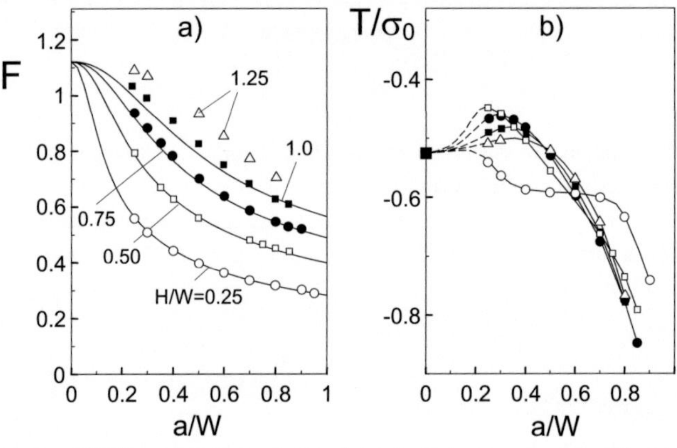

Fig. C10.2 Results of BCM computations; a) stress intensity factor, expressed by F (symbols: BCM results, curves: eq.(C10.1.2)), b) T-stress (symbols as in a)).

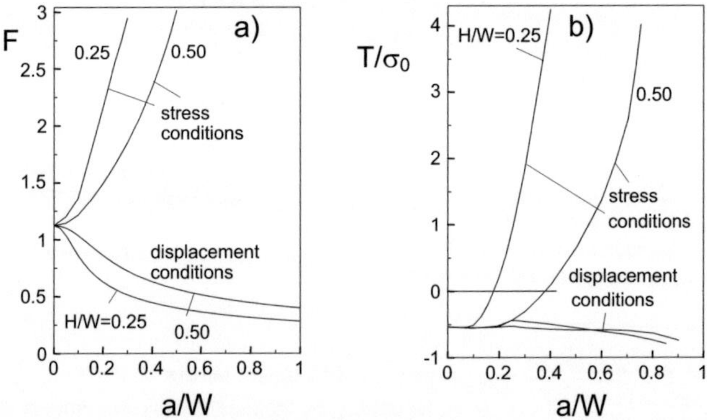

Fig. C10.3 Comparison of solutions for constant normal traction and constant displacements at the plate ends; a) geometric function for stress intensity factor, b) T-stress.

Figure C10.3 gives a comparison of the stress intensity factor and T-stress solutions for the stress conditions of ($\sigma_y = \sigma_0$, $\tau_{xy} = 0$ at $y = H$) and the results obtained with the displacement condition (v=const., $\tau_{xy}=0$ at $y=H$) at $H/W=0.25$ and $H/W=0.5$.

Strong deviations of the results are obvious from Fig. C10.3. Whereas the geometric functions F for the stress boundary conditions increase monotonically with increasing a/W, the geometric function for the displacement boundary conditions decreases with a/W. This result illustrates that application of correct boundary conditions is necessary to compute the fracture mechanics parameters for a given crack problem.

As a second displacement condition, the case of prescribed bending displacements

$$v = \sigma_0 \frac{H}{E}\left(1 - 2\frac{x}{W}\right)$$

(C10.1.3)

is considered with the outer fibre tensile stress σ_0 in the uncracked plate. The results obtained for this type of loading are compiled in Tables C10.8 to C10.10. Higher order coefficients of the Williams stress function are entered into Tables C10.11 and C10.12.

Table C10.1 Geometric function F if the stress intensity factor solution at v=const.

a/W	H/W=0.25	0.50	0.75	1.00	1.25
0.00	1.1215	1.1215	1.1215	1.1215	1.1215
0.25	0.558	0.794	0.938	1.030	1.094
0.3	0.510	0.726	0.883	0.992	1.071
0.4	0.445	0.627	0.782	0.909	1.012
0.5	0.399	0.561	0.701	0.826	0.937
0.6	0.364	0.515	0.638	0.750	0.855
0.7	0.338	0.481	0.588	0.684	0.774
0.8	0.318	0.453	0.548	0.629	0.704

Table C10.2 T-stress data T/σ_0 at v=const.

a/W	H/W=0.25	0.50	0.75	1.00	1.25
0.00	-0.526	-0.526	-0.526	-0.526	-0.526
0.25	-0.536	-0.448	-0.467	-0.490	-0.509
0.3	-0.564	-0.460	-0.462	-0.484	-0.503
0.4	-0.587	-0.505	-0.481	-0.490	-0.498
0.5	-0.592	-0.555	-0.530	-0.525	-0.521
0.6	-0.594	-0.606	-0.596	-0.583	-0.567
0.7	-0.600	-0.662	-0.674	-0.661	-0.641
0.8	-0.634	-0.735	-0.774	-0.776	-0.768

Table C10. 3 Biaxiality ratio β for v=const.

a/W	H/W=0.25	0.50	0.75	1.00	1.25
0.00	-0.469	-0.469	-0.469	-0.469	-0.469
0.25	-0.961	-0.564	-0.498	-0.476	-0.465
0.3	-1.106	-0.634	-0.523	-0.488	-0.470
0.4	-1.319	-0.805	-0.615	-0.539	-0.492
0.5	-1.484	-0.989	-0.756	-0.636	-0.556
0.6	-1.632	-1.177	-0.934	-0.777	-0.663
0.7	-1.775	-1.376	-1.146	-0.966	-0.828
0.8	-1.994	-1.623	-1.412	-1.234	-1.091

Table C10.4 Coefficient A_1 for v=const.

a/W	H/W=0.25	0.50	1.00
0.3	-0.0737	-0.0459	-0.0356
0.4	-0.0744	-0.0489	-0.0296
0.5	-0.0744	-0.0517	-0.0264
0.6	-0.0744	-0.0532	-0.0235
0.7	-0.0748	-0.0532	-0.0186
0.8	-0.0760	-0.050	-0.0098

Table C10.5 Coefficient B_1 for v=const.

a/W	H/W=0.25	0.50	1.00
0.3	0.2775	0.1945	0.1669
0.4	0.2523	0.1752	0.1450
0.5	0.2464	0.1630	0.1364
0.6	0.2468	0.1589	0.1281
0.7	0.2544	0.1613	0.1156
0.8	0.2834	0.1664	0.1024

Table C10.6 Coefficient A_2 for v=const.

a/W	H/W=0.25	0.50	1.00
0.3	-0.1052	-0.0785	-0.0356
0.4	-0.0900	-0.0610	-0.0340
0.5	-0.0886	-0.0468	-0.0166
0.6	-0.0895	-0.0343	0.0123
0.7	-0.0919	-0.0111	0.0649
0.8	-0.0806	0.0590	0.192

Table C10.7 Coefficient B_2 for ν=const.

a/W	H/W=0.25	0.50	1.00
0.3	-0.1880	-0.1082	-0.1501
0.4	-0.1282	-0.0685	-0.0758
0.5	-0.1091	-0.0498	-0.0635
0.6	-0.1017	-0.0394	-0.0870
0.7	-0.0836	-0.0577	-0.153
0.8	-0.0736	-0.1636	-0.380

Table C10.8 Geometric function F for bending displacements, eq.(C10.1.3).

a/W	H/W=0.25	0.50	0.75	1.00	1.25
0.00	1.1215	1.1215	1.1215	1.1215	1.1215
0.2	0.431	0.639	0.740	0.798	0.829
0.3	0.250	0.412	0.531	0.614	0.677
0.4	0.129	0.238	0.344	0.432	0.503
0.5	0.035	0.102	0.186	0.262	0.330
0.6	-0.041	-0.008	0.050	0.109	0.164
0.7	-0.105	-0.103	-0.070	-0.032	0.007
0.8	-0.162	-0.188	-0.183	-0.168	-0.148

Table C10.9 T-stress data T/σ_0 for bending displacements.

a/W	H/W=0.25	0.50	0.75	1.00	1.25
0.00	-0.526	-0.526	-0.526	-0.526	-0.526
0.2	-0.165	-0.121	-0.146	-0.165	-0.182
0.3	-0.072	0.033	0.033	0.016	0.003
0.4	0.040	0.161	0.184	0.176	0.171
0.5	0.158	0.282	0.318	0.323	0.326
0.6	0.276	0.402	0.446	0.462	0.476
0.7	0.396	0.525	0.580	0.608	0.631
0.8	0.525	0.662	0.741	0.790	0.828

Table C10.10 Biaxiality ratio β for bending displacements.

a/W	H/W=0.25	0.50	0.75	1.00	1.25
0.00	-0.469	-0.469	-0.469	-0.469	-0.469
0.2	-0.383	-0.189	-0.197	-0.207	-0.219
0.3	-0.288	0.080	0.062	0.026	0.004
0.4	0.310	0.676	0.535	0.407	0.340
0.5	4.514	2.765	1.710	1.233	0.988
0.6	-6.732	-0.020	8.92	4.238	2.902
0.7	-3.771	-5.097	-8.285	-1.906	90.14
0.8	-3.241	-3.521	-4.050	-4.702	-5.590

Table C10.11 Coefficient A_1 for bending displacements.

a/W	H/W=0.25	0.50	1.00
0.3	0.0170	0.0406	0.0487
0.4	0.0318	0.0534	0.0674
0.5	0.0466	0.0647	0.0822
0.6	0.0615	0.0757	0.0959
0.7	0.0764	0.0870	0.1107
0.8	0.0917	0.0997	0.1304

Table C10.12 Coefficient B_1 for bending displacements.

a/W	H/W=0.25	0.50	1.00
0.3	-0.0206	-0.0843	-0.1074
0.4	-0.0768	-0.1107	-0.1344
0.5	-0.1264	-0.1318	-0.1512
0.6	-0.1754	-0.1518	-0.1681
0.7	-0.2255	-0.1759	-0.1960
0.8	-0.2849	-0.2177	-0.2560

A weight function for the crack problem illustrated in Fig. C10.1 was given in [C10.2] as

$$h = \sqrt{\frac{2}{\pi a}} \left[\frac{1}{\sqrt{1-\rho}} + \sum_{n=1}^{4} C_n (1-\rho)^{n-1/2} \right] \quad , \quad \rho = x/a \qquad \text{(C10.1.4)}$$

with the coefficients C_n compiled in Table C10.13. In order to allow wide range interpolations of the weight function it is of advantage to know the solution for the limit case $H/W{\to}0$ which may be approximated by [C10.1]

226

$$h = \sqrt{\frac{2}{\pi a}} \frac{1}{\sqrt{1-\rho}} \left[1 - 2\left(\frac{a}{H}\right)^2 (1-\rho)^2 \right] \exp\left(-3\frac{a}{H}(1-\rho) - (a/H)^3(1-\rho)^3 \right) \qquad (C10.1.5)$$

Table C10.13 Coefficients for the weight function representation of eq.(C10.1.4).

H/W		a/W=0.3	0.4	0.5	0.6	0.7	0.8
0.25	C_1	-1.6924	-2.3107	-2.9654	-3.6544	-4.3576	-50441
	C_2	0.4181	1.1296	2.3576	4.15225	6.4217	9.0209
	C_3	0.8616	1.0018	0.4213	-1.1047	-3.5700	-6.7893
	C_4	-0.7010	-0.9450	-0.9149	-0.4561	0.4673	1.7795
0.50	C_1	-0.7560	-1.0480	-1.3366	-1.5870	-1.8665	-2.2770
	C_2	0.0813	0.0515	0.1397	0.3347	0.3478	0.0345
	C_3	0.5542	0.6190	0.6893	0.7303	1.3338	3.0820
	C_4	-0.3818	-0.4584	-0.5345	-0.6192	-0.9558	-1.7863
1.00	C_1	0.1158	-0.1735	-0.4305	-0.6369	-0.7176	-0.5953
	C_2	0.1943	0.1825	0.1079	-0.0455	-0.4514	-1.3617
	C_3	0.4413	0.4832	0.5914	0.7634	1.1138	1.8879
	C_4	-0.3196	-0.3369	-0.3962	-0.4931	-0.6423	-0.9200

C10.2 Pure displacement conditions at the plate ends

In the loading situation illustrated in Fig. C10.4 the displacements u are also kept constant. Since a rigid body motion has no influence on the stresses, we restrict the considerations to the case $u=0$. The characteristic stress is chosen as

$$\sigma_0 = \frac{v}{H} E \qquad\qquad (C10.2.1)$$

Geometric functions F for stress intensity factors defined by

$$K = \sigma_0 \, F\sqrt{\pi a} \qquad\qquad (C10.2.2)$$

are represented in Tables C10.14-C10.16. An impression of the influence of the Poisson's ratio on the geometric function is given in Fig. C10.5.

T-stress solutions for several Poisson's ratios v are compiled in Tables C10.17-C10.19.

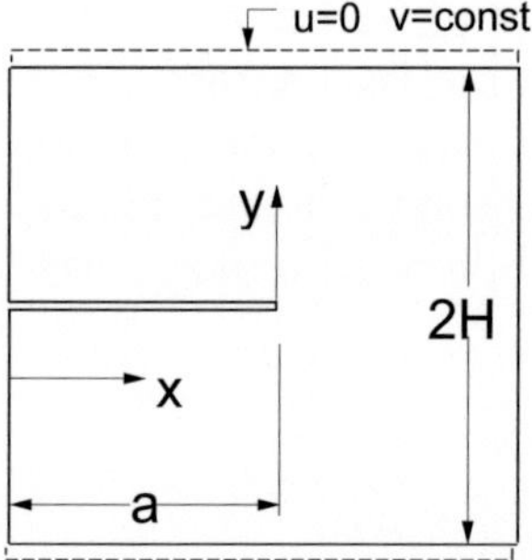

Fig. C10.4 Edge crack under pure displacement boundary conditions.

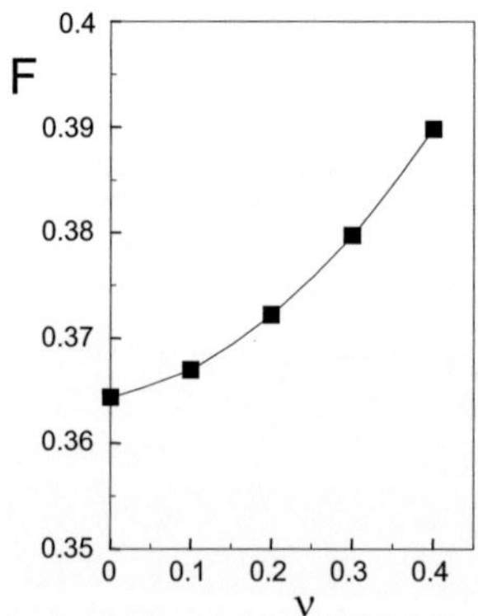

Fig. C10.5 Influence of Poisson's ratio v on the geometric function F in eq.(C10.2.2).

The higher-order stress coefficients A_1 and B_1 (see eq.(A1.1.4)) are compiled in Tables C10.20-C10.25.

Table C10.14 Geometric function for $H/W = 0.25$.

a/W	$\nu=0$	0.1	0.2	0.3	0.4
0	1.1215				
0.3	0.512	0.516	0.524	0.537	0.555
0.4	0.444	0.447	0.455	0.466	0.482
0.5	0.398	0.401	0.407	0.417	0.430
0.6	0.364	0.367	0.372	0.380	0.390
0.7	0.338	0.341	0.345	0.351	0.358
0.8	0.318	0.320	0.322	0.326	0.330

Table C10.15 Geometric function for $H/W = 0.5$.

a/W	$\nu=0$	0.1	0.2	0.3	0.4
0	1.1215				
0.3	0.727	0.730	0.736	0.744	0.754
0.4	0.630	0.636	0.643	0.652	0.664
0.5	0.563	0.568	0.575	0.584	0.595
0.6	0.516	0.520	0.525	0.532	0.540
0.7	0.480	0.482	0.485	0.490	0.496
0.8	0.451	0.452	0.453	0.455	0.458

Table C10.16 Geometric function for $H/W = 1.0$.

a/W	$\nu=0$	0.1	0.2	0.3	0.4
0	1.1215				
0.3	0.993	0.994	0.996	1.000	1.005
0.4	0.909	0.911	0.914	0.918	0.924
0.5	0.827	0.828	0.831	0.835	0.840
0.6	0.751	0.752	0.754	0.757	0.762
0.7	0.684	0.685	0.686	0.688	0.692
0.8	0.629	0.629	0.630	0.632	0.635

Table C10.17 T/σ_0 for $H/W = 0.25$.

a/W	$\nu=0$	0.1	0.2	0.3	0.4
0	-0.526				
0.3	-0.547	-0.522	-0.506	-0.498	-0.499
0.4	-0.577	-0.547	-0.525	-0.511	-0.505
0.5	-0.590	-0.563	-0.544	-0.533	-0.529
0.6	-0.599	-0.579	-0.568	-0.565	-0.570
0.7	-0.614	-0.607	-0.605	-0.608	-0.616
0.8	-0.651	-0.653	-0.659	-0.669	-0.682

Table C10.18 T/σ_0 for $H/W = 0.5$.

a/W	$\nu = 0$	0.1	0.2	0.3	0.4
0	-0.526				
0.3	-0.468	-0.479	-0.494	-0.513	-0.535
0.4	-0.509	-0.518	-0.531	-0.549	-0.571
0.5	-0.557	-0.564	-0.575	-0.591	-0.611
0.6	-0.608	-0.614	-0.623	-0.635	-0.651
0.7	-0.664	-0.668	-0.674	-0.684	-0.696
0.8	-0.740	-0.740	-0.742	-0.747	-0.754

Table C10.19 T/σ_0 for $H/W = 1.0$.

a/W	$\nu = 0$	0.1	0.2	0.3	0.4
0	-0.526				
0.3	-0.484	-0.488	-0.494	-0.501	-0.510
0.4	-0.492	-0.497	-0.504	-0.512	-0.521
0.5	-0.526	-0.531	-0.538	-0.546	-0.555
0.6	-0.583	-0.587	-0.592	-0.599	-0.607
0.7	-0.661	-0.664	-0.668	-0.673	-0.678
0.8	-0.776	-0.776	-0.779	-0.784	-0.791

Table C10.20 Coefficient A_1 for $H/W = 0.25$.

a/W	$\nu = 0$	0.1	0.2	0.3	0.4
0.3	-0.0752	-0.0775	-0.0815	-0.0871	-0.0944
0.4	-0.0761	-0.0782	-0.0817	-0.0868	-0.0933
0.5	-0.0762	-0.0783	-0.0817	-0.0863	-0.0922
0.6	-0.0763	-0.0785	-0.0817	-0.0859	-0.0911
0.7	-0.0767	-0.0787	-0.0815	-0.0850	-0.0891
0.8	-0.0771	-0.0784	-0.0799	-0.0818	-0.0839

Table C10.21 Coefficient A_1 for $H/W = 0.5$.

a/W	$\nu = 0$	0.1	0.2	0.3	0.4
0.3	-0.0489	-0.0518	-0.0551	-0.0589	-0.0632
0.4	-0.0509	-0.0531	-0.0558	-0.0589	-0.0625
0.5	-0.0528	-0.0544	-0.0564	-0.0588	-0.0616
0.6	-0.0538	-0.0549	-0.0562	-0.0578	-0.0596
0.7	-0.0536	-0.0539	-0.0545	-0.0552	-0.0562
0.8	-0.0506	-0.0503	-0.0501	-0.0500	-0.0501

Table C10.22 Coefficient A_1 for $H/W = 1.0$.

a/W	$\nu = 0$	0.1	0.2	0.3	0.4
0.3	-0.0356	-0.0363	-0.0370	-0.0378	-0.0387
0.4	-0.0298	-0.0302	-0.0310	-0.0321	-0.0326
0.5	-0.0265	-0.0269	-0.0274	-0.0280	-0.0286
0.6	-0.0234	-0.0236	-0.0239	-0.0243	-0.0248
0.7	-0.0188	-0.0189	-0.0190	-0.0192	-0.0195
0.8	-0.0106	-0.0105	-0.0106	-0.0108	-0.0112

Table C10.23 Coefficient B_1 for $H/W = 0.25$.

a/W	$\nu = 0$	0.1	0.2	0.3	0.4
0.3	0.2742	0.2626	0.2559	0.2542	0.2574
0.4	0.2532	0.2494	0.2506	0.2568	0.2679
0.5	0.2466	0.2489	0.2561	0.2682	0.2852
0.6	0.2472	0.2555	0.2672	0.2822	0.3006
0.7	0.2552	0.2673	0.2815	0.2978	0.3163
0.8	0.2778	0.2868	0.2951	0.3027	0.3095

Table C10.24 Coefficient B_1 for $H/W = 0.5$.

a/W	$\nu = 0$	0.1	0.2	0.3	0.4
0.3	0.1936	0.1953	0.1993	0.2056	0.2141
0.4	0.1744	0.1759	0.1790	0.1837	0.1899
0.5	0.1647	0.1672	0.1706	0.1749	0.1801
0.6	0.1611	0.1635	0.1663	0.1695	0.1731
0.7	0.1628	0.1632	0.1637	0.1643	0.1649
0.8	0.1726	0.1699	0.1672	0.1645	0.1619

Table C10.25 Coefficient B_1 for $H/W = 1.0$.

a/W	$\nu = 0$	0.1	0.2	0.3	0.4
0.3	0.1684	0.1713	0.1743	0.1775	0.1809
0.4	0.1455	0.1468	0.1486	0.1509	0.1536
0.5	0.1363	0.1363	0.1367	0.1375	0.1386
0.6	0.1280	0.1271	0.1265	0.1263	0.1264
0.7	0.1171	0.1155	0.1144	0.1138	0.1136
0.8	0.1058	0.1066	0.1073	0.1078	0.1081

References C10

[C10.1] Fett, T., Munz, D., Stress intensity factors and weight functions, Computational Mechanics Publications, Southampton, 1997.

[C10.2] Fett, T., Bahr, H.-A., Mode I stress intensity factors and weight functions for short plates under different boundary conditions, Engng. Fract. Mech. **62**(1999), 593-606.

C11

Double-edge-cracked circular disk

The double-edge-cracked circular disk is shown in Fig. C11.1. Different traction and displacement boundary conditions will be considered in the following sections.

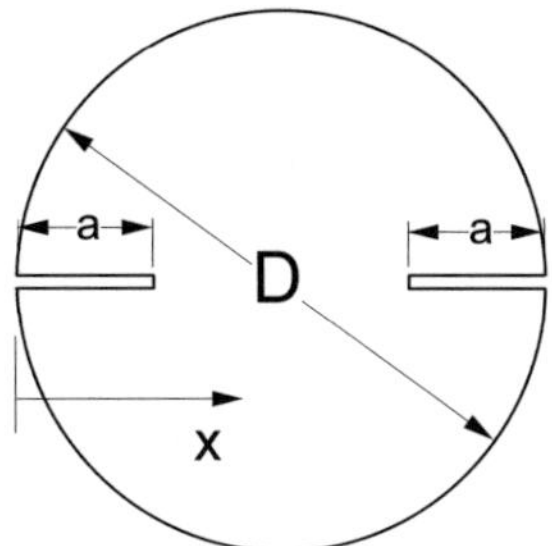

Fig. C11.1 Double-edge-notched disk.

C11.1 Traction boundary conditions

The pure traction loading by σ_n = constant and $\tau_{R\omega}$ = 0 is illustrated in Fig. C11.2.

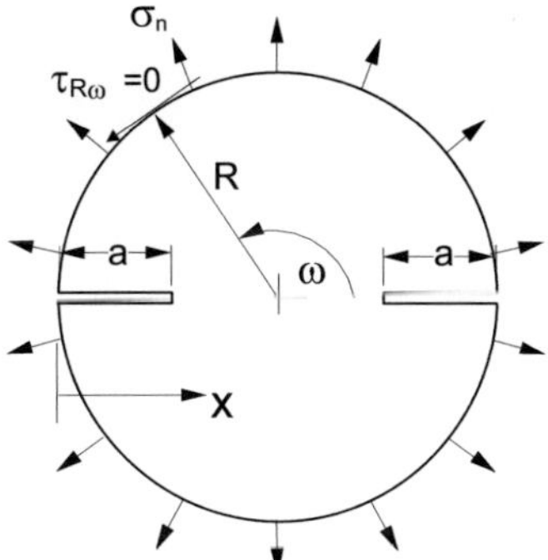

Fig. C11.2 Double-edge-cracked disk under traction boundary conditions σ_n =const., $\tau_{R\omega}$=0.

The geometric function F for the stress intensity factor is

$$K = \sigma_n F \sqrt{\pi a} \, , \quad \alpha = a/R \tag{C11.1.1}$$

as shown in Fig. C11.3 and approximated by

$$F \cong \frac{1.1215 + 0.2746\alpha - 0.7959\alpha^2 - 1.1411\alpha^3 + 1.1776\alpha^4}{\sqrt{1-\alpha}}$$

(C11.1.2)

In contrast to the single-edge-cracked disk, the relative crack length here is defined by $\alpha = a/R$ ($R = D/2$).

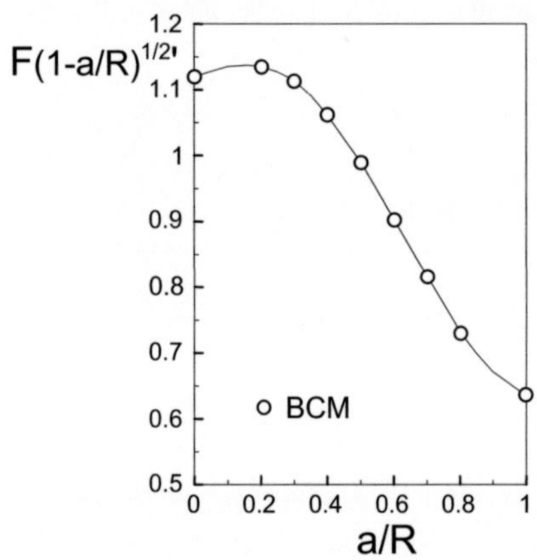

Fig. C11.3 Geometric function F for the double-edge-cracked disk.

The weight function for the double-edge cracked disk under traction boundary conditions reads

$$h = \sqrt{\frac{2}{\pi a}}\left[\frac{1}{\sqrt{1-\rho}} + D_0\sqrt{1-\rho} + D_1(1-\rho)^{3/2} + D_2(1-\rho)^{5/2}\right], \quad \rho = x/a$$

(C11.1.3)

with the coefficients of

$$D_0 = \frac{0.4594 + 2.3454\,\alpha - 1.0205\,\alpha^2 - 7.7547\,\alpha^3 + 9.1403\,\alpha^4}{\sqrt{1-\alpha}}$$

(C11.1.4)

$$D_1 = \frac{0.6833 - 0.1484\,\alpha - 1.8811\,\alpha^2 + 7.0112\,\alpha^3 - 8.9802\,\alpha^4}{\sqrt{1-\alpha}}$$

(C11.1.5)

$$D_2 = \frac{-0.3059 + 0.2829\,\alpha + 0.3552\,\alpha^2 - 1.9646\,\alpha^3 + 2.4682\,\alpha^4}{\sqrt{1-\alpha}}$$

(C11.1.6)

The T-stress under loading by constant normal traction σ_n along the circumference is shown in Fig. C11.4 together with the biaxiality ratio β.

The T-stress can be expressed by

$$\frac{T}{\sigma_n} = \frac{0.474 + 0.282\alpha^2 - 0.857\alpha^4 - 0.385\alpha^6 + 0.486\alpha^8}{1-\alpha} \tag{C11.1.7}$$

the biaxiality ratio by

$$\beta = \frac{0.423 + 0.145\alpha^2 + 1.301\alpha^4 - 3.679\alpha^6 + 1.81\alpha^8}{\sqrt{1-\alpha}} \tag{C11.1.8}$$

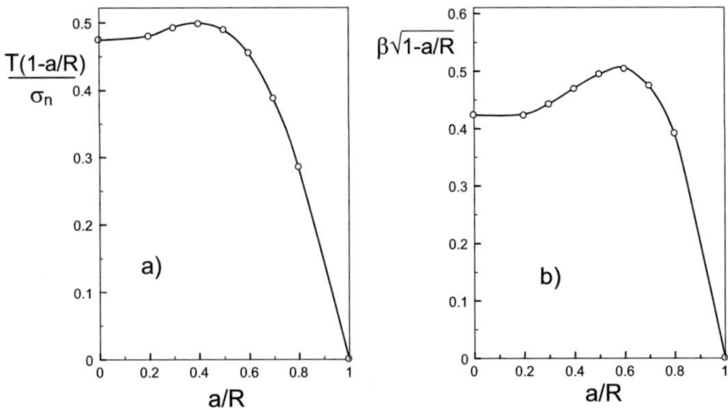

Fig. C11.4 T-stress and biaxiality ratio of the double-edge-cracked circular disk under circumferential normal traction.

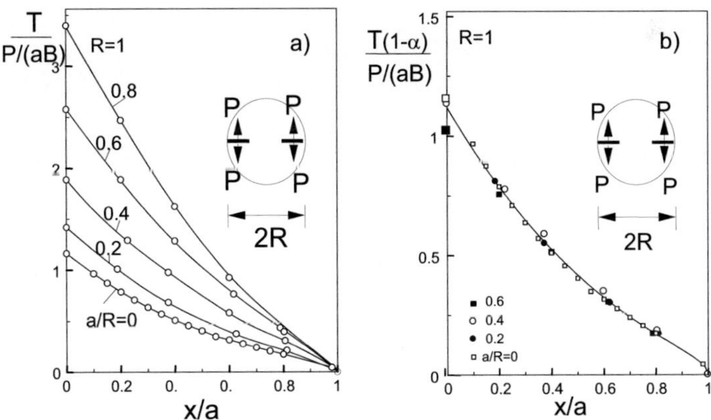

Fig. C11.5 Double-edge-notched disk; a) Green's function from finite element computations, b) normalized representation.

From the data of Fig. C11.5b, the average Green's function results as

$$t \cong \frac{1}{a(1-a/R)}[0.3\sqrt{1-x/a}+0.347(1-x/a)^{3/2}+0.473(1-x/a)^{5/2}] \qquad \text{(C11.9)}$$

For the Green's function under symmetrical loading the same approximate set-up is chosen as used for single-edge-cracked components. It reads

$$t = \tfrac{1}{a}E_0(1-x/a) \qquad \text{(C11.1.10)}$$

with the parameter E_0 entered into Table C11.1 and fitted for $\alpha \le 0.8$ by the polynomial of

$$E_0 = \frac{0.948+0.564\,\alpha^2-1.714\alpha^4-0.770\alpha^6+0.972\alpha^8}{1-\alpha} \qquad \text{(C11.1.11)}$$

Table C11.1 T-stress, biaxiality ratio, and coefficient for the Green's function.
Loading: Constant circumferential normal traction, zero shear traction.

α	T/σ_n	β	E_0
0	0.474	0.423	0.9481
0.2	0.599	0.472	1.199
0.3	0.702	0.528	1.405
0.4	0.829	0.604	1.658
0.5	0.977	0.698	1.954
0.6	1.136	0.795	2.273
0.7	1.290	0.865	2.580
0.8	1.425	0.873	2.850

The higher order coefficients A_1 and B_1 according to eq.(A1.1.4) are compiled in Table C11.2.

Table C11.2 Coefficients A_1 and B_1 according to eq.(A1.1.4).

a/R	A_1	B_1
0.2	-0.039	0.472
0.3	-0.012	0.285
0.4	0.008	0.170
0.5	0.021	0.085
0.6	0.023	0.022
0.7	0.007	-0.016
0.8	-0.051	-0.025

Figure C11.5 represents the displacements under constant normal traction σ_n in the form of

$$\delta = \frac{2a\sigma_n}{E'}\frac{1}{\alpha}\ln\frac{1}{1-\alpha}\lambda(\alpha) , \quad \alpha = a/R \qquad (C11.1.12)$$

The results of boundary collocation computations are represented by the circles. From a least-squares fit we obtain the representation of

$$\lambda = 1.454 + 0.3893\,\alpha + 5.0022\,\alpha^2 - 19.5054\,\alpha^3 + 23.6198\,\alpha^4 - 10.3233\,\alpha^5 \qquad (C11.1.13)$$

The dashed curve in Fig. C11.6 is the solution for the double-edge-cracked endless parallel strip as reported by Tada [C11.1].

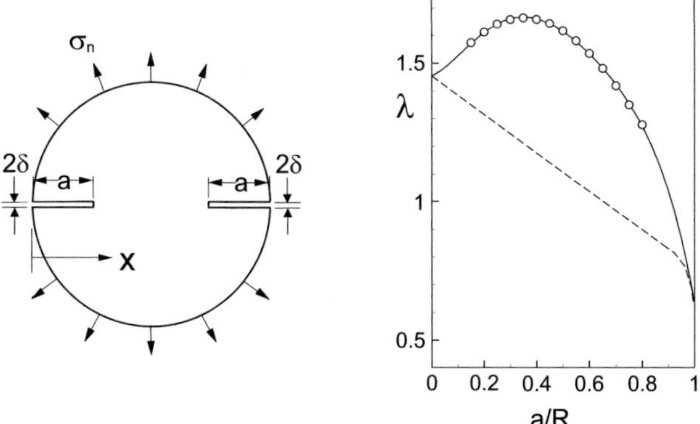

Fig. C11.6 Crack mouth displacements ($x=0$) according to eq.(C11.1.12); circles: Double-edge-cracked disk, dashed curve: Results for the double-edge-cracked endless parallel strip, as reported in [C11.1].

The double-edge-cracked disk under constant shear traction τ_0 on the crack faces is illustrated in Fig. C11.7 together with the stress intensity factor solution represented by

$$K_{II} = \tau_0 F_{II}\sqrt{\pi a} , \quad F'_{II} = F_{II}\sqrt{1-a/R} \qquad (C11.1.14)$$

The data of Fig. C11.7 can be expressed by

$$F_{II} = \frac{1.1215 - 0.5608\alpha + 0.2185\alpha^2 - 0.5007\alpha^3 + 0.3584\alpha^4}{\sqrt{1-\alpha}} , \quad \alpha = a/R \quad (C11.1.15)$$

In addition, Fig. C11.7 contains the mode-II stress intensity factor solution for the double-edge-cracked endless strip [1] as the dashed curve. Only small deviations from this solution are visible in the region of $0.3 < a/R < 0.7$. An approximate weight function can be derived from eq. (C11.1.15) by application of the extended Petroski-Achenbach procedure (see e.g. [C11.2]).

The coefficients for a representation

$$h_{\mathrm{II}} = \sqrt{\frac{2}{\pi a}}\left[\frac{1}{\sqrt{1-\rho}} + D_0\sqrt{1-\rho} + D_1(1-\rho)^{3/2} + D_2(1-\rho)^{5/2}\right], \quad \rho = x/a \quad (C11.1.16)$$

are given by

$$D_0 = \frac{0.4594 - 0.4274\,\alpha + 2.586\,\alpha^2 - 5.42\,\alpha^3 + 5.677\,\alpha^4}{\sqrt{1-\alpha}} \qquad (C11.1.17)$$

$$D_1 = \frac{0.630 - 0.2516\,\alpha + 1.2866\,\alpha^2 - 3.44\,\alpha^3}{\sqrt{1-\alpha}} \qquad (C11.1.18)$$

$$D_2 = \frac{-0.2541 - 0.1658\,\alpha + 0.4396\,\alpha^2 + 0.3636\,\alpha^3}{\sqrt{1-\alpha}} \qquad (C11.1.19)$$

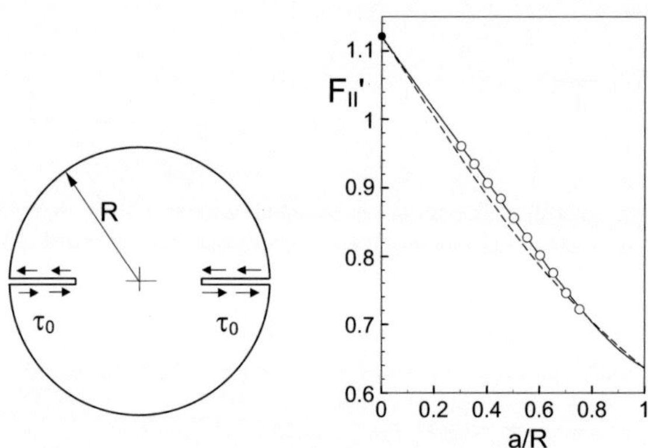

Fig. C11.7 Double-edge-cracked disk under constant shear traction on the crack faces; dashed curve: Solution for the double edge-cracked endless strip (see e.g. [C11.1]).

C11.2 Mixed boundary conditions

Figure C11.8 shows the case of constant normal traction σ_n and disappearing tangential displacements v_ω along the circumference. The stress intensity factor described by

$$K = \sigma_n F(\nu, a/R)\sqrt{\pi a} \qquad (C11.2.1)$$

and the T-stress are plotted in Fig. C11.9. In this loading case the T-stresses are very small. The higher order terms A_1 and B_1 of eq.(A1.1.4) are compiled in Tables C11.3 and C11.4.

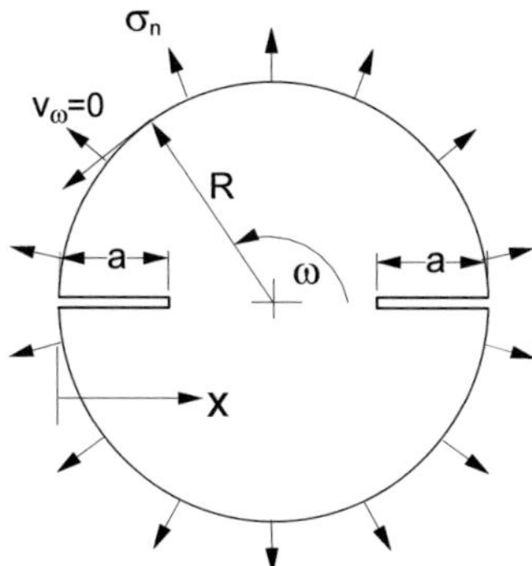

Fig. C11.8 Mixed boundary conditions σ_n=const., v_ω=0.

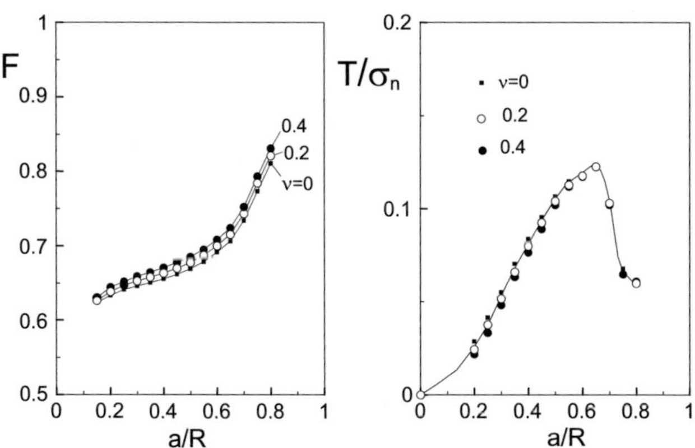

Fig. C11.9 Geometric function F and T-stress as functions of ν and a/R.

For $\nu=0.25$ and $\alpha=a/R \leq 0.75$ the geometric function can be approximated by

$$F = 0.59 + 0.462\,\alpha - 1.171\,\alpha^2 + 1.197\,\alpha^3 \qquad (C11.2.2)$$

and the related T-stress at $\alpha \leq 0.75$ by

$$T / \sigma^* = 0.127\,\alpha^2 + 5.024\,\alpha^3 - 18.468\,\alpha^4 + 26.0173\alpha^5 - 13.7978\alpha^6 \qquad \text{(C11.2.3)}$$

A weight function for $\nu = 0.25$ is given in the form of eq.(C11.1.3) with the coefficients compiled in Table C11.5.

Table C11.3 Coefficient A_1 according to eq.(A1.1.4).

	$\nu=0$	0.2	0.4
a/R=0.2	-0.172	-0.172	-0.171
0.3	-0.137	-0.137	-0.137
0.4	-0.119	-0.118	-0.118
0.5	-0.108	-0.107	-0.107
0.6	-0.104	-0.103	-0.102
0.7	-0.108	-0.107	-0.107
0.8	-0.127	-0.127	-0.126

Table C11.4 Coefficient B_1 according to eq.(A1.1.42).

	$\nu=0$	0.2	0.25	0.4
a/R=0.2	0.011	0.009	-0.008	-0.014
0.3	-0.035	-0.041	-0.043	-0.048
0.4	-0.038	-0.045	-0.046	-0.051
0.5	-0.039	-0.047	-0.049	-0.053
0.6	-0.038	-0.045	-0.047	-0.051
0.7	-0.029	-0.034	-0.035	-0.039
0.8	-0.017	-0.022	-0.023	-0.026

Table C11.5 Coefficients for the weight function according to eq.(C11.1.3) at $\nu = 0.25$.

a/R	D_0	D_1	D_2
0	-2.1306	2.4203	-0.8261
0.1	-1.9737	2.3346	-0.7985
0.2	-1.9052	2.3238	-0.7918
0.3	-1.8619	2.3044	-0.7850
0.4	-1.7853	2.2064	-0.7603
0.5	-1.6213	1.9680	-0.7017
0.6	-1.3276	1.5511	-0.5987
0.7	-0.8859	0.9657	-0.4522
0.75	-0.6128	0.6308	-0.3670

Figure C11.10 shows the case of constant radial displacements u_n and disappearing shear tractions $\tau_{R\omega}$. The stress intensity factor is given by

$$K = \sigma * F(\nu, a/R)\sqrt{\pi a} \ , \quad \sigma* = \frac{u_n E}{R} \tag{C11.2.4}$$

The geometric function F is plotted in Fig. C11.11a. In the form of

$$F* = F(1-\nu) \tag{C11.2.5}$$

the results (which now coincide for $a/R = 0$) are shown in Fig. C11.11b.
For $\nu = 0.25$ the geometric function F in the region of $a/R \leq 0.8$ can be approximated by

$$F \cong \tfrac{4}{3} + 0.8251\alpha + 0.65527\alpha^2 - 12.6637\alpha^3 + 17.6804\alpha^4 - 7.1736\alpha^5 \tag{C11.2.6}$$

and the T-term by

$$T/\sigma* \cong -0.9611\alpha^2 + 11.812\alpha^3 - 23.8847\alpha^4 + 20.5897\alpha^5 - 6.7657\alpha^6 \tag{C11.2.7}$$

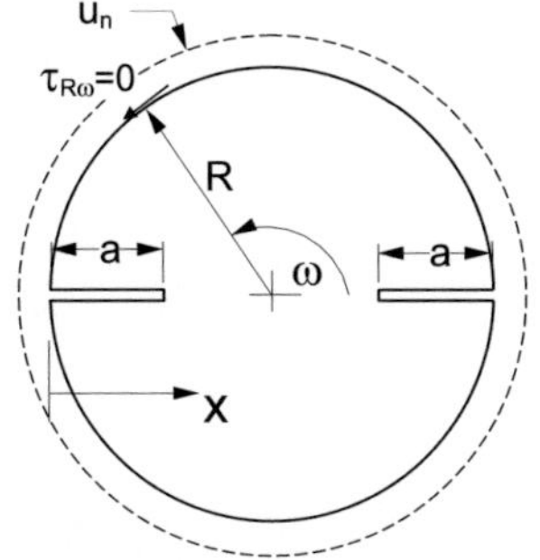

Fig. C11.10 Mixed boundary conditions u_n =const., $\tau_{R\omega}$=0.

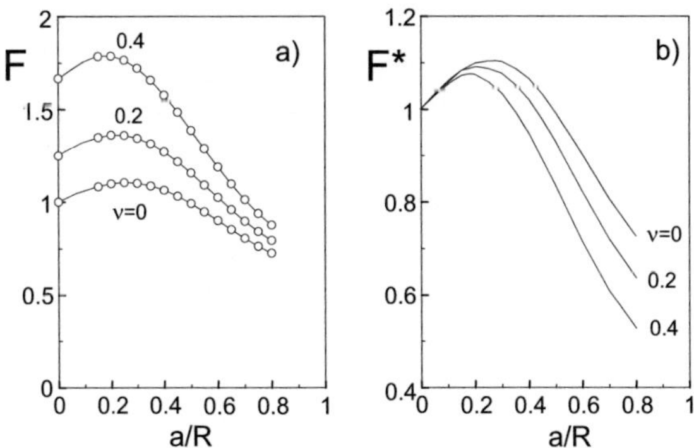

Fig. C11.11 Geometric function according to eq.(C11.2.4).

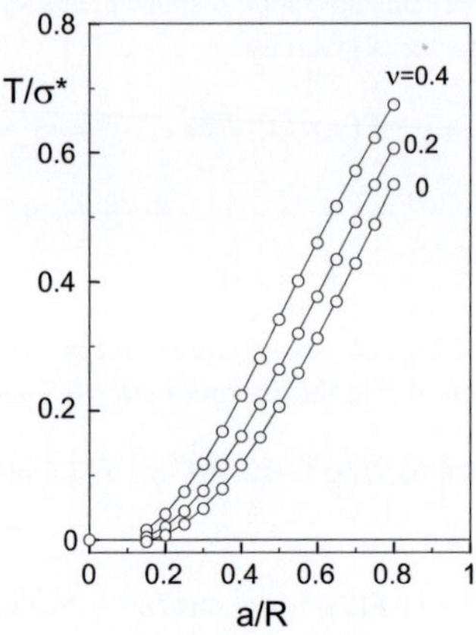

Fig. C11.12 T-stress term under the conditions of u_n =const., $\tau_{R\omega}$=0.

Values of the higher order coefficients A_1 and B_1 of eq.(A1.1.4) are compiled in Tables C11.6 and C11.8.

Table C11.6 Coefficient A_1 according to eq.(A1.1.4).

a/R	$v=0$	0.2	0.25	0.4
0.2	-0.094	-0.116	-0.122	-0.149
0.3	-0.064	-0.075	-0.079	-0.094
0.4	-0.041	-0.047	-0.049	-0.055
0.5	-0.025	-0.027	-0.028	-0.030
0.6	-0.016	-0.016	-0.016	-0.017
0.7	-0.015	-0.015	-0.015	-0.016
0.8	-0.027	-0.030	-0.030	-0.033

Table C11.7 Coefficient B_1 according to eq.(A1.1.4).

a/R	$v=0$	0.2	0.4
0.2	0.003	0.002	0.001
0.3	0.007	0.006	0.004
0.4	0.010	0.008	0.007
0.5	0.011	0.009	0.008
0.6	0.010	0.008	0.007
0.7	0.007	0.006	0.004
0.8	0.003	0.003	0.002

A weight function for $v = 0.25$ is given in the form of eq.(C11.1.3) with the coefficients compiled in Table C11.8.

Table C11.8 Coefficients for the weight function according to eq.(C11.1.3) at $v = 0.25$.

a/R	D_0	D_1	D_2
0	-0.1384	1.0933	-0.4279
0.1	0.0604	1.0713	-0.4102
0.2	0.1270	1.1218	-0.4159
0.3	0.0186	1.2145	-0.4417
0.4	-0.2367	1.3033	-0.4764
0.5	-0.5643	1.3301	-0.5036
0.6	-0.8484	1.1932	-0.4952
0.7	-0.9663	0.7742	-0.4193
0.8	-0.9132	0.1456	-0.2900

Figure C11.13 represents the crack opening displacements δ (for δ see Fig. C11.6) under constant radial displacements and disappearing shear traction at the circumference in the form of

$$\delta = \frac{2a\sigma^*}{E'} \lambda(a/R)$$

(C11.2.8)

with σ^* given by eq.(C11.2.4).

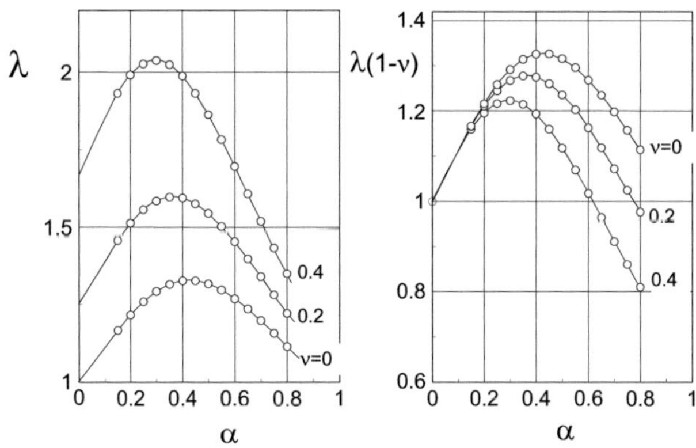

Fig. C11.13 Crack mouth displacement represented by eq.(C11.2.8). Boundary conditions: $u_n = $ const., $\tau_{R\omega} = 0$.

C11.3 Displacement boundary conditions

The case of pure displacement boundary conditions is illustrated in Fig. C11.14. Under these boundary conditions, the geometric functions F for the stress intensity factor defined by eq.(C11.2.4) result as shown in Fig. C11.15a. The T-stress term is given in Fig. C11.15b.

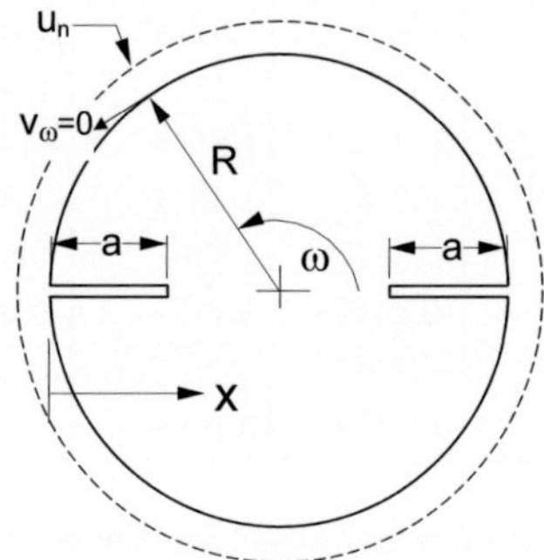

Fig. C11.14 Displacement boundary conditions u_n =const., v_ω=0.

For $\nu = 0.25$ the geometric function F in the range of $a/R \leq 0.8$ can be approximated by

$$F \cong 0.824 - 0.1267\alpha + 0.02799\alpha^2 \qquad (C11.3.1)$$

and the T-stress in the range $0.2 \leq a/R \leq 0.8$ by

$$T/\sigma^* \cong 0.0496 + 0.175\alpha - 0.1016\alpha^2 - 0.2251\alpha^3 + 0.5331\alpha^4 \qquad (C11.3.2)$$

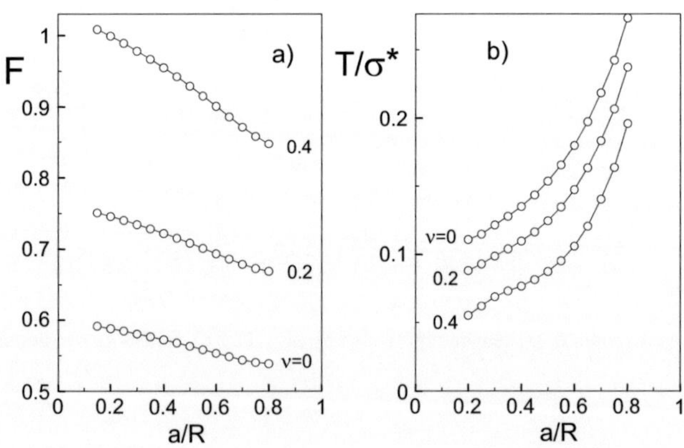

Fig. C11.15 Geometric function F and T-stress as functions of ν and a/R.

Higher-order stress coefficients are compiled in Tables C11.9 and C11.10.

Table C11.9 Coefficient A_1 according to eq.(A1.1.4).

a/R	$\nu=0$	0.2	0.25	0.4
0.2	-0.170	-0.212	-0.226	-0.282
0.3	-0.138	-0.172	-0.183	-0.228
0.4	-0.119	-0.148	-0.158	-0.195
0.5	-0.106	-0.132	-0.140	-0.172
0.6	-0.096	-0.119	-0.126	-0.154
0.7	-0.090	-0.111	-0.117	-0.142
0.8	-0.090	-0.110	-0.117	-0.140

Table C11.10 Coefficient B_1 according to eq.(A1.1.4).

a/R	$\nu=0$	0.2	0.25	0.4
0.2	0.239	0.237	0.232	0.23
0.3	0.167	0.171	0.174	0.188
0.4	0.130	0.138	0.141	0.156
0.5	0.106	0.114	0.117	0.133
0.6	0.085	0.093	0.096	0.110
0.7	0.065	0.072	0.074	0.085
0.8	0.043	0.048	0.050	0.056

A weight function for $\nu = 0.25$ is given in the form of eq.(C11.1.3) with the coefficients compiled in Table C11.11.

Table C11.11 Coefficients for the weight function according to eq.(C11.1.3) for $\nu = 0.25$.

a/R	D_0	D_1	D_2
0	-1.9974	2.3316	-0.7995
0.1	-2.0420	2.3613	-0.8084
0.2	-2.0850	2.3898	-0.8170
0.3	-2.1258	2.4166	-0.8250
0.4	-2.1642	2.4414	-0.8326
0.5	-2.2002	2.4640	-0.8395
0.6	-2.2336	2.4842	-0.8457
0.7	-2.2643	2.5017	-0.8513
0.8	-2.2922	2.5164	-0.8561

C11.4 Double-edge-cracked Brazilian disk

The Brazilian disk test with a double-edge-cracked circular disk is illustrated by Fig. C11.16.

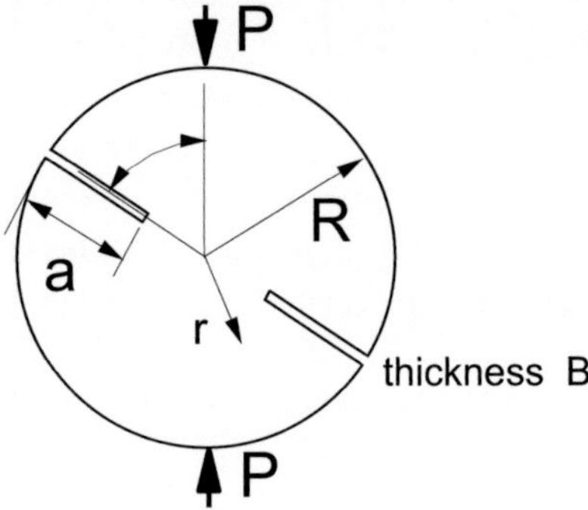

Fig. C11.16 Brazilian disk test with double-edge-cracked specimen.

For the computation of the biaxiality ratio, the mode-I stress intensity factor is necessary. Stress intensity factors K_I computed with the weight function, eqs.(C11.1.3)-(C11.1.6), and expressed by the geometric function F are presented in Table C11.12. The geometric function F is defined by

$$K_I = \sigma^* F\sqrt{\pi a} \,, \quad \sigma^* = P/(\pi RB) \tag{C11.4.1}$$

Table C11.12 Stress intensity factors represented by the geometric function F for the Brazilian disk.

$\alpha = a/R$	$\Theta = \pi/32$	$\pi/16$	$\pi/8$	$\pi/4$	$3\pi/8$	$7\pi/16$	$\pi/2$
0	0	0	0	0	0	0	0
0.1	-6.188	-2.952	-0.970	-0.304	-0.180	-0.160	-0.154
0.2	-4.104	-3.311	-1.709	-0.648	-0.399	-0.357	-0.344
0.3	-2.728	-2.680	-1.989	-0.987	-0.652	-0.590	-0.571
0.4	-1.901	-2.044	-1.927	-1.274	-0.927	-0.854	-0.832
0.5	-1.343	-1.541	-1.713	-1.479	-1.212	-1.145	-1.127
0.6	-0.934	-1.153	-1.469	-1.607	-1.500	-1.459	-1.445
0.7	-0.614	-0.855	-1.263	-1.705	-1.809	-1.817	-1.817

Using the Green's function and the stress distribution given by eqs.(7.5.1) and (7.5.2), the T-stress was computed. Table C11.13 contains the data for several angles Θ (see Fig. C11.16).

The biaxiality ratio resulting from the T- and the K_I-solution is plotted in Fig. C11.17 for several values of the angle Θ.

Table C11.13 T-stress T/σ^* for the Brazilian disk test ($\sigma^*=P/(\pi RB)$).

a/R	$\Theta = \pi/32$	$\pi/16$	$\pi/8$	$\pi/4$	$3\pi/8$	$7\pi/16$	$\pi/2$
0	0	0	0	0	0	0	0
0.1	-3.04	1.88	1.074	0.378	0.228	0.204	0.196
0.2	-8.85	-1.96	1.105	0.763	0.513	0.465	0.450
0.3	-8.75	-4.62	-0.051	1.013	0.835	0.783	0.765
0.4	-8.00	-5.60	-1.516	1.001	1.159	1.141	1.132
0.5	-7.40	-5.90	-2.726	0.688	1.430	1.508	1.527
0.6	-7.01	-6.02	-3.611	0.126	1.577	1.824	1.896
0.7	-6.43	-5.84	-4.11	-0.50	1.61	2.07	2.22

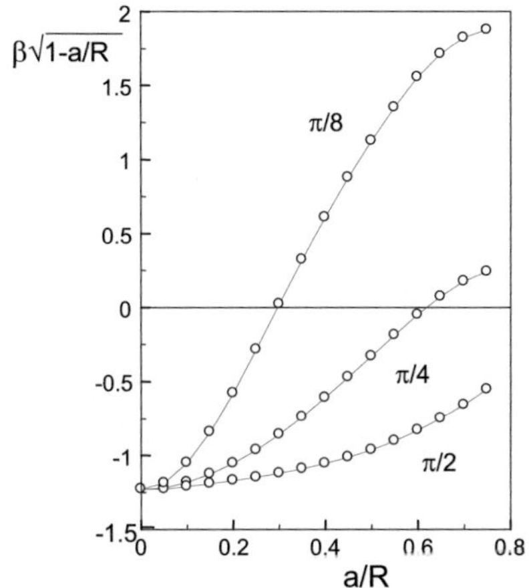

Fig. C11.17 Biaxiality ratio for the double-edge-notched Brazilian disk.

References C11

[C11.1] Tada, H., Paris, P.C., Irwin, G.R., The stress analysis of cracks handbook, Del Research Corporation, 1986.
[C11.2] Fett, T., Munz, D., Stress intensity factors and weight functions, Computational Mechanics Publications, Southampton, 1997.

C12

Double-edge-cracked rectangular plate

C12.1 Double-edge-cracked plate under traction boundary conditions

A double-edge-cracked rectangular plate under pure tensile loading is shown in Fig. C12.1. Stress intensity factors defined by

$$K_\mathrm{I} = \sigma F \sqrt{\pi a} \ , \quad F' = F(1 - a/W)^{1/2} \tag{C12.1.1}$$

are compiled in Table C12.1. A weight function for symmetric loading is given by

$$h = \sqrt{\frac{2}{\pi a}} \left(\frac{1}{\sqrt{1-\rho}} + D_0 \sqrt{1-\rho} + D_1(1-\rho)^{3/2} \right), \quad \rho = x/a \tag{C12.1.2}$$

with the coefficients D_0, D_1 listed in Tables C12.2 and C12.3. T-stresses are compiled in Table C12.4.

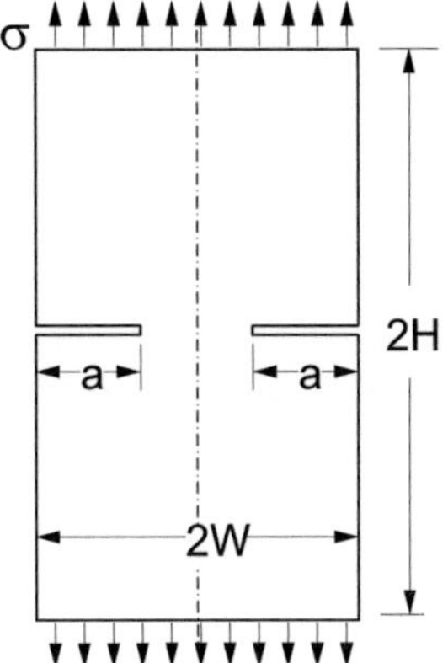

Fig. C12.1 Double-edge-cracked rectangular plate

For a long plate ($H/W = 1.5$) the T-stress term and biaxiality ratio β may be approximated by

$$\frac{T}{\sigma} = \frac{-0.526 + 0.4672\,\alpha + 0.1844\,\alpha^2 - 0.1256\,\alpha^3}{1-\alpha} \tag{C12.1.3}$$

$$\beta = \frac{-0.469 + 0.14067\,\alpha + 0.35646\,\alpha^2 - 0.00986\,\alpha^3}{\sqrt{1-\alpha}} \qquad (C12.1.4)$$

and for the quadratic plate ($H/W=1$) by

$$T/\sigma = -0.526 + 0.1804\,\alpha - 2.7241\,\alpha^2 + 9.5966\,\alpha^3 - 6.3883\,\alpha^4 \qquad (C12.1.5)$$

$$\beta = -0.469 + 0.1229\,\alpha - 1.2256\,\alpha^2 + 6.0628\,\alpha^3 - 4.4983\,\alpha^4 \qquad (C12.1.6)$$

The biaxiality ratios β are given in Table C12.5.

Table C12.1 Geometric function F_1', eq.(C12.1.1).

a/W	L/W=1.5	1.25	1.0	0.75	0.50	0.35
0	1.1215	1.1215	1.1215	1.1215	1.1215	1.1215
0.3	0.94	0.96	1.029	1.18	1.496	1.891
0.4	0.8891	0.9197	0.9946	1.1926	1.646	2.196
0.5	0.8389	0.8659	0.9427	1.1537	1.719	2.437
0.6	0.7900	0.8135	0.8760	1.0597	1.6529	2.535
0.7	0.7420	0.7492	0.8029	0.9297	1.4142	2.46
1.0	0.6366	0.6366	0.6366	0.6366	0.6366	0.6366

Table C12.2 Coefficient D_0 for eq.(C12.1.2).

a/W	L/W=0.35	0.50	0.75	1.00	1.50
0	0.585	0.584	0.584	0.584	0.584
0.3	3.75	2.43	1.403	0.932	0.614
0.4	4.91	3.26	1.777	1.085	0.720
0.5	6.46	3.93	2.004	1.252	0.879
0.6	8.14	4.29	2.12	1.478	1.160
0.7	9.62	4.05	2.33	1.88	1.494

Table C12.3 Coefficient D_1 for eq.(C12.1.2).

a/W	L/W=0.35	0.50	0.75	1.00	1.50
0	0.256	0.256	0.256	0.256	0.256
0.3	1.303	0.953	0.552	0.302	0.216
0.4	2.56	1.48	0.624	0.335	0.178
0.5	3.37	2.05	0.739	0.325	0.134
0.6	3.71	2.43	0.787	0.243	0.01
0.7	3.95	2.83	0.557	0.024	0.034

In Fig. C12. 2 results of Table C12.5 are compared with data from literature (Kfouri [C12.1]). Differences of less than 0.01 were found, i.e. an excellent agreement can be stated. Further coefficients of the Williams stress function, eq.(A1.1.4), are listed in Tables C12.6 - C12.9.

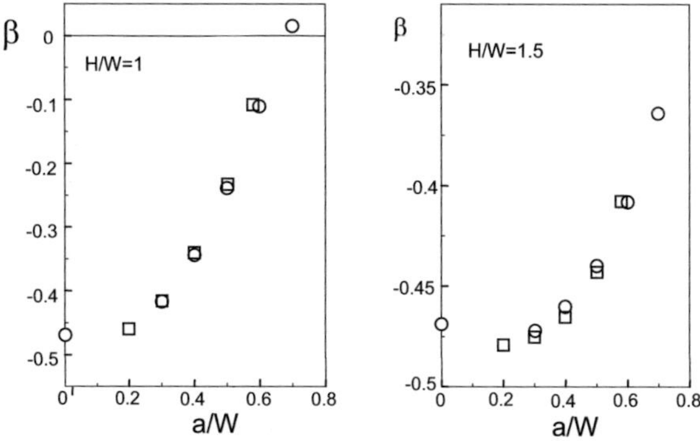

Fig. C12. 2 Comparison with data available from literature. Circles: Table C12.5, squares: Kfouri [C12.1].

Table C12.4 T-stress T/σ of the double-edge-cracked plate in tension.

$\alpha = a/W$	H/W=1.5	1.25	1.00	0.75	0.50	0.35
0.0	-0.526	-0.526	-0.526	-0.526	-0.526	-0.526
0.1	-0.530	-0.530	-0.530			
0.2	-0.532	-0.528	-0.527			
0.3	-0.532	-0.520	-0.512	-0.473	-0.257	0.293
0.4	-0.528	-0.504	-0.440	-0.282	0.256	1.546
0.5	-0.522	-0.464	-0.316	0.045	1.058	3.135
0.6	-0.510	-0.409	-0.153	0.483	2.202	5.24
0.7	-0.4932	-0.32	0.023	0.969	3.68	8.13

Table C12.5 Biaxiality ratio β of the double-edge-cracked plate in tension.

$\alpha = a/W$	H/W=1.5	1.25	1.00	0.75	0.50	0.35
0.0	-0.469	-0.469	-0.469	-0.469	-0.469	-0.469
0.1	-0.475	-0.470	-0.464			
0.2	-0.476	-0.465	-0.451			
0.3	-0.472	-0.453	-0.416	-0.336	-0.144	0.174
0.4	-0.460	-0.425	-0.343	-0.183	0.120	0.545
0.5	-0.440	-0.379	-0.237	0.028	0.435	0.910
0.6	-0.408	-0.318	-0.110	0.288	0.842	1.307
0.7	-0.364	-0.228	0.016	0.571	1.424	1.903

Table C12.6 Coefficient A_1 of the double-edge-cracked plate in tension.

a/W	$H/W=1.5$	1.25	1.00	0.75	0.50
0.3	-0.045	-0.043	-0.0362	-0.0192	0.0441
0.4	-0.0416	-0.0371	-0.0237	0.0147	0.1395
0.5	-0.0414	-0.0339	-0.0118	0.0522	0.2591
0.6	-0.0454	-0.0277	-0.0053	0.0840	0.3936
0.7	-0.0591	-0.0457	-0.0110	0.0956	0.5074

Table C12.7 Coefficient B_1 of the double-edge-cracked plate in tension.

a/W	$H/W=1.5$	1.25	1.00	0.75	0.50
0.3	0.1555	0.148	0.1208	0.0771	-0.0509
0.4	0.1086	0.0911	0.0489	-0.0382	-0.1991
0.5	0.0759	0.0505	-0.0099	-0.1384	-0.3478
0.6	0.0515	0.0014	-0.0496	-0.2157	-0.5472
0.7	0.0356	0.0039	-0.0671	-0.2510	-0.7722

Table C12.8 Coefficient A_2.

a/W	0.25	0.50	1.00
0.3	0.541	0.0447	-0.0173
0.4	-1.867	0.007	0.0026
0.5	-3.24	-0.061	0.0023
0.6	-4.43	-0.158	-0.022
0.7	-5.54	-0.372	-0.083

Table C12.9 Coefficient B_2.

a/W	0.25	0.50	1.00
0.3	3.37	-0.096	-0.244
0.4	5.90	0.203	-0.142
0.5	8.50	0.390	-0.075
0.6	10.48	0.497	-0.017
0.7	11.45	0.661	0.036

C12.2 Mixed boundary conditions at the ends

The double-edge-cracked plate under mixed boundary conditions is shown in Fig. C12.3. Results for stress intensity factors (expressed by the geometric function F according to eq.(C12.1.1)) are illustrated in Fig. C12.4a and compiled in Table C12.10. Also in this case, the curves introduced are described by eq.(C12.2.1). The numerical data are represented well up to $H/W=0.5$ by

$$F = \sqrt{\frac{H}{\pi a}}\, \tanh^{1/\gamma}\!\left(1.1215\sqrt{\frac{\pi a}{H}}\right)^{\gamma} \quad , \quad \gamma = 2.2 \tag{C12.2.1}$$

with a maximum deviation of less than 3%. For the characteristic stress $\sigma=\sigma_0$ it holds

$$\sigma_0 = \frac{v}{H} E \tag{C12.2.2}$$

Figure C12.4b and Table C12.11 represent the T-stress. Values for β are compiled in Table C12.12. Higher order coefficients A_1, B_1, A_2, and B_2 according to eq.(A1.1.4) are given in Tables C12.13 - C12.16.

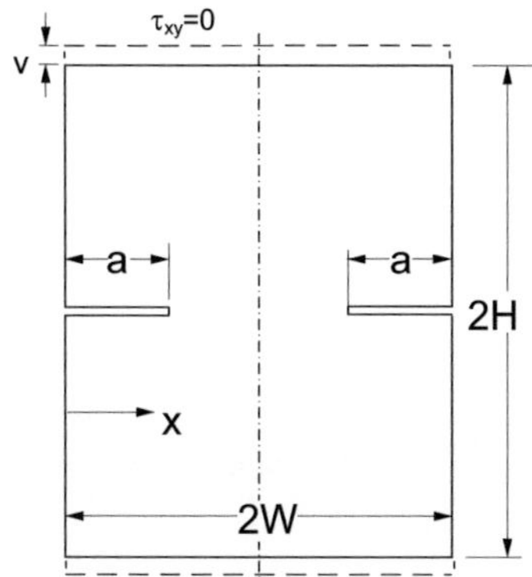

Fig. C12.3 Double-edge-cracked plate under mixed boundary conditions.

Table C12.10 Geometric function F for the double-edge-cracked plate.

a/W	H/W=0.25	0.50	0.75	1.00	1.25
0.00	1.1215	1.1215	1.1215	1.1215	1.1215
0.3	0.5104	0.726	0.868	0.940	0.976
0.4	0.4446	0.625	0.764	0.853	0.905
0.5	0.3987	0.557	0.680	0.772	0.834
0.6	0.3641	0.508	0.614	0.703	0.772
0.7	0.337	0.468	0.563	0.648	0.722
0.8	0.314	0.480	0.527	0.612	0.693

Table C12.11 T-stress data T/σ_0 for the double-edge-cracked plate.

a/W	H/W=0.25	0.50	0.75	1.00	1.25
0.00	-0.526	-0.526	-0.526	-0.526	-0.526
0.3	-0.5632	-0.456	-0.443	-0.455	-0.471
0.4	-0.5872	-0.494	-0.434	-0.423	-0.433
0.5	-0.5919	-0.530	-0.437	-0.396	-0.396
0.6	-0.5922	-0.546	-0.436	-0.369	-0.359
0.7	-0.5903	-0.534	-0.417	-0.336	-0.315
0.8	-0.5740	-0.480	-0.370	-0.290	-0.290

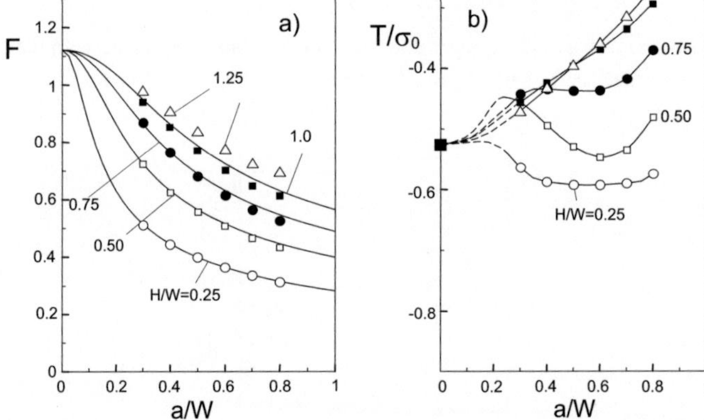

Fig. C12.4 Results of BCM computations for the double-edge-cracked plate; a) stress intensity factor expressed by the geometric function F (symbols: BCM results, curves: Eq.(C12.2.1)), b) T-stress (symbols as in a)).

Table C12.12 Biaxiality ratio β for the double-edge-cracked plate.

a/W	H/W=0.25	0.50	0.75	1.00	1.25
0.0	-0.469	-0.469	-0.469	-0.469	-0.469
0.3	-1.103	-0.628	-0.510	-0.484	-0.483
0.4	-1.321	-0.790	-0.568	-0.496	-0.478
0.5	-1.485	-0.952	-0.643	-0.513	-0.475
0.6	-1.626	-1.075	-0.710	-0.525	-0.465
0.7	-1.752	-1.141	-0.741	-0.519	-0.436
0.8	-1.828	-1.00	-0.702	-0.474	-0.418

Table C12.13 Coefficient A_1 for the double-edge-cracked plate.

a/W	H/W=0.25	0.50	0.75	1.00	1.25
0.3	-0.0737	-0.0457	-0.0387	-0.0386	-0.0397
0.4	-0.0744	-0.0490	-0.0364	-0.0335	-0.0342
0.5	-0.0743	-0.0504	-0.0366	-0.0314	-0.0315
0.6	-0.0742	-0.0509	-0.0372	-0.0313	-0.0311
0.7	-0.0740	-0.0501	-0.0383	-0.0334	-0.0337
0.8	-0.0726	-0.0495	-0.0424	-0.0409	-0.0433

Table C12.14 Coefficient B_1 for the double-edge-cracked plate.

a/W	H/W=0.25	0.50	0.75	1.00	1.25
0.3	0.2776	0.1913	0.1543	0.1426	0.1368
0.4	0.2522	0.172	0.1245	0.1021	0.0960
0.5	0.2461	0.1470	0.1044	0.0772	0.0676
0.6	0.2449	0.1266	0.0841	0.0573	0.0465
0.7	0.2420	0.1027	0.0610	0.0394	0.0303
0.8	0.2220	0.0697	0.0371	0.0236	0.0200

Table C12.15 Coefficient A_2.

a/W	H/W=0.25	0.50	1.00
0.3		-0.0773	-0.0416
0.4	-0.0899	-0.0600	-0.0291
0.5	-0.0885	-0.0432	-0.0242
0.6	-0.0884	-0.0326	-0.0233
0.7	-0.0866	-0.0264	-0.0312
0.8	-0.0766	-0.0362	-0.0694

Table C12.16 Coefficient B_2.

a/W	H/W=0.25	0.50	1.00
0.3	-0.188	-0.113	-0.159
0.4	-0.128	-0.09	-0.088
0.5	-0.110	-0.067	-0.058
0.6	-0.108	-0.065	-0.047
0.7	-0.117	-0.074	-0.041
0.8	-0.176	-0.091	-0.032

C12.3 Displacement boundary conditions at the ends

The geometric function F, the T-stress, and the higher-order coefficients A_1 and B_1 of the double-edge-cracked rectangular plate under pure displacement conditions at the plate ends (see Fig. C12.5) are given in Tables C12.17-C12.30 (for σ_0 see eq.(C12.2.2)).

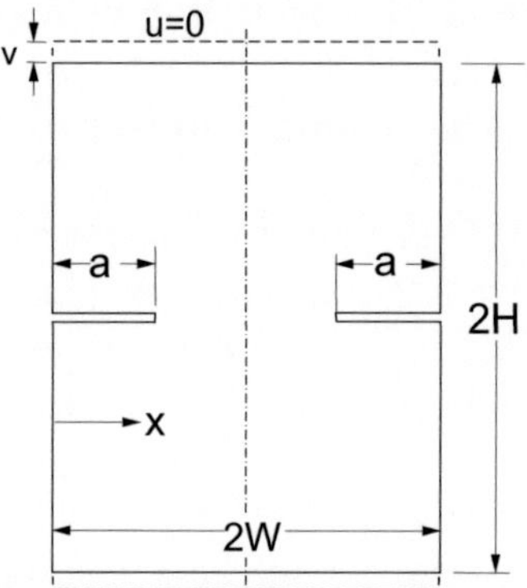

Fig. C12.5 Double-edge-cracked plate under pure displacement boundary conditions.

Table C12.17 Geometric function F for $H/W=0.25$.

a/W	$\nu = 0$	0.1	0.2	0.3	0.4
0	1.1215				
0.3	0.5119	0.5156	0.5243	0.5381	0.557
0.4	0.4443	0.4471	0.4549	0.4677	0.4854
0.5	0.3982	0.4003	0.4071	0.4185	0.4346
0.6	0.3637	0.3656	0.3717	0.3821	0.3967
0.7	0.3365	0.3384	0.3441	0.3536	0.3670
0.8	0.3137	0.3159	0.3214	0.3301	0.3420

Table C12.18 T-stress T/σ_0 for $H/W=0.25$.

a/W	$\nu=0$	0.1	0.2	0.3	0.4
0	-0.526				
0.3	-0.5460	-0.5152	-0.4915	-0.4749	-0.4654
0.4	-0.5744	-0.5337	-0.4997	-0.4724	-0.4517
0.5	-0.5845	-0.5404	-0.5024	-0.4705	-0.4448
0.6	-0.5856	-0.5412	-0.5030	-0.4709	-0.4449
0.7	-0.5794	-0.5375	-0.5021	-0.4732	-0.4507
0.8	-0.5578	-0.5232	-0.4953	-0.4741	-0.4596

Table C12.19 Coefficient A_1 for $H/W = 0.25$.

a/W	$\nu = 0$	0.1	0.2	0.3	0.4
0.3	-0.0752	-0.0774	-0.0816	-0.0879	-0.0960
0.4	-0.0760	-0.0778	-0.0816	-0.0873	-0.0950
0.5	-0.0760	-0.0778	-0.0815	-0.0872	-0.0948
0.6	-0.0759	-0.0778	-0.0815	-0.0871	-0.0947
0.7	-0.0757	-0.0777	-0.0815	-0.0871	-0.0944
0.8	-0.0747	-0.0770	-0.0809	-0.0863	-0.0932

Table C12.20 Coefficient B_1 for $H/W = 0.25$.

a/W	$\nu = 0$	0.1	0.2	0.3	0.4
0.3	0.2737	0.2568	0.2442	0.2359	0.2318
0.4	0.2518	0.2412	0.2355	0.2347	0.2387
0.5	0.2432	0.2354	0.2331	0.2361	0.2446
0.6	0.2388	0.2327	0.2322	0.2374	0.2483
0.7	0.2330	0.2292	0.2311	0.2386	0.2517
0.8	0.2149	0.2156	0.2214	0.2324	0.2485

Table C12.21 Geometric function F for $H/W = 0.5$.

a/W	$\nu = 0$	0.1	0.2	0.3	0.4
0	1.1215				
0.3	0.722	0.722	0.725	0.732	0.742
0.4	0.625	0.629	0.637	0.649	0.666
0.5	0.558	0.563	0.573	0.587	0.605
0.6	0.509	0.515	0.524	0.538	0.555
0.7	0.469	0.475	0.484	0.496	0.512
0.8	0.437	0.441	0.449	0.460	0.474

Table C12.22 T-stress T/σ_0 for $H/W = 0.5$.

a/W	$\nu = 0$	0.1	0.2	0.3	0.4
0	-0.526				
0.3	-0.456	-0.458	-0.466	-0.481	-0.502
0.4	-0.479	-0.472	-0.473	-0.481	-0.496
0.5	-0.502	-0.488	-0.481	-0.482	-0.491
0.6	-0.512	-0.494	-0.483	-0.480	-0.485
0.7	-0.500	-0.482	-0.472	-0.496	-0.473
0.8	-0.455	-0.441	-0.433	-0.460	-0.436

Table C12.23 Coefficient A_1 for $H/W = 0.50$.

a/W	$\nu = 0$	0.1	0.2	0.3	0.4
0.3	-0.0439	-0.0529	-0.0578	-0.0638	-0.0711
0.4	-0.0506	-0.0534	-0.0574	-0.0626	-0.0690
0.5	-0.0519	-0.0541	-0.0575	-0.0620	-0.0676
0.6	-0.0523	-0.0542	-0.0572	-0.0613	-0.0664
0.7	-0.0518	-0.0539	-0.0564	-0.060	-0.0646
0.8	-0.0515	-0.0532	-0.0556	-0.0587	-0.0624

Table C12.24 Coefficient B_1 for $H/W = 0.50$.

a/W	$\nu = 0$	0.1	0.2	0.3	0.4
0.3	0.1808	0.1726	0.1678	0.1664	0.1683
0.4	0.1550	0.1460	0.1398	0.1364	0.1357
0.5	0.1368	0.1302	0.1261	0.1245	0.1254
0.6	0.1207	0.1173	0.1162	0.1174	0.1210
0.7	0.1012	0.1007	0.1022	0.1058	0.1114
0.8	0.0716	0.0727	0.0753	0.0793	0.0847

Table C12.25 Geometric function F for $H/W = 1.0$.

a/W	$\nu = 0$	0.1	0.2	0.3	0.4
0	1.1215				
0.3	0.925	0.918	0.913	0.911	0.912
0.4	0.841	0.839	0.840	0.844	0.851
0.5	0.767	0.769	0.774	0.781	0.791
0.6	0.703	0.708	0.715	0.724	0.736
0.7	0.653	0.658	0.666	0.676	0.688
0.8	0.619	0.627	0.634	0.642	0.649

Table C12.26 T-stress T/σ_0 for $H/W = 1.0$.

a/W	$\nu = 0$	0.1	0.2	0.3	0.4
0	-0.526				
0.3	-0.460	-0.473	-0.486	-0.498	-0.509
0.4	-0.434	-0.446	-0.460	-0.477	-0.497
0.5	-0.411	-0.425	-0.441	-0.460	-0.482
0.6	-0.385	-0.399	-0.416	-0.436	-0.459
0.7	-0.351	-0.364	-0.379	-0.398	-0.419
0.8	-0.302	-0.316	-0.329	-0.342	-0.354

Table C12.27 Coefficient A_1 for $H/W = 1.0$.

a/W	$\nu = 0$	0.1	0.2	0.3	0.4
0.3	-0.0409	-0.0433	-0.0459	-0.0488	-0.0520
0.4	-0.0371	-0.0395	-0.0422	-0.0452	-0.0484
0.5	-0.0354	-0.0377	-0.0403	-0.0432	-0.0463
0.6	-0.0351	-0.0371	-0.0395	-0.0422	-0.0452
0.7	-0.0367	-0.0384	-0.0404	-0.0426	-0.0451
0.8	-0.0433	-0.0452	-0.0469	-0.0483	-0.0493

Table C12.28 Coefficient B_1 for $H/W = 1.0$.

a/W	$\nu = 0$	0.1	0.2	0.3	0.4
0.3	0.1456	0.1477	0.1521	0.1589	0.1681
0.4	0.1047	0.1082	0.1128	0.1185	0.1252
0.5	0.0783	0.0806	0.0840	0.0883	0.0937
0.6	0.0575	0.0592	0.0616	0.0647	0.0686
0.7	0.0391	0.0402	0.0418	0.0439	0.0465
0.8	0.0239	0.0245	0.0252	0.0259	0.0266

Table C12.29 Geometric function F for $H/W = 1.25$.

a/W	$\nu = 0$	0.2	0.4
0	1.1215		
0.3	0.964	0.954	0.958
0.4	0.895	0.894	0.904
0.5	0.829	0.833	0.847
0.6	0.770	0.779	0.795
0.7	0.724	0.733	0.752

Table C12.30 T-stress T/σ_0 for $H/W = 1.25$.

a/W	$\nu = 0$	0.2	0.4
0	-0.526		
0.3	-0.477	-0.490	-0.509
0.4	-0.442	-0.462	-0.488
0.5	-0.411	-0.436	-0.465
0.6	-0.377	-0.404	-0.432
0.7	-0.338	-0.356	-0.399

C12.4 Transverse loading

A double-edge-cracked plate under transverse traction σ_x is illustrated in Fig. C12.6. Under this loading, the stress intensity factor is defined by

$$K = \sigma_x F \sqrt{\pi a} \qquad (C12.4.1)$$

The geometric function F is plotted in Fig. C12.7 for several values of a/W, H/W, and d/W. Figure C12.8 represents the T-stresses.

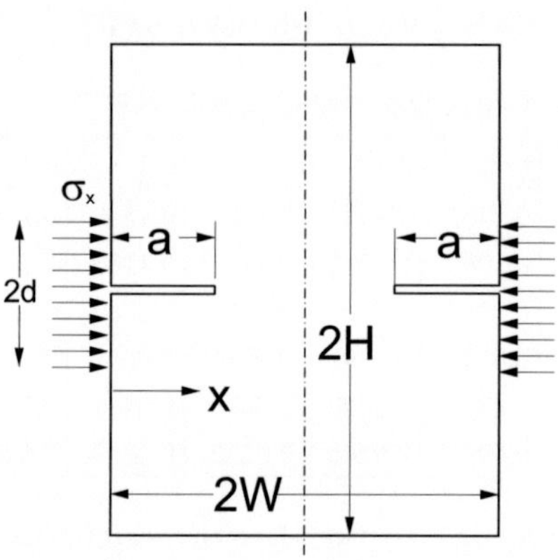

Fig. C12.6 Double-edge-cracked plate under transverse loading.

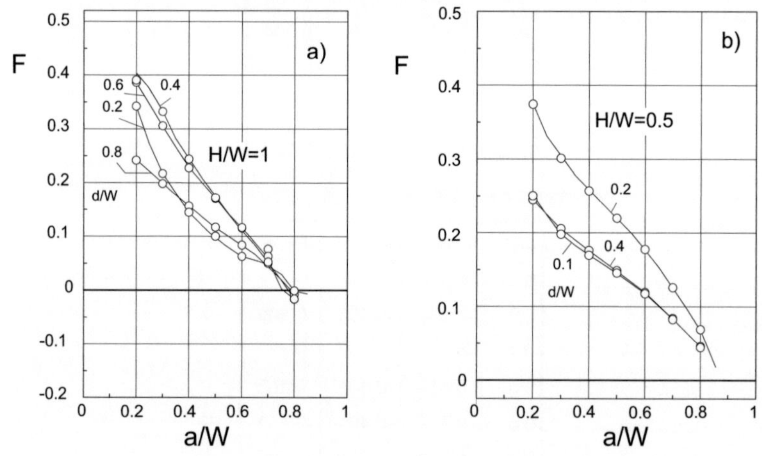

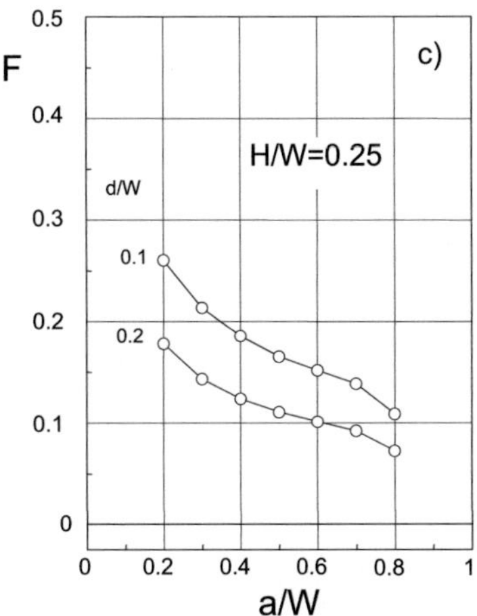

Fig. C12.7 Geometric function *F* according to eq.(C12.4.1).

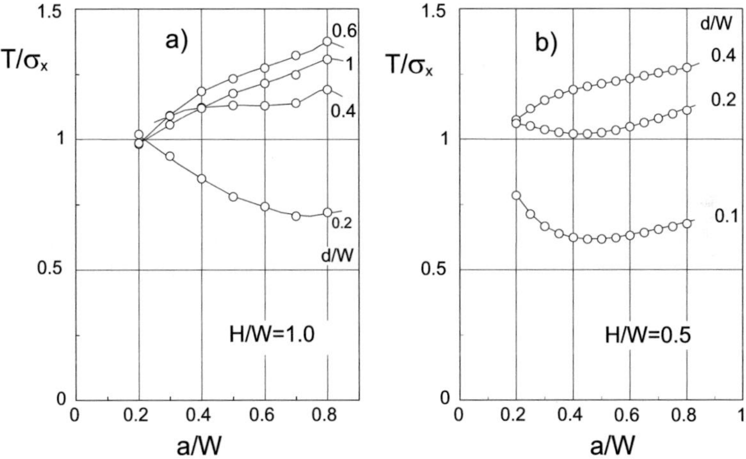

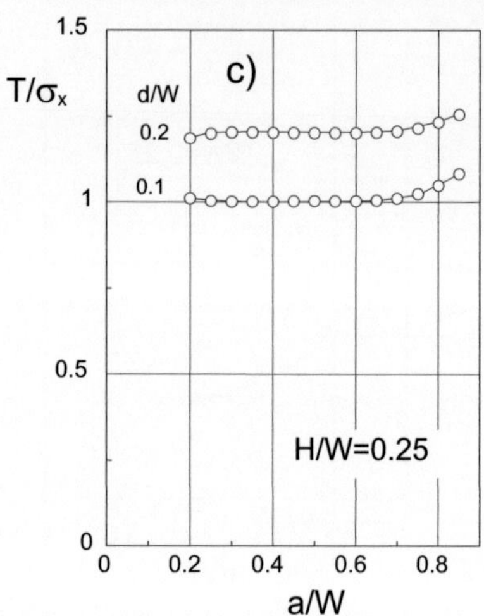

Fig. C12.8 T-stress represented as T/σ_x.

Reference C12

[C12.1] Kfouri, A.P., Some evaluations of the elastic T-term using Eshelby's method, Int. J. Fract. **30**(1986), 301-315.

C13

Edge-cracked bar in 3-point bending

C13.1 Symmetric loading

An edge-cracked bending bar loaded by a concentrated force P is shown in Fig. C13.1. Under symmetric loading (crack and force P on the symmetry line $y=0$), only mode-I stress intensity factors K_I occur. These stress intensity factors are expressed by the geometric function which are defined as

$$K_I = \sigma_0 F_I \sqrt{\pi a}, \quad \sigma_0 = \frac{3PL}{W^2 B}, \quad F_I = F_I'/(1 - a/W)^{3/2} \qquad (\text{C13.1.1})$$

The geometric function is given in Table C13.1.

The T-stresses for the 3-point bending test were computed by application of the Green's function method using an expansion with two terms. T is entered into Table C13.2 and the related biaxiality ratios are compiled into Table C13.3 and plotted in Fig. C13.2.

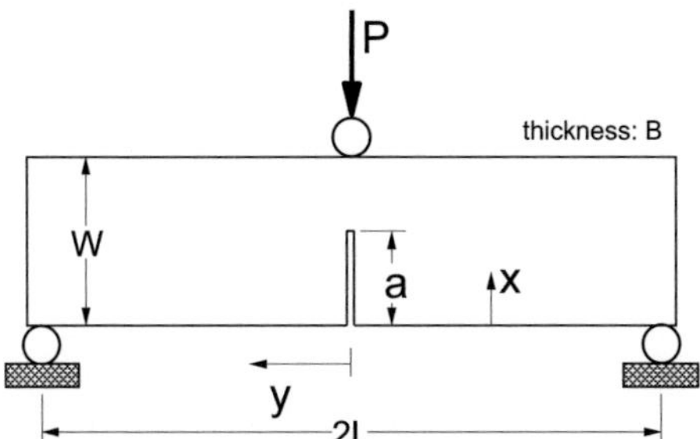

Fig. C13.1 3-point bending test.

Table C13.1 Geometric function $F(1-a/W)^{3/2}$.

a/W	$L/W=10$	5	4	3	2.5	2
0.1	0.8964	0.8849	0.8791	0.8694	0.8616	0.8504
0.2	0.7493	0.7381	0.7325	0.7231	0.7156	0.7046
0.3	0.6485	0.6387	0.6337	0.6255	0.6188	0.6091
0.4	0.5774	0.5690	0.5651	0.5582	0.5527	0.5447
0.5	0.5242	0.5177	0.5145	0.5091	0.5048	0.4985
0.6	0.4816	0.4770	0.4744	0.4704	0.4672	0.4626
0.7	0.4458	0.4430	0.4408	0.4381	0.4359	0.4328
0.8	0.4154	0.4140	0.4124	0.4108	0.4094	0.4076

Table C13.2 T-stress in the form of $T/\sigma^*(1-a/W)^2$.

a/W	$L/W=10$	5	4	3	2.5	2
0	-0.526	-0.526	-0.526	-0.526	-0.526	-0.526
0.1	-0.291	-0.292	-0.291	-0.290	-0.289	-0.288
0.2	-0.150	-0.149	-0.149	-0.149	-0.149	-0.149
0.3	-0.044	-0.049	-0.054	-0.056	-0.058	-0.063
0.4	0.035	0.026	0.022	0.014	0.008	-0.001
0.5	0.088	0.077	0.071	0.061	0.054	0.044
0.6	0.122	0.111	0.105	0.096	0.088	0.077
0.7	0.141	0.132	0.127	0.119	0.113	0.103
0.8	0.143	0.137	0.132	0.125	0.120	0.112
0.9	0.132	0.128	0.126	0.122	0.119	0.115
1	0.113	0.113	0.113	0.113	0.113	0.113

Table C13.3 Biaxiality ratio in the form of $\beta(1-a/W)^{1/2}$.

a/W	$L/W=10$	5	4	3	2.5	2
0	-0.469	-0.469	-0.469	-0.469	-0.469	-0.469
0.1	-0.325	-0.330	-0.331	-0.334	-0.335	-0.339
0.2	-0.200	-0.202	-0.203	-0.206	-0.208	-0.211
0.3	-0.068	-0.077	-0.085	-0.090	-0.094	-0.103
0.4	0.061	0.046	0.039	0.025	0.014	-0.002
0.5	0.168	0.149	0.138	0.120	0.107	0.088
0.6	0.253	0.233	0.221	0.204	0.188	0.166
0.7	0.316	0.298	0.288	0.272	0.259	0.238
0.8	0.344	0.331	0.320	0.304	0.293	0.259
0.9	0.332	0.327	0.321	0.314	0.309	0.301
1	0.302	0.302	0.302	0.302	0.302	0.302

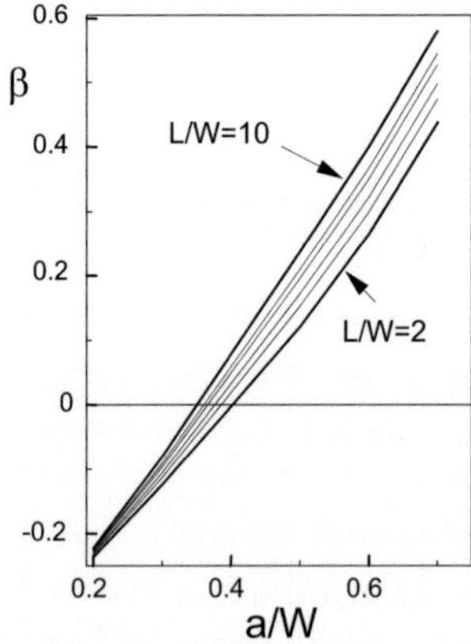

Fig. C13.2 Biaxiality ratio β for edge-cracked 3-point bending specimens with different ratios L/W.

C13.2 Misalignment in loading and crack location

If the loading application point or the crack location are out of the symmetry plane, the stress distribution also reveals shear stresses in the uncracked body at the location of the crack. Based on the normal stresses σ_x and the shear stresses τ_{xy}, the stress intensity factors for bending tests with edge cracks can be determined by application of the weight function technique [C13.1]. Figure C13.3 shows the relevant geometric data. The mixed-mode stress intensity factors are represented in Tables C13.4-C13.25 in terms of the geometric function F_{I}, eq.(C13.1.1), and the F_{II} by

$$K_{\mathrm{II}} = \sigma_0 F_{\mathrm{II}} \sqrt{\pi a} \qquad (C13.2.1)$$

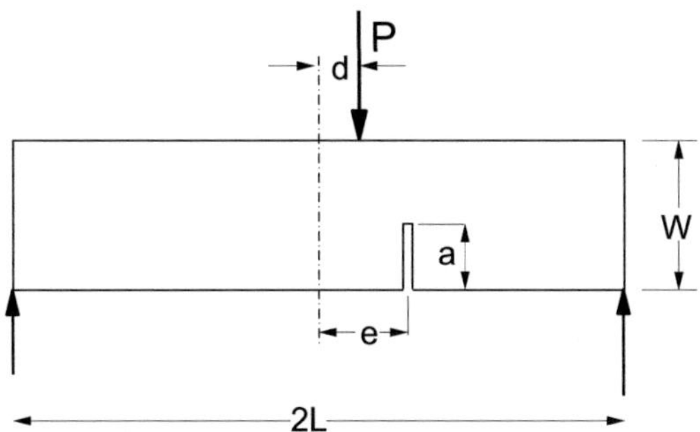

Fig. C13.3 Edge-cracked bar under 3-point loading.

The mode-I stress intensity factors for L/W=2.5 and 5 are shown in Fig. C13.4 and the mode-II solutions in Fig. C13.5.

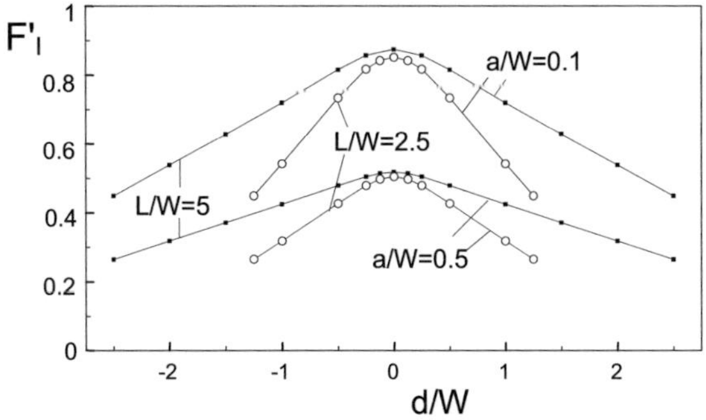

Fig. C13.4 Geometric function F'_{I} as a function of eccentricity and crack depth.

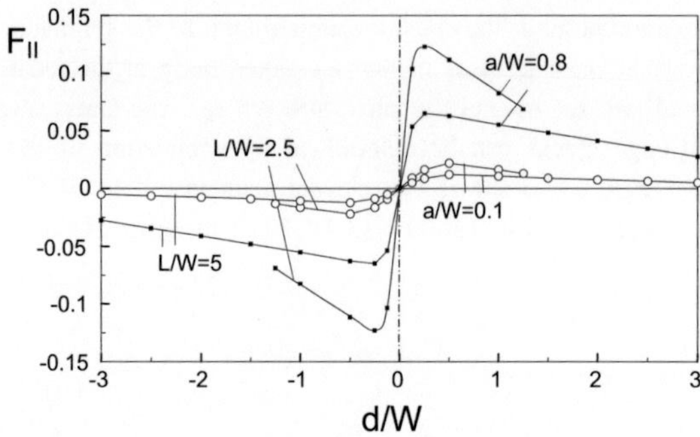

Fig. C13.5 Geometric function F_{II} as a function of eccentricity and crack depth.

For the crack depth $a/W = 0.5$ the influences of the misalignments e/W and d/W are illustrated in normalised form. In Fig. C13.6a the geometric functions F_I are plotted versus e/W-d/W, i.e. versus the relative distance between the crack and the inner load application point. The data points obtained for different L/W ratios are nearly symmetrical to the axis e/W-$d/W=0$. In Fig. C13.6b the data points of Fig. C13.6a are plotted in the form of

$$F'_{e/W-d/W} - F'_{e/W-d/W=0} = f\left(\frac{e/W - d/W}{\sqrt{L/W}}\right). \tag{C13.2.2}$$

In this representation all data points can be represented by the same curve. The function f in eq.(C13.2.2) can be approximated by

$$f\left(\frac{e/W - d/W}{\sqrt{L/W}}\right) \cong -1.075\frac{(e/W - d/W)^2}{L/W}. \tag{C13.2.3}$$

The geometric function F' can be written as

$$F'(e/W - d/W) = F'_{e/W-d/W=0} - 1.075\frac{(e/W - d/W)^2}{L/W} \tag{C13.2.4}$$

with

$$F'(0) = F'_{e/W-d/W=0} = \begin{cases} 0.4980 \text{ for } L/W = 2.0 \\ 0.5048 \text{ for } L/W = 2.5 \\ 0.5177 \text{ for } L/W = 5.0 \end{cases}. \tag{C13.2.5}$$

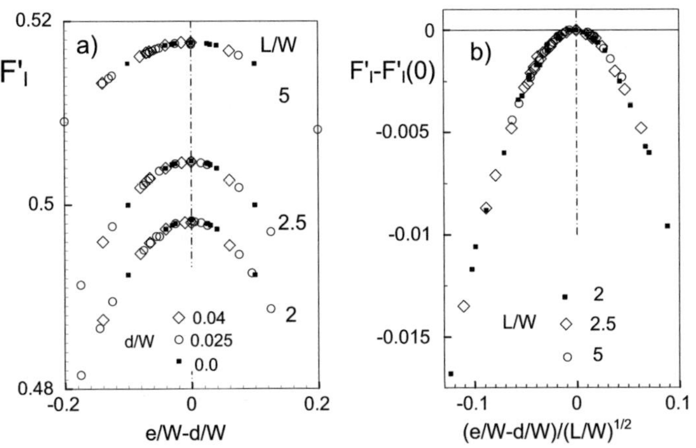

Fig. C13.6 Influence of e/W and d/W on the mode-I geometric function in normalised representation.

Table C13.4 Geometric function F'_1 for $d/W=0$, $L/W=2.0$.

e/W	$a/W=0.1$	0.2	0.3	0.4	0.5	0.6	0.7	0.8
0.000	0.8500	0.7043	0.6089	0.5445	0.4984	0.4625	0.4327	0.4075
0.025	0.8495	0.7040	0.6086	0.5442	0.4980	0.4620	0.4321	0.4067
0.040	0.8488	0.7033	0.6080	0.5436	0.4974	0.4614	0.4313	0.4056
0.100	0.8425	0.6982	0.6033	0.5389	0.4924	0.4557	0.4246	0.3971
0.300	0.7882	0.6536	0.5632	0.5003	0.4536	0.4157	0.3832	0.3553
0.500	0.7013	0.5827	0.5014	0.4439	0.4009	0.3663	0.3373	0.3131
1.000	0.4624	0.3860	0.3327	0.2946	0.2661	0.2434	0.2245	0.2087
-0.025	0.8495	0.7040	0.6086	0.5442	0.4980	0.4620	0.4321	0.4067
-0.040	0.8488	0.7033	0.6080	0.5436	0.4974	0.4614	0.4313	0.4056
-0.100	0.8425	0.6982	0.6033	0.5389	0.4924	0.4557	0.4246	0.3971
-0.300	0.7882	0.6536	0.5632	0.5003	0.4536	0.4157	0.3832	0.3553
-0.500	0.7013	0.5827	0.5014	0.4439	0.4009	0.3663	0.3373	0.3131
-1.000	0.4624	0.3860	0.3327	0.2946	0.2661	0.2434	0.2245	0.2087

Table C13.5 Geometric function F_{II} for $d/W=0$, $L/W=2.0$.

e/W	$a/W=0.1$	0.2	0.3	0.4	0.5	0.6	0.7	0.8
0.000	0	0	0	0	0	0	0	0
0.025	-0.0027	-0.0048	-0.0067	-0.0086	-0.0110	-0.0147	-0.0213	-0.0366
0.040	-0.0043	-0.0077	-0.0107	-0.0137	-0.0176	-0.0233	-0.0337	-0.0573
0.100	-0.0105	-0.0188	-0.0261	-0.0335	-0.0425	-0.0556	-0.0777	-0.1212
0.300	-0.0267	-0.0477	-0.0653	-0.0816	-0.0984	-0.1177	-0.1416	-0.1733
0.500	-0.0336	-0.0603	-0.0818	-0.0999	-0.1159	-0.1314	-0.1488	-0.1739
1.000	-0.0314	-0.0578	-0.0798	-0.0980	-0.1137	-0.1287	-0.1460	-0.1719
-0.025	0.0027	0.0048	0.0067	0.0086	0.0110	0.0147	0.0213	0.0366
-0.040	0.0043	0.0077	0.0107	0.0137	0.0176	0.0233	0.0337	0.0573
-0.100	0.0105	0.0188	0.0261	0.0335	0.0425	0.0556	0.0777	0.1212
-0.300	0.0267	0.0477	0.0653	0.0816	0.0984	0.1177	0.1416	0.1733
-0.500	0.0336	0.0603	0.0818	0.0999	0.1159	0.1314	0.1488	0.1739
-1.000	0.0314	0.0578	0.0798	0.0980	0.1137	0.1287	0.1460	0.1719

Table C13.6 Geometric function F'_{I} for $d/W=0.00$, $L/W=2.5$.

e/W	$a/W=0.1$	0.2	0.3	0.4	0.5	0.6	0.7	0.8
0.000	0.8617	0.7156	0.6189	0.5528	0.5048	0.4673	0.4359	0.4095
0.025	0.8613	0.7153	0.6186	0.5525	0.5045	0.4669	0.4355	0.4089
0.040	0.8607	0.7148	0.6181	0.5520	0.5040	0.4664	0.4348	0.4079
0.100	0.8557	0.7107	0.6143	0.5483	0.5000	0.4619	0.4295	0.4011
0.300	0.8123	0.6750	0.5822	0.5174	0.4690	0.4299	0.3964	0.3677
-0.025	0.8613	0.7153	0.6186	0.5525	0.5045	0.4669	0.4355	0.4089
-0.040	0.8607	0.7148	0.6181	0.5520	0.5040	0.4664	0.4348	0.4079
-0.100	0.8557	0.7107	0.6143	0.5483	0.5000	0.4619	0.4295	0.4011
-0.300	0.8123	0.6750	0.5822	0.5174	0.4690	0.4299	0.3964	0.3677

Table C13.7 Geometric function F_{II} for $d/W=0$, $L/W=2.5$.

e/W	$a/W=0.1$	0.2	0.3	0.4	0.5	0.6	0.7	0.8
0.000	0	0	0	0	0	0	0	0
0.025	-0.0022	-0.0038	-0.0053	-0.0069	-0.0088	-0.0117	-0.0170	-0.0293
0.040	-0.0034	-0.0061	-0.0085	-0.0110	-0.0141	-0.0187	-0.0270	-0.0458
0.100	-0.0084	-0.0151	-0.0209	-0.0268	-0.0340	-0.0445	-0.0621	-0.0970
0.300	-0.0214	-0.0382	-0.0523	-0.0653	-0.0788	-0.0942	-0.1133	-0.1387
-0.025	0.0022	0.0038	0.0053	0.0069	0.0088	0.0117	0.0170	0.0293
-0.040	0.0034	0.0061	0.0085	0.0110	0.0141	0.0187	0.0270	0.0458
-0.100	0.0084	0.0151	0.0209	0.0268	0.0340	0.0445	0.0621	0.0970
-0.300	0.0214	0.0382	0.0523	0.0653	0.0788	0.0942	0.1133	0.1387

Table C13.8 Geometric function F'_I for $d/W=0$, $L/W=5.0$.

e/W	$a/W=0.1$	0.2	0.3	0.4	0.5	0.6	0.7	0.8
0.000	0.8848	0.7381	0.6387	0.5692	0.5177	0.4768	0.4425	0.4134
0.025	0.8846	0.7379	0.6385	0.5690	0.5176	0.4767	0.4423	0.4131
0.040	0.8844	0.7377	0.6383	0.5688	0.5174	0.4764	0.4419	0.4126
0.100	0.8819	0.7356	0.6364	0.5669	0.5154	0.4742	0.4393	0.4092
0.300	0.8602	0.7178	0.6203	0.5515	0.4998	0.4582	0.4227	0.3925
-0.025	0.8846	0.7379	0.6385	0.5690	0.5176	0.4767	0.4423	0.4131
-0.040	0.8844	0.7377	0.6383	0.5688	0.5174	0.4764	0.4419	0.4126
-0.100	0.8819	0.7356	0.6364	0.5669	0.5154	0.4742	0.4393	0.4092
-0.300	0.8602	0.7178	0.6203	0.5515	0.4998	0.4582	0.4227	0.3925

Table C13.9 Geometric function F_II for $d/W=0$, $L/W=5.0$.

e/W	$a/W=0.1$	0.2	0.3	0.4	0.5	0.6	0.7	0.8
0.000	0	0	0	0	0	0	0	0
0.025	-0.0011	-0.0019	-0.0027	-0.0035	-0.0044	-0.0059	-0.0085	-0.0146
0.040	-0.0017	-0.0031	-0.0043	-0.0055	-0.0070	-0.0093	-0.0135	-0.0229
0.100	-0.0042	-0.0075	-0.0104	-0.0134	-0.0170	-0.0222	-0.0311	-0.0485
0.300	-0.0107	-0.0191	-0.0262	-0.0327	-0.0394	-0.0471	-0.0567	-0.0693
-0.025	0.0011	0.0019	0.0027	0.0035	0.0044	0.0059	0.0085	0.0146
-0.040	0.0017	0.0031	0.0043	0.0055	0.0070	0.0093	0.0135	0.0229
-0.100	0.0042	0.0075	0.0104	0.0134	0.0170	0.0222	0.0311	0.0485
-0.300	0.0107	0.0191	0.0262	0.0327	0.0394	0.0471	0.0567	0.0693

Reference C13

[C13 1] Baratta, F.I., Fett, T., The effect of load and crack misalignment on stress intensity factors for bend-type fracture toughness specimens, J of Testing and Evaluation, **28**(2000), 96-102.

C14

Four-point bending test with edge-cracked bars

The 4-point bending test on an edge-cracked bar is shown in Fig. C14.1 for the most general case with misalignments in the load application and in the crack location. Both influences were studied in [C14.1]. The case of an offset e in the crack location is addressed here exclusively.

The mixed-mode stress intensity factors are given by

$$K_\mathrm{I} = \sigma_0 F_\mathrm{I} \sqrt{\pi a} \ , \quad F_\mathrm{I} = F_\mathrm{I}'/(1 - a/W)^{3/2} \tag{C14.1}$$

and
$$K_\mathrm{II} = \sigma_0 F_\mathrm{II} \sqrt{\pi a} \tag{C14.2}$$

with the bending stress

$$\sigma_0 = \frac{6P(L-d)}{W^2 B} \tag{C14.3}$$

Results are compiled in Tables C14.1-C14.4.

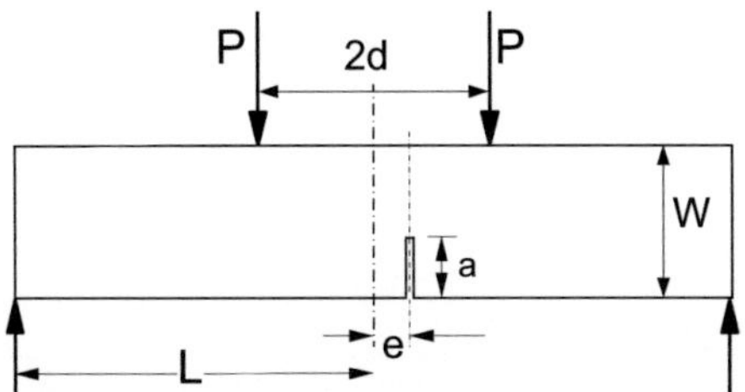

Fig. C14.1 Four-point bending test with edge-cracked specimen.

Table C14.1 Geometric function F'_I for d/W=1.25, L/W=2.5 (for e/W<0 the same geometric function F'_I results as in case of e/W>0).

e/W	a/W=0.1	0.2	0.3	0.4	0.5	0.6	0.7	0.8
0.000	0.9096	0.7617	0.6592	0.5860	0.5308	0.4864	0.4490	0.4173
0.125	0.9100	0.7620	0.6594	0.5861	0.5309	0.4865	0.4490	0.4173
0.250	0.9113	0.7628	0.6599	0.5864	0.5310	0.4865	0.4490	0.4173
0.500	0.9169	0.7663	0.6620	0.5876	0.5317	0.4868	0.4491	0.4173
0.800	0.9247	0.7703	0.6644	0.5895	0.5332	0.4879	0.4496	0.4175
1.000	0.9173	0.7628	0.6585	0.5857	0.5316	0.4879	0.4504	0.4180
1.250	0.8633	0.7168	0.6196	0.5532	0.5050	0.4673	0.4359	0.4095
1.500	0.7425	0.6152	0.5294	0.4699	0.4260	0.3908	0.3606	0.3345

Table C14.2 Geometric function F_{II} for d/W=1.25, L/W=2.5 (for e/W<0 the geometric function is the same, but with a changed sign).

e/W	a/W=0.1	0.2	0.3	0.4	0.5	0.6	0.7	0.8
0.000	0	0	0	0	0	0	0	0
0.125	0.0005	0.0008	0.0009	0.0009	0.0007	0.0005	0.0003	0.0001
0.250	0.0011	0.0017	0.0019	0.0018	0.0015	0.0011	0.0006	0.0002
0.500	0.0023	0.0035	0.0039	0.0037	0.0033	0.0025	0.0016	0.0007
0.800	0.0010	0.0005	-0.0002	-0.0003	0.0003	0.0013	0.0021	0.0019
1.000	-0.0063	-0.0125	-0.0174	-0.0199	-0.0195	-0.0158	-0.0088	-0.0008
1.250	-0.0245	-0.0451	-0.0624	-0.0770	-0.0897	-0.1021	-0.1163	-0.1374
1.500	-0.0415	-0.0760	-0.1057	-0.1325	-0.1588	-0.1876	-0.2234	-0.2737

Table C14.3 Geometric function F'_{I} for d/W=2.50, L/W=5.0.

e/W	a/W=0.1	0.2	0.3	0.4	0.5	0.6	0.7	0.8
0.000	0.9080	0.7605	0.6584	0.5856	0.5307	0.4864	0.4490	0.4173
0.125	0.9080	0.7605	0.6584	0.5856	0.5307	0.4864	0.4490	0.4173
0.250	0.9080	0.7605	0.6584	0.5856	0.5307	0.4864	0.4490	0.4173
0.500	0.9080	0.7605	0.6584	0.5856	0.5307	0.4864	0.4490	0.4173
1.000	0.9081	0.7606	0.6585	0.5856	0.5307	0.4864	0.4490	0.4173
1.250	0.9084	0.7608	0.6586	0.5857	0.5307	0.4864	0.4490	0.4173
1.500	0.9097	0.7617	0.6591	0.5860	0.5308	0.4865	0.4491	0.4173
2.000	0.9161	0.7654	0.6614	0.5875	0.5318	0.4870	0.4493	0.4174

Table C14.4 Geometric function F_{II} for d/W=2.500, L/W=5.0.

e/W	a/W=0.1	0.2	0.3	0.4	0.5	0.6	0.7	0.8
0.000	0.0000	0.0000	0.0000	0.0000	0.0000	0.0000	0.0000	0.0000
0.125	0.0000	0.0000	0.0000	0.0000	0.0000	0.0000	0.0000	0.0000
0.250	0.0000	0.0000	0.0000	0.0000	0.0000	0.0000	0.0000	0.0000
0.500	0.0000	0.0000	0.0000	0.0000	0.0000	0.0000	0.0000	0.0000
1.000	0.0000	0.0001	0.0001	0.0001	0.0001	0.0000	0.0000	0.0000
1.250	0.0002	0.0003	0.0004	0.0004	0.0003	0.0002	0.0001	0.0000
1.500	0.0006	0.0009	0.0011	0.0010	0.0008	0.0006	0.0003	0.0001
2.000	0.0009	0.0010	0.0009	0.0008	0.0010	0.0012	0.0012	0.0008

Reference C14

[C14.1] Baratta, F.I., Fett, T., The effect of load and crack misalignment on stress intensity factors for bend-type fracture toughness specimens, J. of Testing and Evaluation, **28**(2000), 96-102.

C15

DCDC test specimen

C15.1   Symmetric specimen with a central hole

The "double cleavage drilled compression" (DCDC) specimen is a rectangular bar with a circular hole in the centre (Fig. C15.1) [C15.1]. The specimen is loaded by compressive tractions p at the ends. It is used for the determination of crack growth behaviour of brittle materials [C15.2-C15.5].

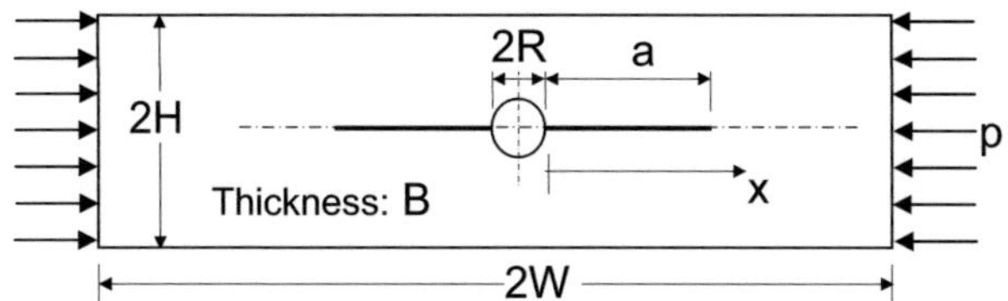

Fig. C15.1 Standard DCDC specimen.

Stress intensity factor solutions for the DCDC specimen are available in literature. The stress intensity factor of the symmetric test specimen, $b/R=0$, was determined by He et al. [C15.2], who proposed

$$\frac{|p|\sqrt{\pi R}}{K_I} = \frac{1}{F} = \frac{H}{R} + \left[0.235\frac{H}{R} - 0.259\right]\frac{a}{R} \tag{C15.1.1}$$

(see also [C15.6]). This relation is shown by the dashed curves in Fig. 2.1a.

Based on the finite element results of Fig. C15.2a (circles), the geometric function for the stress intensity factors was fitted as (see [C15.7], C15.8)

$$\frac{1}{F} = -0.37 + 1.116\frac{H}{R} + \left[0.216\frac{H}{R} - 0.1575\right]\frac{a}{R} \tag{C15.1.2}$$

This relation is plotted in Fig. C15.2a as the solid curves.

The T-stress is shown in Fig. C15.2b. It can be expressed by [C15.8]

$$\frac{1}{(T/|p|)+1} = -1.39 + 1.18\frac{H}{R} + \left[0.2\frac{H}{R} - 0.24\right]\frac{a}{R} \tag{C15.1.3}$$

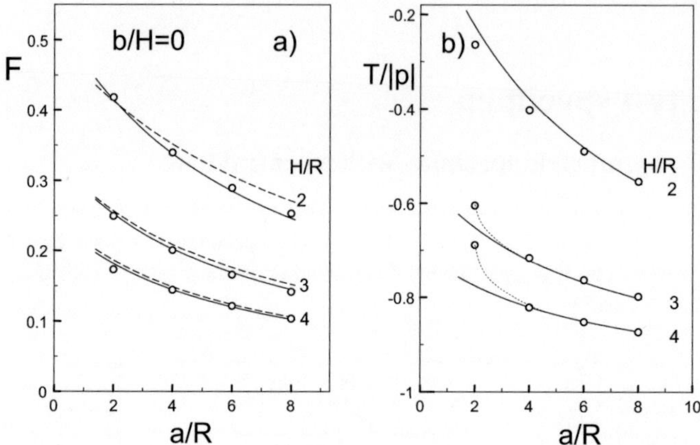

Fig. C15.2 a) Geometric function for the stress intensity factor (symbols: finite element results, solid curves: fit relation eq.(C15.1.2), dashed curves: eq.(C15.1.1) from [C15.2], b) T-stress (symbols: finite element results, solid curves: fit relation eq.(C15.1.3).

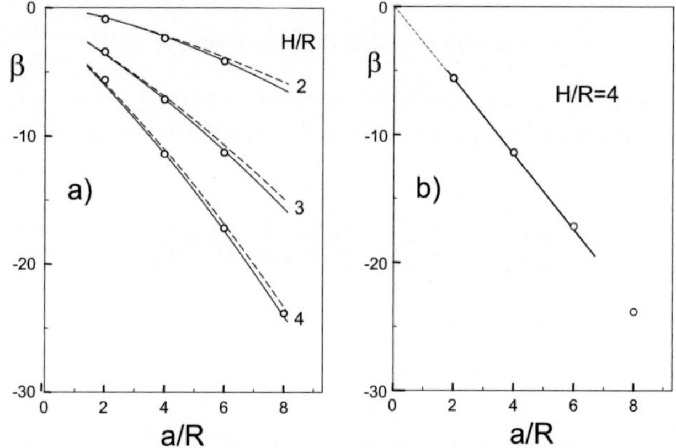

Fig. C15.3 Biaxiality ratio obtained from the results of Figs. C15.2a and C15.2b; a) symbols: finite element results [C15.7], solid curves: based on data fit relation eq.(C15.1.2), dashed curves: eq.(C15.1.1), from [C15.2], b) simplified straight-line fit according to eq.(C15.1.5).

The biaxiality ratio β according to Leevers and Radon [C15.9]

$$\beta = \frac{T\sqrt{\pi a}}{K_I} \tag{C15.1.4}$$

is plotted in Fig. C15.3a. The symbols represent the finite element results. The dashed curves were computed from eq.(C15.1.1) and eq.(C15.1.3), the solid curves from eq.(C15.1.2) and eq.(C15.1.3). For the mostly chosen geometry $H/R=4$, the biaxiality ratio may be expressed for $2\leq a/R\leq 7$ by a simple straight-line approximation

$$\beta = -3\frac{a}{R} \qquad (C15.1.5)$$

as introduced in Fig. C15.3b.

C15.2 Asymmetric specimen with hole offset

For mixed-mode crack loading, the asymmetric DCDC specimen was applied [C15.3]. This specimen with an offset of the hole is shown in Fig. C15.4 (see [C15.6]).

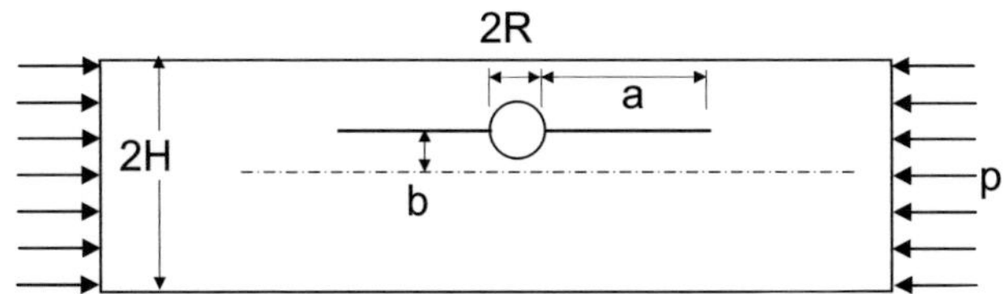

Fig. C15.4 DCDC specimen with hole offset for mixed-mode tests.

The mode-I stress intensity factors, Fig. C15.5a, were fitted for $a/R\geq 4$ [C15.7] according to

$$\frac{1}{F_{\mathrm{I}}} = c_0 + c_1\frac{H}{R} + \left[c_2\frac{H}{R} + c_3\right]\frac{a}{R} \qquad (C15.2.1a)$$

with the coefficients

$$c_0 = -0.3703 - 0.2706|\,b/R\,| - 0.2716(b/R)^2 \qquad (C15.2.1b)$$

$$c_1 = 1.1163 + 0.1864|\,b/R\,| - 0.0140(b/R)^2 \qquad (C15.2.1c)$$

$$c_2 = 0.2160 - 0.0326(b/R) + 0.0040(b/R)^2 \qquad (C15.2.1d)$$

$$c_3 = -0.1575 + 0.0176|\,b/R\,| + 0.0040(b/R)^2 \qquad (C15.2.1e)$$

For the commonly used geometry $H/R=4$ and small misalignments $|b|/R\leq 0.2$, it holds

$$F_{\mathrm{I}} \cong F_{\mathrm{I},0}\left[1 + \left(0.122 - 0.003\frac{a}{R}\right)\left(\frac{b}{R}\right)^2\right] \qquad (C15.2.2)$$

Mode-II stress intensity factors are given in Fig. C15.5b and C15.5c in the form of the mixed-mode ratio $K_{\mathrm{II}}/K_{\mathrm{I}}$.

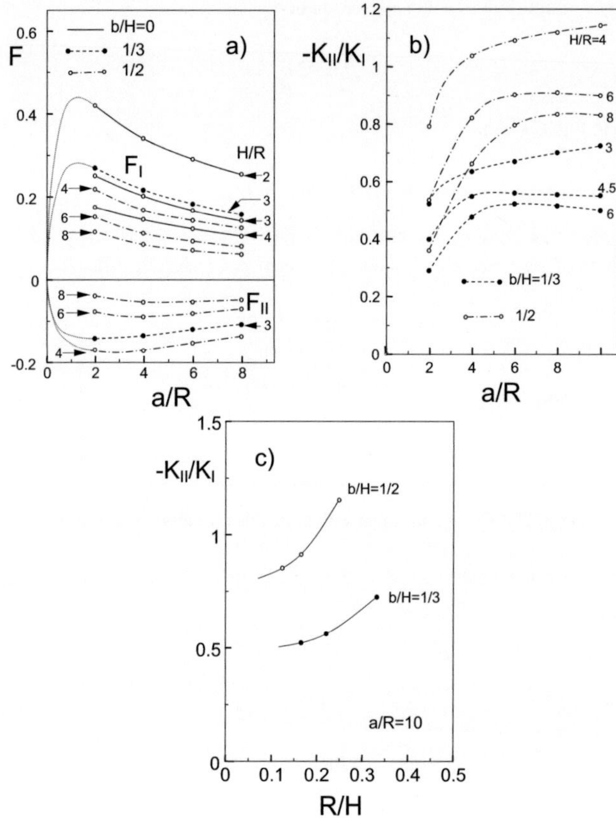

Fig. C15.5 a) Mixed-mode stress intensity factors, b) mode mixity, c) mode mixity at $a/R=10$ [C15.8].

Figure C15.6a represents the T-stress results, and in Fig. C15.6b the biaxiality ratio β is plotted. Since β is strongly negative, growing cracks must exhibit a very high path stability [C15.10,C15.11,C15.12].

The data of Fig. C15.6a were fitted for $a/R \geq 4$ by the equation of

$$\frac{1}{(T/|p|)+1} \cong f_0 + f_1\frac{H}{R} + \left[f_2\frac{H}{R} + f_3 \right]\frac{a}{R} \qquad \text{(C15.2.3a)}$$

with the coefficients

$$f_0 = -1.3903 + 0.9310|b/R| - 0.1038(b/R)^2 \qquad \text{(C15.2.3b)}$$

$$f_1 = 1.1801 - 0.3575|b/R| + 0.0222(b/R)^2 \qquad \text{(C15.2.3c)}$$

$$f_2 = 0.2003 + 0.1061|b/R| - 0.0039(b/R)^2 \qquad \text{(C15.2.3d)}$$

$$f_3 = -0.2397 - 0.3471|b/R| - 0.0827(b/R)^2 \qquad \text{(C15.2.3e)}$$

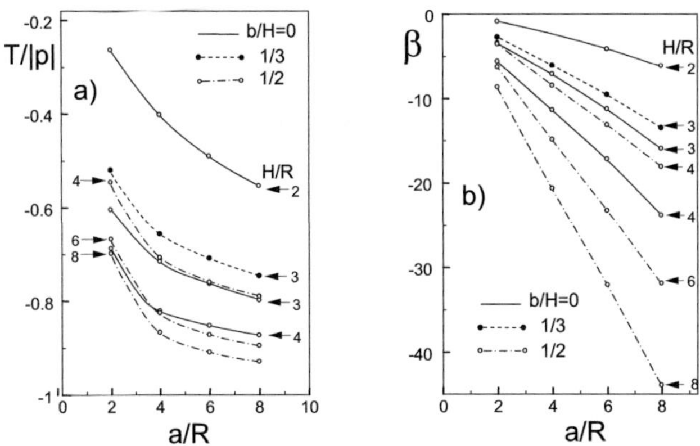

Fig. C15.6 a) T-stress of an asymmetric DCDC specimen, b) biaxiality ratio β.

C15.3 Weight functions

Weight functions were determined in [C15.8] by finite element computations. Figure C15.7 shows stress intensity factors for pairs of concentrated normal forces P and shear forces Q acting symmetrically on the faces of the crack at distance x. These stress intensity factors are identical with the weight functions. It becomes obvious that for shear loading the total weight function differs only slightly from the limit cases represented by the singular term (dashed curves).

The stress intensity factors of Fig. C15.7 were fitted by the relation

$$h_{\mathrm{I,II}} = \sqrt{\frac{2}{\pi a}}\left(\frac{1}{\sqrt{1-x/a}} + D_1^{(\mathrm{I,II})}\sqrt{1-x/a} + D_2^{(\mathrm{I,II})}(1-x/a)^{3/2} \right) \qquad \text{(C15.3.1)}$$

The coefficients for several crack lengths are compiled in Table C15.1. Values of intermediate lengths can be obtained by parabolic interpolation. Table C15.2 presents the coefficients for the case of "one-side loading" as illustrated in Fig. C15.8.

Stress intensity factors for antisymmetrically applied shear forces (Fig. C15.9) are given in Table C15.3.

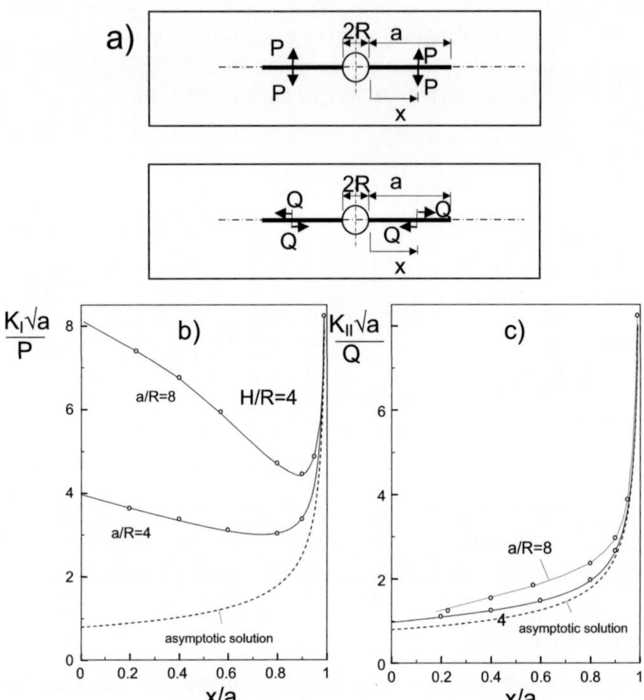

Fig. C15.7 a) Symmetric loading by point forces acting on the crack, b) stress intensity factors for pairs of normal stresses, c) stress intensity factor for shear forces.

Table C15.1 Coefficients for the weight function representation (C15.3.1) under symmetric loading.

Crack length a/R	Mode	$D_1^{(I)}$	$D_2^{(I)}$
2	I	1.465	0.208
4	I	3.379	0.591
6	I	5.561	1.184
8	I	8.222	1.482
Crack length a/R	Mode	$D_1^{(II)}$	$D_2^{(II)}$
2	II	0.1460	0.1843
4	II	0.5687	-0.3493
6	II	1.266	-1.132
8	II	2.034	-2.020

The coefficients in Table C15.1 for mode-I were fitted by

$$D_1^{(I)} = -3.97 + 1.46604\,\tfrac{a}{R} + 4.533\exp[-0.291\tfrac{a}{R}] \qquad\text{(C15.3.2)}$$

$$D_2^{(I)} = -0.246 + 0.222\,\tfrac{a}{R} + 0.5288\exp[-2.08\tfrac{a}{R}] \qquad\text{(C15.3.3)}$$

and for mode-II by

$$D_1^{(II)} = 0.568 - 0.4544\,\tfrac{a}{R} + 0.2134\left(\tfrac{a}{R}\right)^2 - 0.0293\left(\tfrac{a}{R}\right)^3 + 0.0014\left(\tfrac{a}{R}\right)^4 \qquad\text{(C15.3.4)}$$

$$D_2^{(II)} = 0.283 + 0.0956\,\tfrac{a}{R} - 0.0824\left(\tfrac{a}{R}\right)^2 + 0.00518\left(\tfrac{a}{R}\right)^3 - 0.00011\left(\tfrac{a}{R}\right)^4 \qquad\text{(C15.3.5)}$$

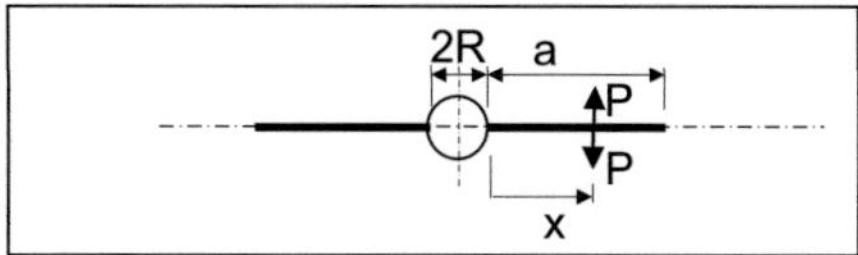

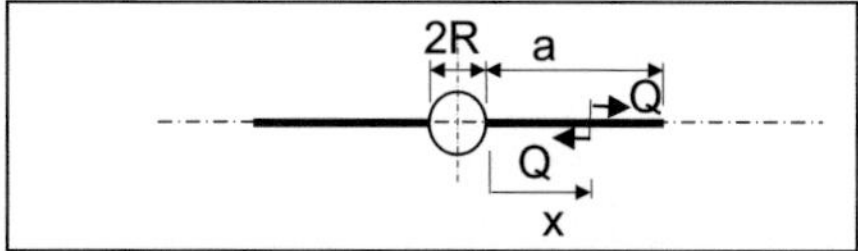

Fig. C15.8 One-side loading by point forces acting on the crack.

Table C15.2 Coefficients for the weight function representation (C15.3.1) under one-side loading.

Crack length a/R	Mode	$D_1^{(I)}$	$D_2^{(I)}$
2	I	0.6988	-0.0445
4	I	2.242	-0.379
6	I	4.226	-0.8553
8	I	6.740	-1.816
Crack length a/R	Mode	$D_1^{(II)}$	$D_2^{(II)}$
2	II	0.4529	0.1358
4	II	1.0752	-0.1948
6	II	1.8564	-0.7194
8	II	2.704	-1.413

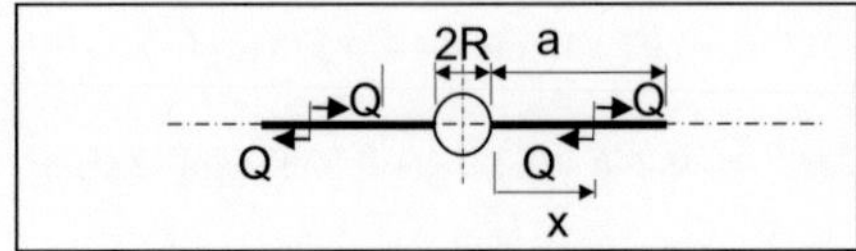

Fig. C15.9 Antisymmetric shear loading.

Table C15.3 Coefficients for the weight function representation (C15.3.1) under anti-symmetric shear loading.

Crack length a/R	Mode	$D_1^{(II)}$	$D_2^{(II)}$
2	II	0.7594	0.0879
4	II	1.600	-0.0674
6	II	2.449	-0.3095
8	II	3.373	-0.8041

C15.4 Eccentricity of loading

A possibility of misalignment is an offset y_P between the externally applied force P and the symmetry axis of the specimen (see Fig. C15.10). This eccentricity gives rise to a moment

$$M_b = P \times y_P \qquad\qquad (C15.4.1)$$

and bending stresses in the specimen. Whereas for small y_P/H the mode-I stress intensity factor is hardly affected, a mode-II stress intensity factor contribution is created, as shown in Fig. C15.11 for $H/R=4$ in normalised form. This effect does not only occur for point forces. Also in the case of a non-symetrically distributed load, a moment results as

$$M_b = \int_{-H}^{H} p(y)\,y\,dy \qquad\qquad (C15.4.2)$$

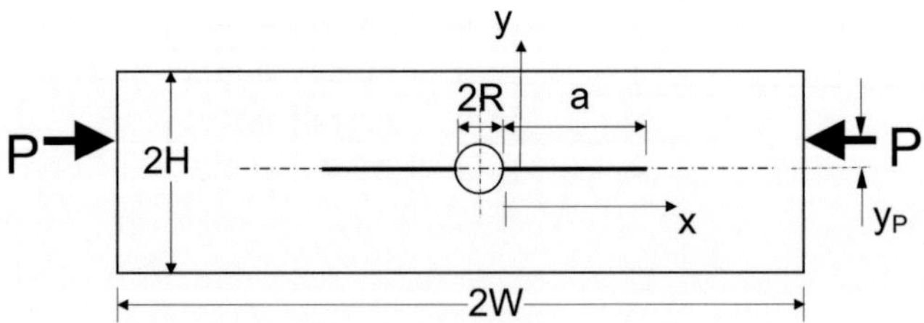

Fig. C15.10 DCDC specimen under eccentric loading.

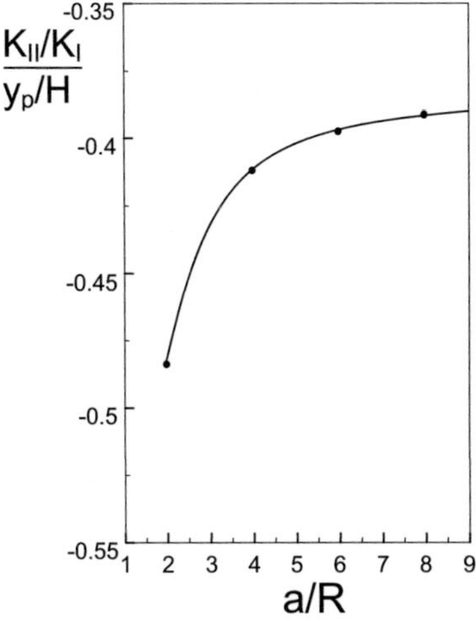

Fig. C15.11 Mode-II stress intensity factor caused by a misalignment y_P of the externally applied load P.

C15.5 Non-symmetrically extending cracks

Cracks may be generated during loading which are not exactly symmetric with different lengths a_0 and a_1. At the longer crack, the stress intensity factor is reduced. Finite element results for such different cracks are shown in Fig. C15.12 as the squares.

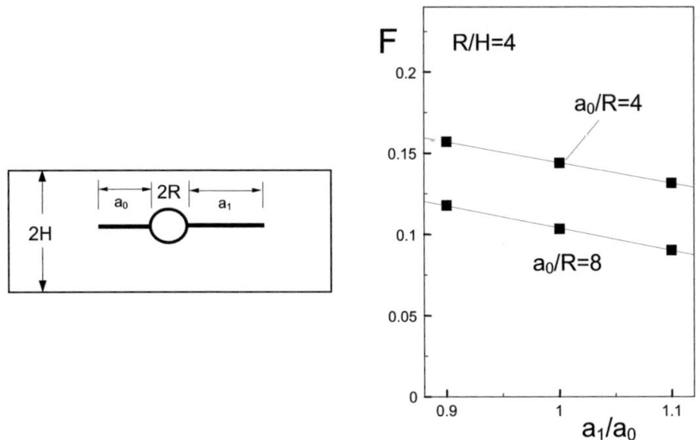

Fig. C15.12 Influence of non-symmetric cracks on the mode-I stress intensity factor.

A fit of these data yields

$$F(a_1) \cong F(a_0) - 0.13(\tfrac{a_1}{a_0} - 1)$$ (C15.5.1)

where $F(a_0)$ is the geometric function for two cracks with identical length of a_0, given for instance by eqs.(C15.1.1) or (C15.1.2).

References C15:

[C15.1] Janssen, C., Specimen for fracture mechanics studies on glass, in Proceedings Xth International Congress on Glass, (1974), p. 23, Kyoto, Japan.

[C15.2] He, M.Y., Turner, M.R., Evans, A.G., Analysis of the double cleavage drilled compression specimen for interface fracture energy measurements over a range of mode mixities, Acta metall. mater. **43**(1995), 3453-3458.

[C15.3] Ritter, J.E., Huseinovic, A., Chakravarthy, S., Lardner, T.J., Subcritical crack growth in soda-lime glass under mixed-mode loading, J. Amer. Ceram. Soc. **83**(2000), 2109-2111.

[C15.4] F. Célarié, S Prades, D. Bonamy, L. Ferrero, E. Bouchaud, C. Guillot, and C. Marlière, Glass breaks like metals, but at the nanometer scale, *Phys. Rev. Let.*, **90** [7] 075504 (2003).

[C15.5] Maliere, C., Despetis, F., Phalippou, J., Crack path instabilities in DCDC experiments in the low speed regime, J. of Non-Crystalline Solids 316(2003), 21-27.

[C15.6] Lardner, T.J., Chakravarthy, S., Quinn, J.D., Ritter, J.E., Further analysis of the DCDC specimen with an offset hole, Int. J. Fract. **109**(2001), 227-237.

[C15.7] T. Fett, G. Rizzi, D. Munz, T-stress solution for DCDC specimens, Engng. Fract. Mech. **72**(2005), 145-149.

[C15.8] Fett, T., Rizzi, G., A fracture mechanics analysis of the DCDC specimen, Report FZKA 7094, Forschungszentrum Karlsruhe, 2005.

[C15.9] Leevers, P.S., Radon, J.C., Inherent stress biaxiality in various fracture specimen geometries, Int. J. Fract. **19**(1982), 311-325.

[C15.10] Cotterell, B. and Rice, J.R., Slightly curved or kinked cracks, *International Journal of Fracture* **16**(1980), 155-169.

[C15.11] Melin, S., On the directional stability of wedging, Int. J. Fract. **50**(1991), 293-300.

[C15.12] Melin, S., The influence of the T-stress on the directional stability of cracks, Int. J. Fract. **114**(2002), 259-265.

<u>C16</u>

Compact tensile (CT) specimen

C16.1 Rectangular CT specimen

Several analyses of the CT specimen (see Fig. C16.1) are available in literature [C16.1-C16.4]. Under slightly modified boundary conditions, rather strong differences were found especially for short crack lengths.

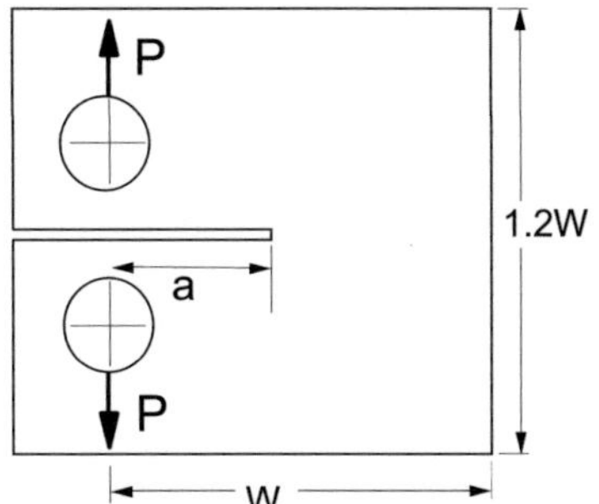

Fig. C16.1 Compact tension specimen.

A stress intensity factor solution for the standard CT specimens was proposed by Newman [C16.5] and Srawley [C16.6]

$$K_1 = \frac{P}{B\sqrt{W}} \frac{(2+\alpha)(0.886 + 4.64\alpha - 13.32\alpha^2 + 14.72\alpha^3 - 5.6\alpha^4)}{(1-\alpha)^{3/2}} \tag{C16.1.1}$$

with $\alpha = a/W$.
A weight function was proposed in [C16.7] for $0.2 \leq \alpha \leq 0.8$ by the polynomial of

$$h = \sqrt{\frac{2}{\pi a}} \frac{1}{\sqrt{1-x/a}(1-\alpha)^{3/2}} [(1-\alpha)^{3/2} + \sum D_{nm}(1-x/a)^{m+1}\alpha^n] \tag{C16.1.2}$$

with the coefficients listed in Table C16.1.
Literature results for the biaxiality ratio β

$$\beta = \frac{T\sqrt{\pi a}}{K} \tag{C16.1.3}$$

proposed by Leevers and Radon [C16.2] are entered in Fig. C16.2a. Figure C162b compiles own FE results (solid symbols) and data from Knesl and Bednar [C16.8] (open symbols) for the ASTM E399 standard specimen together with the limit case $\alpha \to 1$ taken from Tables C8.5 and C8.12. The curve plotted in Fig. C16.2b can be described by

$$\beta \cong \frac{0.7702 - 6.572\alpha + 26.665\alpha^2 - 43.446\alpha^3 + 29.695\alpha^4 - 6.6886\alpha^5}{\sqrt{1-\alpha}} \qquad (C16.1.4)$$

Table C16.1 Coefficients D_{nm} for eq.(C16.1.2)

m	n=0	1	2	3	4
0	2.673	-8.604	20.621	-14.635	0.477
1	-3.557	24.973	-53.398	50.707	-11.837
2	1.230	-8.411	16.957	-12.157	-0.940
3	-0.157	0.954	-1.284	-0.393	1.655

The T-stress term results from eqs.(C16.1.1), (C16.1.3), and (C16.1.4). In this context, it has to be noted that the results in Fig. C16.2a were not derived for the standard CT specimen with large holes. In reference [C16.2], the T-stress was determined by applying of shear tractions along the loading line and by application of point forces in the centres of fictitious holes. In Fig. C16.2b the test specimen according to ASTM E399 was modelled with point forces to be active at the contact points. Based on these results, it is recommended to use cracks with α>0.25.

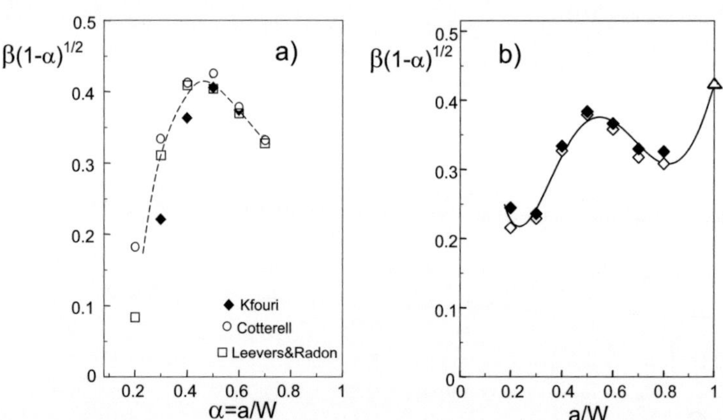

Fig. C16.2 a) Biaxiality ratio of the CT specimen from literature; curve: eq.(C16.1.4), squares: Leevers and Radon [C16.2], circles: Cotterell [C16.3], diamond squares: Kfouri [C16.1], b) biaxiality ratio for a standard CT specimen loaded by point forces: full symbols: FE results, open symbols: Knesl and Bednar [C16.8], triangle: limit case from Tables C8.5 and C8.12.

C16.2   Round-CT specimen

The RCT specimen is identical with the single-edge-cracked circular disk, if the load application holes are neglected. Figure C16.3 shows this fracture mechanics test specimen.

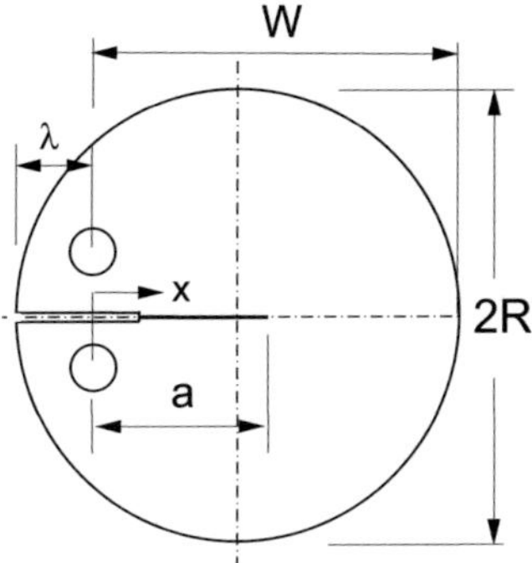

Fig. C16.3 Geometric data of the RCT specimen.

The stress intensity factor solution was derived by Newman [C16.9] as

$$K_{\mathrm{I}} = \frac{P}{B\sqrt{W}}Y^*(\alpha) , \quad \alpha = a/W \tag{C16.2.1a}$$

$$Y^* = \frac{(2+\alpha)(0.76 + 4.8\,\alpha - 11.58\,\alpha^2 + 11.43\,\alpha^3 - 4.08\,\alpha^4)}{(1-\alpha)^{3/2}} \tag{C16.2.1b}$$

valid for $\alpha \geq 0.2$. This stress intensity factor solution deviates by less than 6% from the solution for the rectangular CT specimen addressed in Section C16.1.

The biaxiality ratio can be approximated by

$$\beta \cong \frac{0.5473 - 5.5305\alpha + 21.478\alpha^2 - 30.305\,\alpha^3 + 17.265\alpha^4 - 3.0312\alpha^5}{\sqrt{1-\alpha}} \tag{C16.2.2}$$

This relation is plotted in Fig. C16.4 together with eq.(C16.1.2) for the CT specimen.

The weight function can be expressed for $0.2 \leq \alpha \leq 0.8$ by the polynomial [C16.10] of

$$h = \sqrt{\frac{2}{\pi a}}\,\frac{1}{\sqrt{1-x/a}(1-\alpha)^{3/2}}[(1-\alpha)^{3/2} + \sum D_{nm}(1-x/a)^{m+1}\alpha^n] \tag{C16.2.3}$$

with the coefficients listed in Table C16.2.

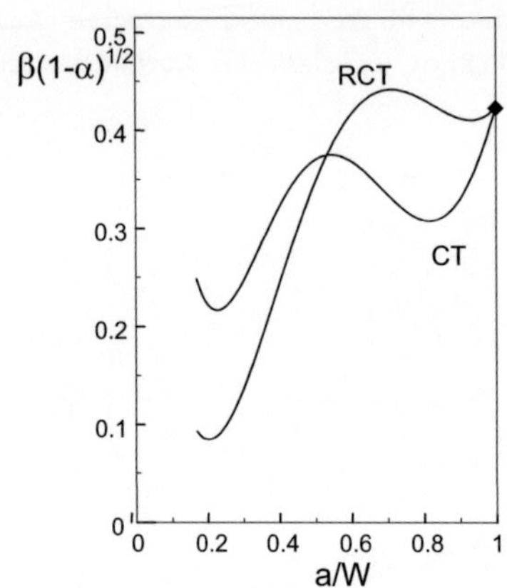

Fig. C16.4 Comparison of biaxiality ratios for the CT and RCT specimens.

Table C16.2 Coefficients D_{nm} for eq.(C16.2.3)

n	m=0	1	2	3	4
0	2.826	-5.865	0.8007	-0.2584	0.6856
1	-10.948	48.095	-3.839	1.280	-6.734
2	35.278	-143.789	6.684	-5.248	25.188
3	-41.438	196.012	-4.836	11.435	-40.140
4	15.191	-92.787	-0.7274	-7.328	22.047

References C16:

[C16.1] Kfouri, A.P., Some evaluations of the elastic T-term using Eshelby's method, Int. J. Fract. **30**(1986), 301-315.

[C16.2] Leevers, P.S., Radon, J.C., Inherent stress biaxiality in various fracture specimen geometries, Int. J. Fract. **19**(1982), 311-325.

[C16.3] Cotterell, B., On the fracture path stability in the compact tension test, Int. J. Fract. Mech. **6**(1970), 189-192.

[C16.4] Sherry, A.H., France, C.C., Goldthorpe, M.R., Compendium of T-stress solutions for two and three- dimensional cracked geometries, Engng. Fract. Mech. **18**(1995), 141-155.

[C16.5] Newman, J.C., Stress analysis of compact specimens including the effects of pin loading, ASTM STP 560, 1974, 105.

[C16.6] Srawley, J.E., Wide range stress intensity factor expressions for ASTM E399 standard fracture toughness specimens, Int. J. Fract. Mech. **12**(1976), 475-476.

[C16.7] Fett, T., Munz, D., Stress intensity factors and weight functions, Computational Mechanics Publications, Southampton, 1997.

[C16.8] Knesl, Z., Bednar, K., Two parameter fracture mechanics: Calculation of parameters and their values, Report of the Institute of Physics of Materials, Brno, 1998.

[C16.9] Newman, J.C., Stress intensity factors and crack opening displacements for round compact specimens, Int. J. Fract. **17**(1981) 567-578.

[C16.10] Fett, T., A weight function for the RCT-specimen, Int. J. Fract. **63**(1993) R81-R85.

C17

Double Cantilever Beam

The Double-Cantilever-Beam (DCB) specimen is shown in Fig. C17.1. A line load P/B (B= specimen thickness, here chosen as B=1) is applied in distance (a-x) from the crack tip.

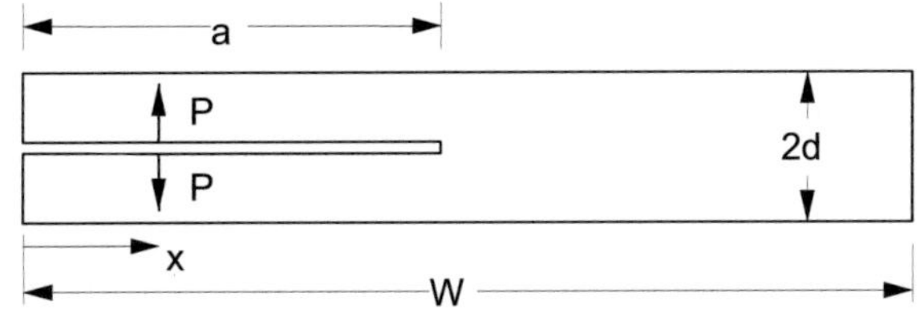

Fig. C17.1 Double Cantilever Beam specimen under crack-face loading by line load P.

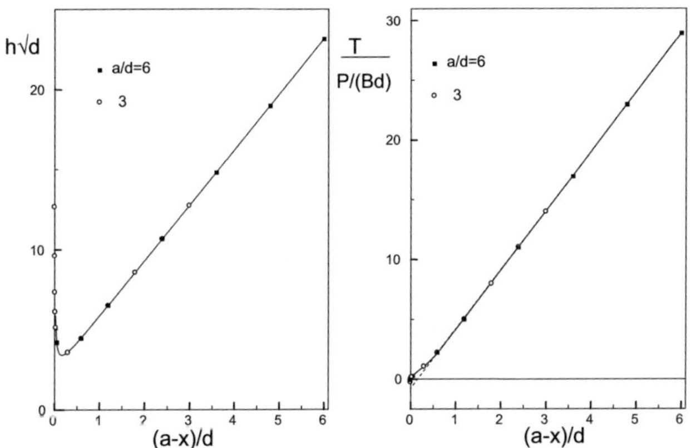

Fig. C17.2 a) Weight function for stress intensity factors, b) Green's function for T-stress (W>6d).

The weight function is given in [C17.1] as

$$h = \sqrt{\frac{12}{d}}\left(\frac{a}{d}+0.68\right) - \sqrt{\frac{12}{d}}\frac{x}{d} + \sqrt{\frac{2}{\pi(a-x)}}\exp\left(-\sqrt{12\frac{a-x}{d}}\right) \tag{C17.1a}$$

and introduced in Fig. C17.2a as the curve. Its agreement with the FE results is excellent. The stress intensity factor for the specimen loaded at the crack mouth x=0 results from eq.(C17.1) as

$$\frac{K_1}{P/B} = \sqrt{\frac{12}{d}}\left(\frac{a}{d}+0.68\right) + \sqrt{\frac{2}{\pi a}}\exp\left(-\sqrt{12\frac{a}{d}}\right) \tag{C17.1b}$$

The T-stress results are shown in Fig. C17.2b. The asymptotic solution for $(a-x)/d>1$ can be expressed by

$$\frac{T}{P/(Bd)} = \begin{cases} 4.95\dfrac{a-x}{d} - 0.9 & \textit{for } (a-x)/d > 1 \\[3mm] 0.3\sqrt{\dfrac{a-x}{d}} & \textit{for } (a-x)/d \to 0 \end{cases} \qquad (C17.2)$$

This relation is represented by the dashed line. A relation for any $(a-x)/d$ reads

$$\frac{T}{P/(Bd)} = \lambda\left(\tfrac{9}{8}\sqrt{\frac{a-x}{d}} + \tfrac{5}{2}\left(\frac{a-x}{d}\right)^{3/2} \right) + (1-\lambda)\left(4.95\frac{a-x}{d} - 0.9 \right) \qquad (C17.3)$$

with
$$\lambda = \exp\left(-2\left(\frac{a-x}{d}\right)^2 \right) \qquad (C17.4)$$

represented by the (solid curve).
Figure C17.3 shows the biaxiality ratio as the curve. An approximation from [C17.1] for $0.1<d/a<0.6$

$$\frac{1}{\beta} \cong 0.681\frac{d}{a} + 0.0685 \qquad (C17.5)$$

is entered as the dash-dotted line.

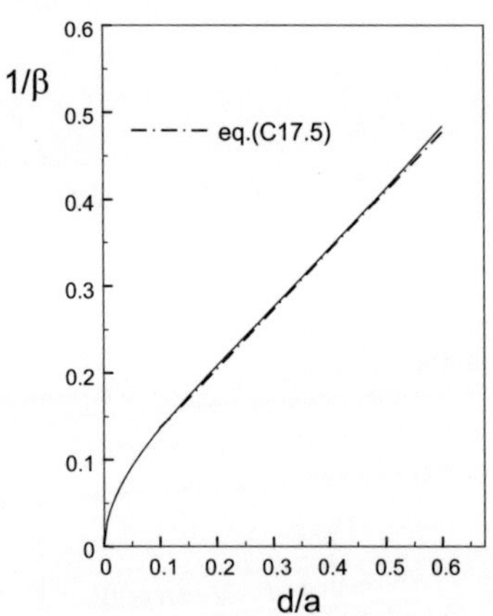

Fig. C17.3 Biaxiality ratio.

Reference C17:

[C17.1] Fett, T., Munz, D., Stress intensity factors and weight functions, Computational Mechanics Publications, Southampton, 1997.

C18

Cracked bars under opposed forces

C18.1 Stresses by a single pair of concentrated opposite line forces

Figure C18.1 shows an edge-cracked parallel strip under loading by a pair of opposite concentrated forces at the distance x from an edge crack.

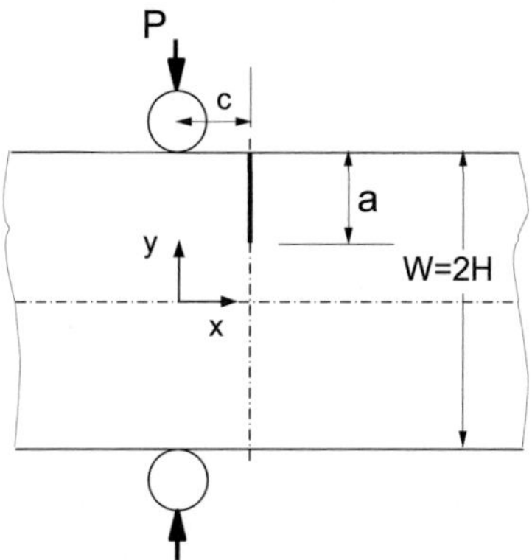

Fig. C18.1 Edge-cracked plate loaded by a single pair of opposite forces.

The stresses in the strip of width $W = 2H$ and thickness B under the loading of opposite concentrated line forces P have been computed by Filon [C18.1]. With the geometric data shown in Fig. C18.1 the stresses can be expressed by

$$\sigma_x = -\frac{2P}{\pi HB}\int_0^\infty \frac{\sinh u - u\cosh u}{\sinh 2u + 2u}\cos\frac{ux}{H}\cosh\frac{uy}{H}\,du\, -$$

$$-\frac{2P}{\pi HB}\int_0^\infty \frac{uy}{H}\frac{\sinh u}{\sinh 2u + 2u}\cos\frac{ux}{H}\sinh\frac{uy}{H}\,du \tag{C18.1.1}$$

$$\sigma_y = -\frac{2P}{\pi HB}\int_0^\infty \frac{\sinh u + u\cosh u}{\sinh 2u + 2u}\cos\frac{ux}{H}\cosh\frac{uy}{H}\,du$$

$$+\frac{2P}{\pi HB}\int_0^\infty \frac{uy}{H}\frac{\sinh u}{\sinh 2u + 2u}\cos\frac{ux}{H}\sinh\frac{uy}{H}\,du \tag{C18.1.2}$$

$$\tau_{xy} = \frac{2P}{\pi HB} \int_0^\infty \frac{u \cosh u}{\sinh 2u + 2u} \sin \frac{ux}{H} \sinh \frac{uy}{H} du$$

$$- \frac{2P}{\pi HB} \int_0^\infty \frac{uy}{H} \frac{\sinh u}{\sinh 2u + 2u} \sin \frac{ux}{H} \cosh \frac{uy}{H} du \qquad (C18.1.3)$$

The geometric functions for the mode-I and mode-II stress intensity factors, here denoted as Y_I and Y_{II}, are defined by

$$K_I = \sigma * Y_I \sqrt{W} \qquad (C18.1.4a)$$

$$K_{II} = \sigma * Y_{II} \sqrt{W} \qquad (C18.1.4b)$$

with the characteristic stress

$$\sigma* = \frac{P}{HB} \qquad (C18.1.5)$$

Figures C18.2 and C18.3 show the results which are also compiled in Tables C18.1 and C18.2 [C18.2].

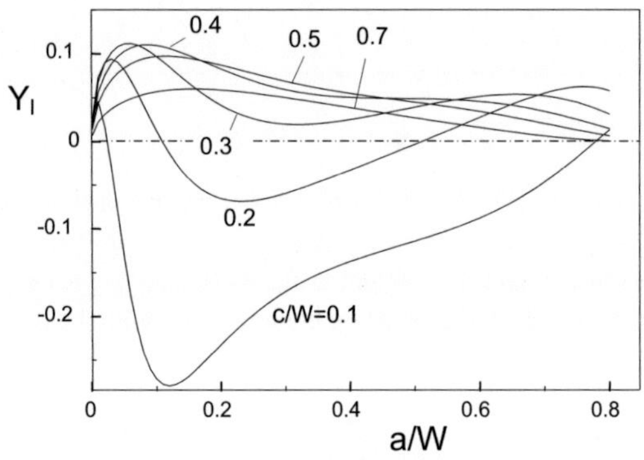

Fig. C18.2 Geometric function Y according to eq.(C18.1.4a) for 2-point loading.

Table C18.1 Mode-I stress intensity factors represented by the geometric function Y_I, eq.(C18.1.4a).

a/W	x/W=0.1	0.2	0.3	0.4	0.5
0.1	-0.2718	0.0123	0.097	0.1096	0.0970
0.2	-0.2324	-0.0654	0.0415	0.0830	0.0872
0.3	-0.1717	-0.0596	0.020	0.0591	0.0687
0.4	-0.1375	-0.0329	0.025	0.0503	0.0549
0.5	-0.1142	-0.0033	0.039	0.0490	0.0449
0.6	-0.0879	0.0273	0.051	0.0457	0.0343
0.7	-0.047	0.055	0.052	0.0323	0.0205

Table C18.2 Mode-II stress intensity factors represented by the geometric function Y_{II}, eq.(C18.1.4b).

a/W	$x/W=0.05$	0.1	0.15	0.2	0.25	0.3	0.4	0.5
0.1	1.433	0.660	0.296	0.134	0.0556	0.0147	-0.021	-0.031
0.2	1.340	0.920	0.634	0.392	0.226	0.117	0.0005	-0.044
0.3	1.185	0.942	0.712	0.506	0.337	0.205	0.0396	-0.037
0.4	1.083	0.876	0.686	0.510	0.358	0.232	0.0598	-0.028
0.5	1.029	0.814	0.625	0.459	0.318	0.205	0.0511	-0.027
0.6	1.012	0.754	0.542	0.371	0.238	0.140	0.0214	-0.032
0.7	1.020	0.673	0.418	0.242	0.128	0.0568	-0.012	-0.034

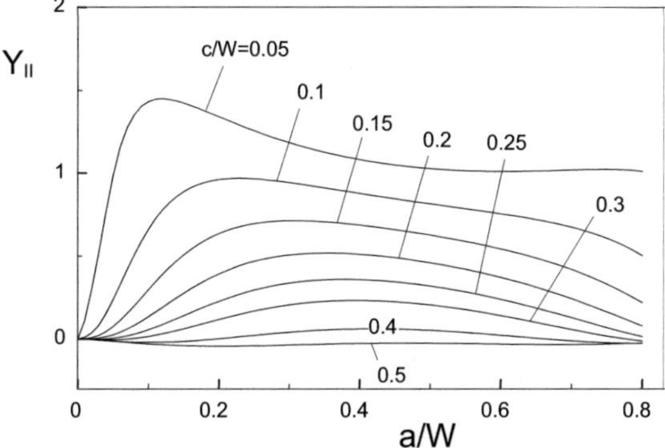

Fig. C18.3 Geometric function Y_{II} according to eq.(C18.1.4b).

T-stresses were computed in [C18.3]. Normalised values T/σ^* are given in Table C18.3.

Table C18.3 T/σ^* for the edge-cracked strip under a pair of opposite concentrated forces.

a/W	$x/W=0.1$	0.2	0.5	0.7	1	1.5
0.2	-1.24	-0.292	0.0522	0.0356	0.013	0.001
0.3	-1.22	-0.585	0.0532	0.0693	0.032	0.003
0.4	-1.14	-0.695	0.0330	0.0879	0.045	0.004
0.5	-1.11	-0.724	0.0219	0.0930	0.050	0.05
0.6	-1.14	-0.701	0.0325	0.0884	0.045	0.004
0.7	-1.23	-0.594	0.0590	0.0733	0.033	0.003

In Fig. C18.4, the biaxiality ratio β is given by the ratio of T-stress and stress intensity factor according to

$$\beta = \frac{T\sqrt{\pi a}}{K_{\mathrm{I}}} \tag{C18.1.6}$$

The curve in Fig. C18.4 can be described by

$$\beta = \frac{-0.469 + 1.8589\,\alpha + 34.527\,\alpha^2 - 133.477\,\alpha^3 + 127.994\,\alpha^4}{\sqrt{1-\alpha}} \tag{C18.1.7}$$

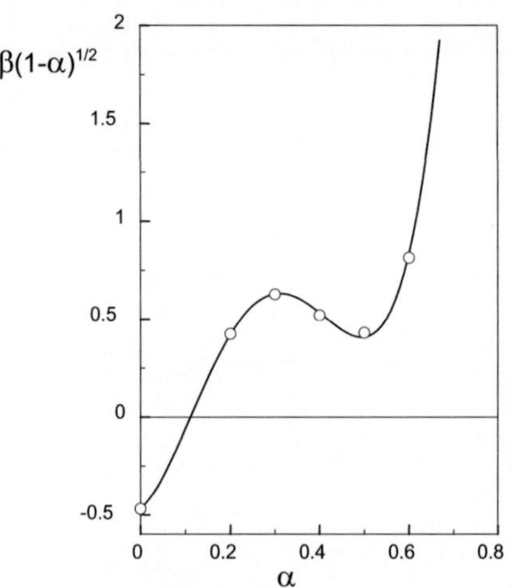

Fig. C18.4 Biaxiality ratio β for $c/W = 1$.

C18.2 Stresses and stress intensity factors for two pairs of forces

Superposition of the results obtained for one pair of concentrated forces (Section C18.1) allows computing the loading problem illustrated in Fig. C18.5.

In the uncracked bar we obtain for the symmetry line ($x = 0$)

$$\sigma_x = -\frac{8P}{\pi WB}\int_0^\infty \frac{\sinh u - u\cosh u}{\sinh 2u + 2u}\cos\frac{ud}{W}\cosh\frac{2uy}{W}\,du -$$

$$-\frac{8P}{\pi WB}\int_0^\infty \frac{2uy}{W}\frac{\sinh u}{\sinh 2u + 2u}\cos\frac{ud}{W}\sinh\frac{2uy}{W}\,du \tag{C18.2.1}$$

The stress σ_x is plotted in Fig. C18.6, normalised to the characteristic stress

$$\sigma^* = \frac{2P}{WB} \tag{C18.2.2}$$

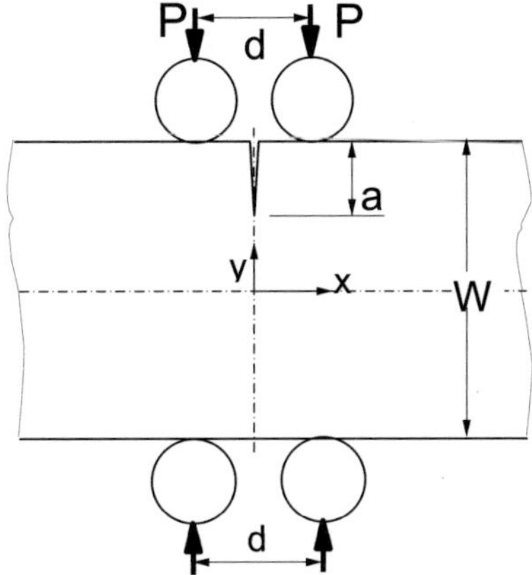

Fig. C18.5 Controlled fracture test device with load application via four symmetrical rollers.

The stress intensity factors for the edge-cracked specimen were computed from the stresses in the uncracked specimen by use of the weight function (Section C8.3). The geometric function for the mode-I stress intensity factor, Y_I, is defined by

$$K_I = \sigma * Y_I \sqrt{W} . \tag{C18.2.3}$$

and plotted in Fig. C18.7 [C18.2] as a function of a/W with d/W as a parameter. It is also entered into Table C18.4.

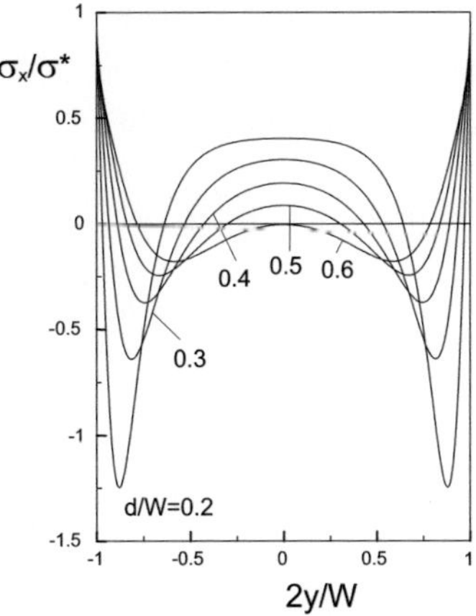

Fig. C18.6 Axial stresses σ_x along the symmetry line $x=0$ in the absence of the crack.

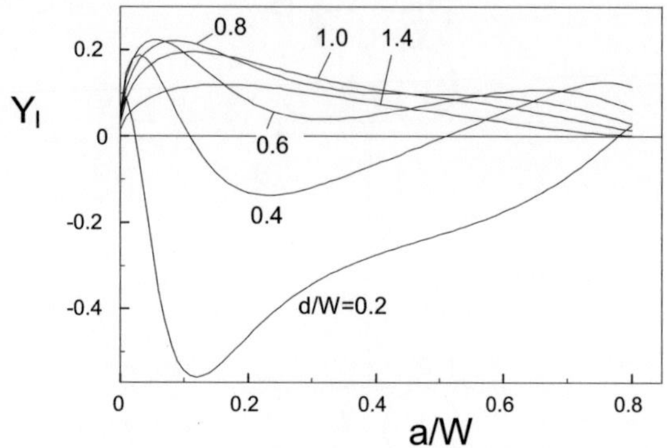

Fig. C18.7 Geometric function Y_I according to eq.(C18.2.3) for 4-point loading.

Table C18.4 Stress intensity factors for 4-point loading, represented by the geometric function Y_I, eq.(C18.2.3).

a/W	d/W							
	0.2	0.4	0.6	0.8	1.0	1.2	1.4	1.6
0.05	-0.26	0.171	0.2234	0.2066	0.1686	0.1279	0.0916	0.0624
0.1	-0.537	0.034	0.1997	0.2196	0.1942	0.1538	0.1136	0.0792
0.15	-0.531	-0.034	0.1346	0.1971	0.1903	0.1578	0.1199	0.0854
0.2	-0.458	-0.077	0.0836	0.1660	0.1746	0.1510	0.1176	0.0853
0.25	-0.389	-0.1331	0.0526	0.1384	0.1556	0.1389	0.1103	0.0811
0.3		-0.117	0.0403	0.1186	0.1377	0.1250	0.1003	0.0743
0.35		-0.0915	0.0414	0.1068	0.1224	0.1110	0.0891	0.0660
0.4			0.0506	0.1011	0.1101	0.0977	0.0774	0.0569
0.45			0.0641	0.0991	0.0998	0.0850	0.0658	0.0419
0.5			0.0791	0.0984	0.0902	0.0727	0.0542	0.0382
0.55			0.0931	0.0968	0.0804	0.0606	0.0431	0.0291
0.6			0.1041	0.0924	0.0693	0.0483	0.0322	0.0207
0.65			0.1094	0.0838	0.0564	0.0359	0.0221	0.0131
0.7			0.1064	0.0703	0.0422	0.0239	0.0132	0.0068

For the special case of $d/W = 1$, Y_I is fitted for $\alpha = a/W \leq 0.6$ by

$$Y_I = 0.905\,\alpha^{1/2} - 3.358\,\alpha^{3/2} + 3.857\,\alpha^{5/2} + 1.4425\,\alpha^{7/2} - 3.873\,\alpha^{9/2} \qquad (C18.2.4)$$

T-stresses were computed in [C18.3]. The normalised T-stresses T/σ^* are given in Table C18.5. The biaxiality ratio β is identical with that for a single pair of opposite forces (see Fig. C18.4) and can be described for $d/W = 1$ by

$$\beta = \frac{-0.469 + 1.8589\,\alpha + 34.527\,\alpha^2 - 133.477\,\alpha^3 + 127.994\,\alpha^4}{\sqrt{1-\alpha}} \qquad (C18.2.5)$$

i.e. by the same relation as given in eq.(C18.1.7).

Table C18.5 T-stress T/σ^* for the edge-cracked strip under two pairs of opposite forces, Fig.C18.5.

a/W	$d/W=0.2$	0.4	1	1.4	2	3
0.2	-2.48	-0.584	0.1044	0.0713	0.026	0.002
0.3	-2.44	-1.169	0.1064	0.1386	0.063	0.006
0.4	-2.28	-1.390	0.0660	0.1758	0.090	0.008
0.5	-2.22	-1.448	0.0438	0.1859	0.100	0.010
0.6	-2.28	-1.401	0.0650	0.1768	0.090	0.008
0.7	-2.47	-1.188	0.1804	0.1466	0.066	0.006

C18.3 Double-edge-cracked bars

In this section the mode-I stress intensity factors are reported for a double-edge-cracked bar under 4-roller loading [C18.2]. The specimen and the load application are illustrated in Fig. C18.8.

The stress intensity factor is plotted in Fig. C18.9 and compiled for a number of crack depths in Table C18.6. Since in case of double-edge-cracked bars, $H = W/2$ is the characteristic width dimension for normalising the crack length a (i.e. a is limited by $a < H$), K_I is defined by

$$K_I = \sigma^* Y_I \sqrt{H} \qquad (C18.3.1)$$

with σ^* defined by eq.(C18.1.5).

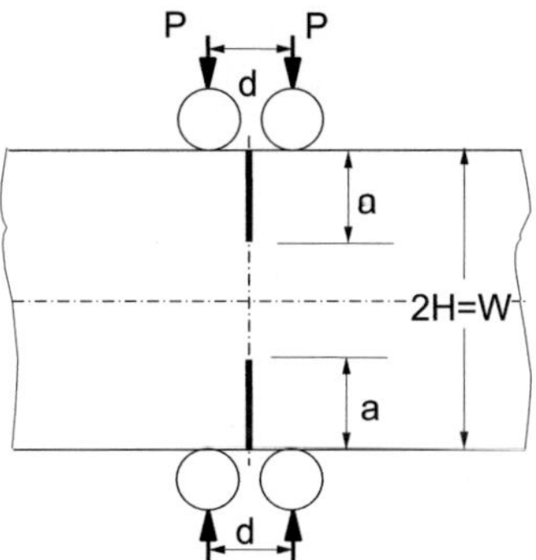

Fig. C18.8 Double-edge-cracked bar loaded by two pairs of opposite forces.

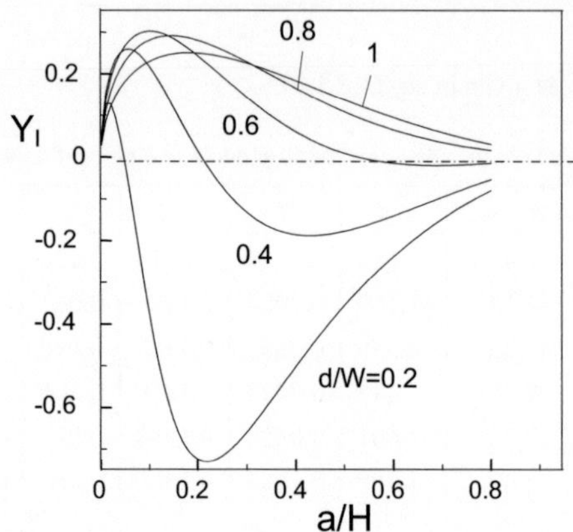

Fig. C18.9 Geometric function Y_{I} according to eq.(C18.3.1) for different roller distances.

Table C18.6 Geometric function Y_{I} according to eq.(C18.3.1).

a/W	d/W=0.2	0.4	0.6	0.8	1.0
0.05	-0.016	0.258	0.267	0.230	0.182
0.1	-0.374	0.218	0.302	0.280	0.229
0.15	-0.627	0.121	0.285	0.291	0.248
0.2		0.017	0.244	0.281	0.250
0.3		-0.132	0.144	0.229	0.226
0.4		-0.185	0.061	0.165	0.184
0.5		-0.177	0.009	0.109	0.138
0.6		-0.140	-0.015	0.065	0.095
0.7		-0.095	-0.020	0.035	0.059
0.8		-0.054	-0.015	0.016	0.031

References C18

[C18.1] Filon, L.N.G., On an approximate solution for the bending of a beam of rectangular cross-section under any system of load, with special reference to points of concentrated or discontinuous loading, Phil. Trans., A, **201**(1903), 63-155.

[C18.2] Fett, T., Munz, D., Thun, A toughness test device with opposite roller loading, Engng. Fract. Mech. **68**(2001), 29-38

[C18.3] Fett, T., T-stresses in rectangular plates and circular disks, Engng. Fract. Mech. **60**(1998), 631-652.

C19

Cracks ahead of notches

C19.1 Stress intensity factor

Many test specimens contain narrow notches which are introduced in order to simulate a starter crack. A specimen containing a slender edge notch of depth a_0 with the notch root the radius R is considered (Fig. C19.1). A small crack of length ℓ is assumed to occur directly at the notch root.

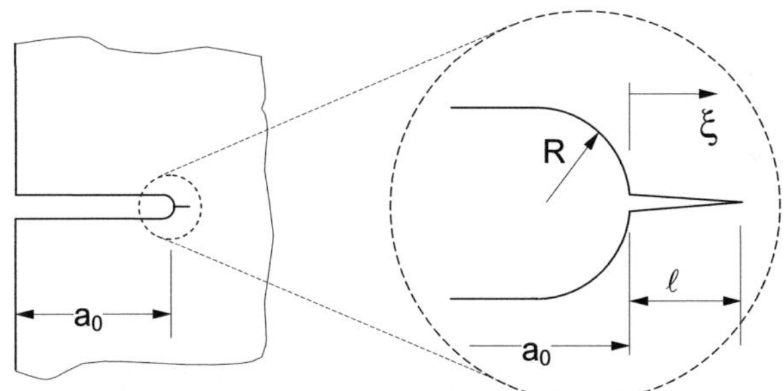

Fig. C19.1 A small crack emanating from the root of a notch.

In the absence of a crack, the stresses near the notch root are given by

$$\sigma_y = \frac{2K(a_0)}{\sqrt{\pi(R+2\xi)}}\frac{R+\xi}{R+2\xi} \tag{C19.1.1}$$

$$\sigma_x = \frac{2K(a_0)}{\sqrt{\pi(R+2\xi)}}\frac{\xi}{R+2\xi} \tag{C19.1.2}$$

(for ξ see Fig. C19.1) as shown by Creager and Paris [C19.1]. The quantity $K(a_0)$ is the stress intensity factor of a crack having the same length a_0 as the notch under identical external load

$$K(a_0) = \sigma * F(a_0)\sqrt{\pi a_0} \tag{C19.1.3}$$

with the characteristic stress $\sigma*$ (e.g. remote tensile stress, outer fibre bending stress) and the geometric function F. The stresses resulting from eqs.(C19.1.1) and (C19.1.2) are plotted in Fig. C19.2. The solid parts of the curves represent the region ($0 \leq \xi \leq R/2$) where higher-order terms in the stress approximation are negligible.

For cracks ahead of slender notches the stress intensity factor can be represented as [C19.2]

$$K \cong K * \tanh(2.243\sqrt{\ell / R}) \tag{C19.1.4}$$

with the stress intensity factor K^* formally computed for a fictitious crack of total length $a_0+\ell$ according to

$$K^* = \sigma * F(a_0 + \ell)\sqrt{\pi(a_0 + \ell)} \tag{C19.1.5}$$

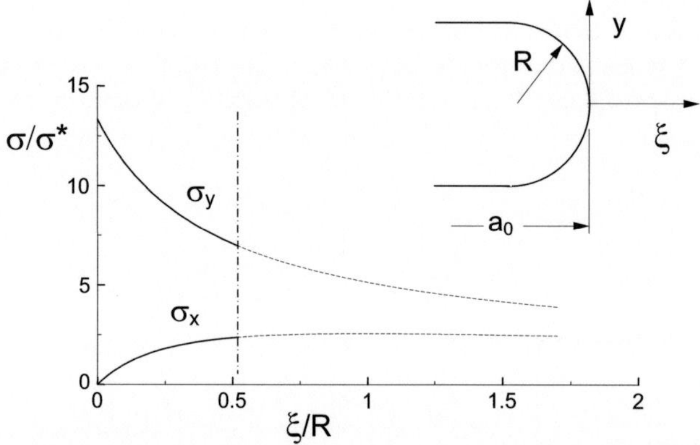

Fig. C19.2 Stresses ahead of a slender notch in bending computed according to Creager and Paris [C19.1] for $a_0/W = 0.5$ and $R/W = 0.025$; W=width of the bending bar.

C19.2 T-stress

In Fig. C19.3 the T-stress for bending is plotted versus a/W for several notch depths a_0. The "long-crack solution" given by eq.(C8.2.4) is introduced as the solid curve. This curve represents the T-stress for an edge crack of the total length $a=a_0+\ell$.

Results obtained under tensile loading are plotted in Fig. C19.4. In this case, the characteristic stress is identical with the remote tensile stress $\sigma^* = \sigma_0$. In this representation the solid line is described by eq.(C8.1.4).

For the limit case $\ell/R \to 0$ the T-stress can be determined from the solution for a crack in a semi-infinite plate under a tensile stress σ_{max} as occurring directly at the notch root

$$\sigma_{max} = 2\sigma * F(a_0)\sqrt{\frac{a_0}{R}} \tag{C19.2.1}$$

Directly at the free surface ($\xi = 0$) it holds $\sigma_x = 0$. It can be concluded

$$T_0 = T_{\ell/R \to 0} = \left.\frac{T_{finite\ plate}}{\sigma *}\right|_{a \to 0} \sigma_{max} \tag{C19.2.2}$$

$$\left.\frac{T_{finite\ plate}}{\sigma^*}\right|_{a/W \to 0} = -0.526 \tag{C19.2.3}$$

and, consequently,

$$\frac{T_0}{\sigma^*} = -1.052\, F(a_0)\sqrt{\frac{a_0}{R}} \tag{C19.2.4}$$

It becomes obvious from eq.(C19.2.4) that for slender notches very strong compressive T-stresses occur in the limit case $\ell/R \to 0$. The limit values T_0 for tension and bending (location indicated by the arrows in Figs. C19.3 and C19.4) are compiled in Table C19.1.

An approximate description for the T-stress is given by

$$T \approx -0.526\,\sigma_{max} + (T^* + 0.526\,\sigma_{max})\tanh^{4/3}(5\ell/R)^{3/4} \tag{C19.2.5}$$

$$T \approx T_0 + (T^* - T_0)\tanh^{4/3}(5\ell/R)^{3/4} \tag{C19.2.6}$$

where T^* is the T-stress term for the "long-crack solution", i.e. the T-stress for a crack of total length $a = \ell + a_0$ according to eqs. (C8.1.4) and (C8.2.4).

The T-stress approximation by eq.(C19.2.6) is plotted in Fig. C19.5 for bending together with the data of Fig. C19.3.

Table C19.1 Limit values for the T-stress term ($\ell/R \to 0$), R/W=0.025.

a/W	T_0/σ^* (bending)	T_0/σ^* (tension)
0.3	-4.11	-6.05
0.4	-5.28	-8.91
0.5	-7.01	-13.31
0.6	-9.86	-20.74

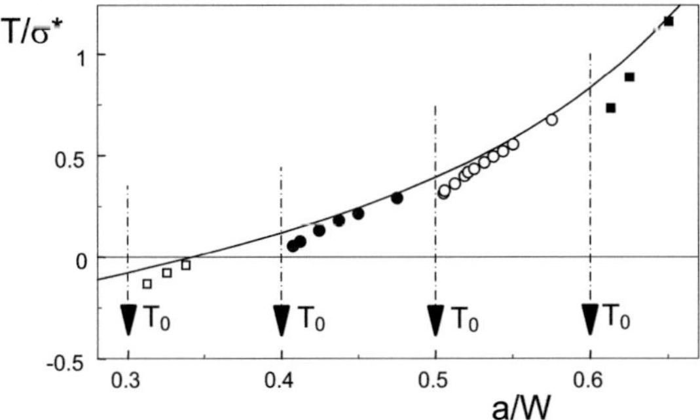

Fig. C19.3 T-stress for a small crack ahead of a slender notch in bending, computed with the Boundary Collocation Method for $R/W = 0.025$. Solid line: Long-crack solution.

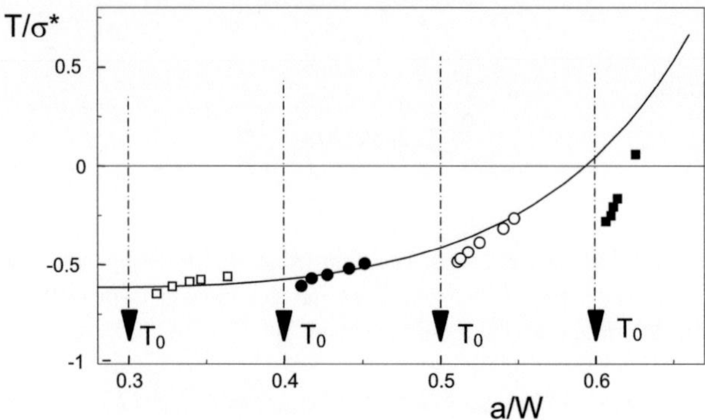

Fig. C19.4 T-stress for a small crack ahead of a slender notch in tension, computed with the Boundary Collocation Method for $R/W = 0.025$. Solid line: Long-crack solution.

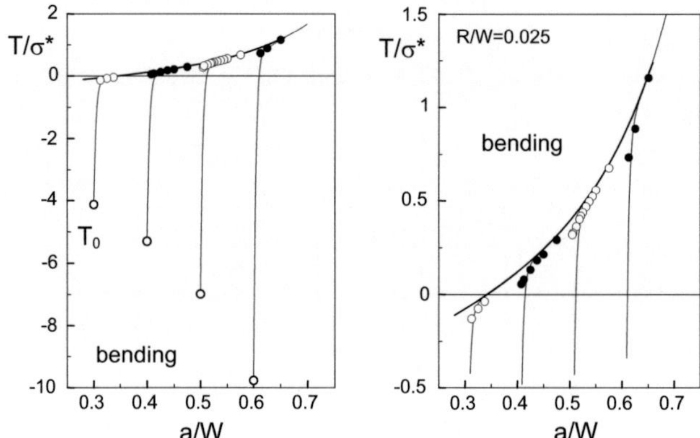

Fig. C19.5 Comparison of the approximation eq.(C19.2.6) with the results of Fig. C19.3, R/W=0.025.

Figure C19.6 shows the influence of the crack length ℓ on the T-stress for a bending bar of width W=4mm and two different notch root radii. From this diagram, it can be concluded that the first few micrometers of crack extension from a notch are automatically stable, since $T<0$ over about $a-a_0$=6-12 µm for notch radii of R=10 and 20 µm [C19.3]. The crack lengths ℓ_{stable} for which path stability occurs even at a/W>0.35 can be roughly expressed as

$$\ell_{stable} = a_{stable} - a_0 \approx 0.6\,R \tag{C19.2.7}$$

For a/W<0.35, the condition $T<0$ is trivially fulfilled, since the "long-crack solution" $T(a)$ is negative in this case, see Fig. C19.5.

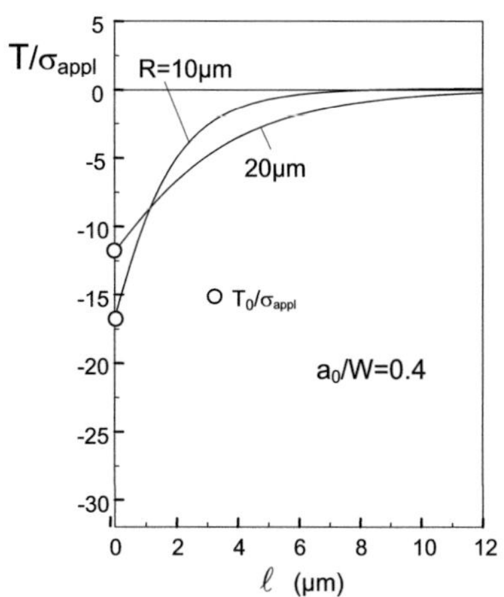

Fig. C19.6 T-stress for small cracks [C19.3] ahead of a narrow notch (bending).

References C19

[C19.1] Creager, M., Paris, P.C., Elastic field equations for blunt cracks with reference to stress corrosion cracking, Int. J. Fract. **3**(1967), 247-252.

[C19.2] Fett, T., Munz, D., Stress intensity factors and weight functions, Computational Mechanics Publications, Southampton, 1997.

[C19.3] Fett, T., Munz, D., Influence of narrow starter notches on the initial crack growth resistance curve of ceramics, Arch. Appl. Mech. **76**(2006), 667-679.

C20

Array of edge cracks

Figure C20.1 shows an array of periodical edge cracks. BCM computations were performed for an element of periodicity. The boundary conditions are given by a constant displacements v (defining the characteristic stress σ) and disappearing shear stresses along the symmetry lines (dash-dotted lines), i.e.

$$v = \frac{\sigma}{E'} \frac{d}{2}; \quad \tau_{xy} = 0 \text{ for } y = \pm d/2 \tag{C20.1}$$

($E' = E$ for plane stress and $E' = E/(1-v^2)$ for plane strain, $E =$ Young's modulus, $v =$ Poisson's ratio) as illustrated in Fig. C20.2. The coefficient B_0 is shown in Fig. C20.3a as a function of the ratio d/a for different relative crack lengths a/W. The result can be summarised as

$$B_0 = 0.148 \quad , d/a \leq 1.5 \tag{C20.2}$$

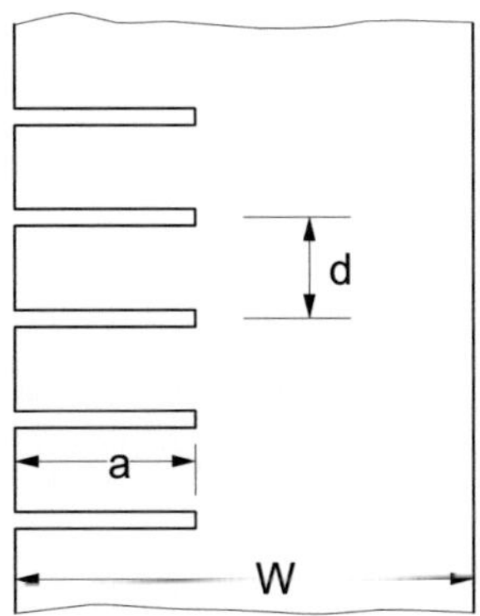

Fig. C20.1 Periodical edge cracks in an endless strip.

The coefficient A_0 is plotted in Fig. C20.3b in the normalised form of

$$A_0^* = 6A_0 \sqrt{\pi W / d} \tag{C20.3}$$

For all values of a/W investigated it was found

$$A_0^* = 1.000 \pm 0.002 \tag{C20.4}$$

resulting in the stress intensity factor solution of

$$K_I = \sigma\sqrt{d/2}$$ (C20.5)

(see e.g. [C20.1]). The T-stress term is

$$T = -4B_0 = -0.592\sigma$$ (C20.6)

and the biaxiality ratio β according to eq.(A1.3.12) results as

$$\beta = -1.484\sqrt{a/d}$$ (C20.7)

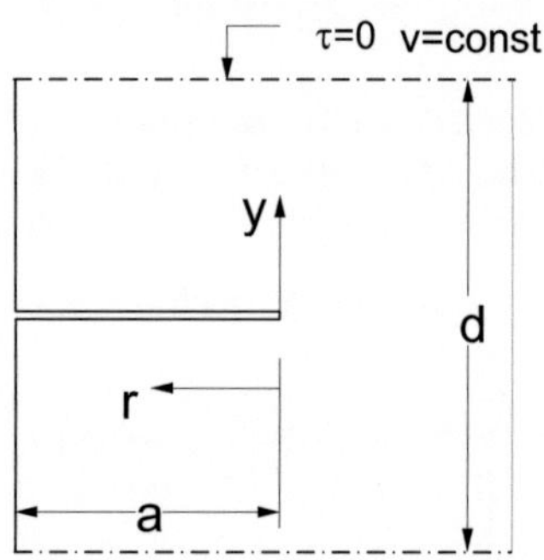

Fig. C20.2 Boundary conditions representing an endless strip with periodical cracks.

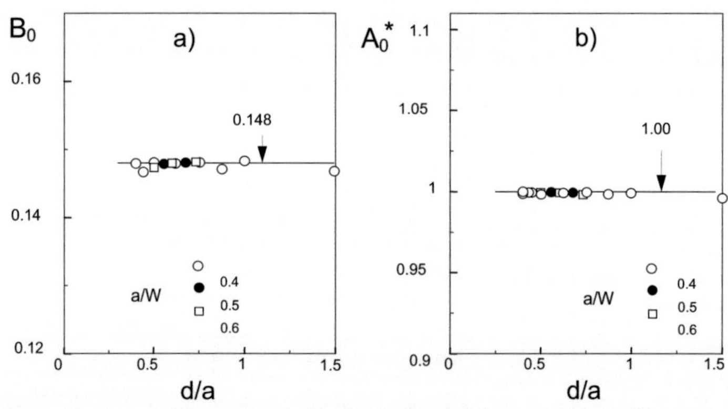

Fig. C20.3 a) Influence of the geometric data on the first regular term of the Williams stress function B_0, b) Coefficient A_0 in the normalisation $A_0^* = 6A_0\sqrt{\pi W/d}$.

[C20.1] Tada, H., Paris, P.C., Irwin, G.R., The stress analysis of cracks handbook, Del Research Corporation, 1986.

C21

Special problems

C21.1 The first derivative of the weight function

For the application of the Petroski-Achenbach procedure (Section A3.2.1) and the direct adjustment method (Section A3.2.2) disappearing second and third derivatives of the crack displacement field and the weight function allowed determining higher-order coefficients. As a further condition, a disappearing first derivative of the crack opening displacements at the crack mouth was proposed [C21.1]

$$\left.\frac{\partial h}{\partial x}\right|_{x=0} = 0 \qquad\qquad (C21.1.1)$$

The validity of this condition can be proved for edge-cracked semi-infinite bodies [C21.2] and infinitely long double-edge cracked plates [C21.3]. In this Section, the second possibility is addressed.

C21.1.1 Double edge-cracked strip of infinite length

In order to compute the first derivative of the weight function for the double-edge-cracked strip, first this specimen is loaded by pairs of forces P acting directly at the crack faces (Fig. C21.1).

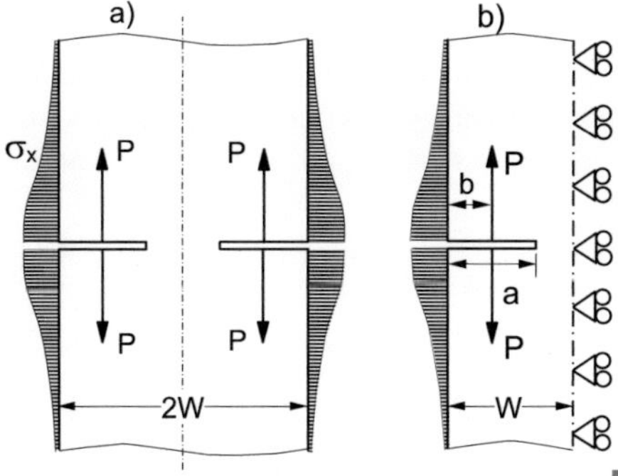

Fig. C21.1 a) Double-edge-cracked plate, b) symmetry conditions (shaded areas: surface tractions which reproduce the stress state of the internal cracks for the same load, see Fig. C21.2).

The specimen is assumed to be cut out of an infinite array of collinear cracks in an infinite body (see Fig. C21.2), which are also loaded by concentrated forces at the crack faces. The cutting lines are the dash-dotted lines of Fig. C21.2. The total stress state in such a structure can be derived from the Westergaard stress function (see e.g. [C21.4])

$$\Phi = \frac{2P}{W} \frac{\cos(\pi b / W)\sqrt{\sin^2(\pi a / W) - \sin^2(\pi b / W)}}{[\sin^2(\pi z / W) - \sin^2(\pi b / W)]\sqrt{1 - \sin^2(\pi a / W)/\sin^2(\pi z / W)}} \qquad (C21.13.2)$$

with the geometric data a, b, and W as illustrated in Fig. C21.2. The complex variable z is given by $z = x + iy$ with the origin of x, y in the crack centre.
The σ_x-stresses are given by

$$\sigma_x(x, y, b) = \operatorname{Re}\Phi - y\operatorname{Im}\frac{d}{dz}\Phi \qquad (C21.1.3)$$

The tractions, which have to be applied at the surface $x=0$ of the double-edge-cracked plate to satisfy the displacement conditions in the case of the internal crack are

$$\sigma_{appl} = -\sigma_x(0, y, b) \qquad (C21.1.4)$$

These surface tractions (illustrated schematically by the shaded areas in Fig. C21.1) are responsible for a stress intensity factor contribution ΔK_{I} which can be expressed by the weight function formulation

$$\Delta K_{\mathrm{I}} = \int_0^{\infty} \sigma_x(0, y, b)\, h_{x,DE}(a, y)\, dy \qquad (C21.1.5)$$

where $h_{x,DE}$ is the weight function for surface tractions acting in x-direction. The subscript DE stands for "Double Edge". On the other hand, the stress intensity factor caused by the free surface condition can be expressed by

$$\Delta K_{\mathrm{I}} = K_{\mathrm{I},DE} - K_{\mathrm{I},INT} = \int_0^{a}[h_{y,DE}(a, x) - h_{y,INT}(a, x)]\sigma_y(y = 0, x)\, dx \qquad (C21.1.6)$$

with the crack surface tractions σ_y acting normal to the crack and the subscript INT denoting the "Internal Crack". The crack-face weight functions are $h_{y,DE}$ for the double-edge crack and $h_{y,INT}$ for the internal crack.

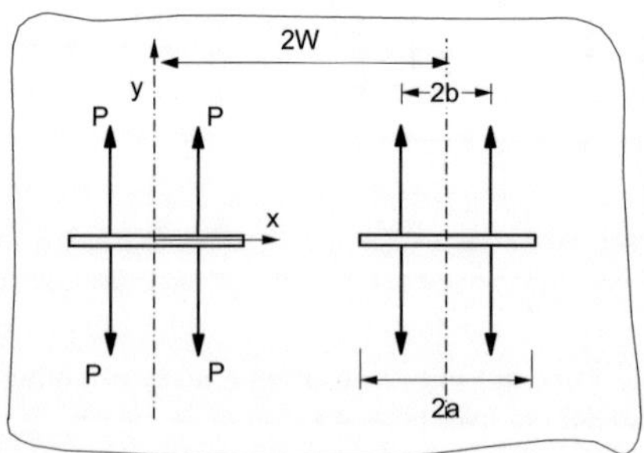

Fig. C21.2 Array of collinear cracks loaded by point forces.

The stress distribution of a pair of forces $\pm P$ acting on the crack surface at $x = b$ is expressed in terms of the Dirac δ function as

$$\sigma_y(x) = P\delta(x - b) \tag{C21.1.7}$$

(assuming the plate thickness $B=1$). Introducing this into eq.(C21.1.6) gives

$$\Delta K_I = Ph_{y,DE}(a,b) - Ph_{y,INT}(a,b) \tag{C21.1.8}$$

and, finally,

$$h_{y,DE}(a,b) = h_{y,INT} + \frac{1}{P}\int_0^\infty \sigma_x(0,y,b)\,h_{x,DE}(a,y)\,dy \tag{C21.1.9}$$

We are now interested in the first derivative of the weight function directly at the free surface. Since the weight function for the internal crack is symmetric at the crack centre,

$$\left.\frac{\partial h_{y,INT}(a,b)}{\partial b}\right|_{b=0} = 0, \tag{C21.1.10}$$

we find

$$\left.\frac{\partial h_{y,DE}(a,b)}{\partial b}\right|_{b=0} = \frac{1}{P}\int_0^\infty h_{x,DE}(a,y)\left.\frac{\partial \sigma_{appl}}{\partial b}\right|_{b=0} dy \tag{C21.1.11}$$

From eq.(C21.13.2) we can see that the Westergaard stress function is symmetric with respect to the real variable b. This is the case because b occurs in the terms of

$$\cos(\pi b/W), \quad \sin^2(\pi b/W)$$

only. Consequently, it holds $\partial\Phi/\partial b = 0$ and we obtain for the derivatives

$$\left.\frac{\partial \sigma_x}{\partial b}\right|_{b=0} = \mathrm{Re}\left.\frac{\partial \Phi}{\partial b}\right|_{b=0} - y\,\mathrm{Im}\left.\frac{d}{dz}\frac{\partial \Phi}{\partial b}\right|_{b=0} = 0 \tag{C21.1.12}$$

From eq.(C21.1.11) it results

$$\left.\frac{\partial h_{y,DE}}{\partial b}\right|_{b=0} = 0 \tag{C21.1.13}$$

i.e. the first derivative of the crack-face weight function of an infinitely long double-edge-cracked strip must disappear at the crack mouth.

C21.1.2 Finite edge-cracked plate

It has to be emphasized that the condition (C21.1.1) is fulfilled only for special problems. In most crack problems, eq.(21.1.1) is invalid. This can be shown for the case of a single edge-cracked strip of infinite length. Figure C21.3a shows the well-established weight function derived by Kaya and Erdoan [C21.5]. A monotonically increasing non-disappearing slope at x/a=0 is clearly visible. Figure C21.3b represents the first derivative of h at x=0 as a function of the relative crack length a/W. The first derivative h' vanishes only for a/W=0, i.e. for the edge crack in the half space.

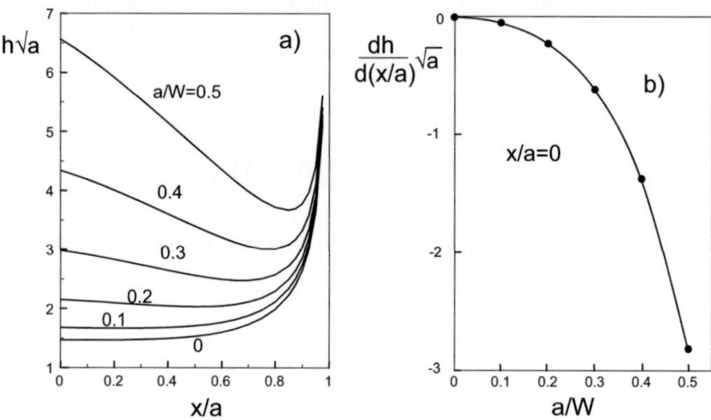

Fig. C21.3 a) Weight function solution by Kaya and Erdogan [C21.5], b) first derivative of the weight function.

C21.2 Limit values for stress intensity factor and T-stress

Limit case a/W=0

The limit values of stress intensity factor and T-stress can be computed according to the analysis made by Wigglesworth [C21.6]. The exact value of the geometric function F for the semi-infinite body under constant crack-face pressure p is given by

$$K = pF\sqrt{\pi a}, \quad F = \exp\left[\frac{1}{\pi}\int_0^\infty \ln\left(\frac{1}{1-u^2\cosech^2(\pi u/2)}\right)\frac{du}{1+u^2}\right] \tag{C21.2.1}$$

resulting in [C21.7]

$$F = 1.12152225523... \tag{C21.2.2}$$

The T-stress under remote tensile loading σ reads (see Section A2)

$$T = -4\sigma_\infty B_0 = -0.5259676026\,\sigma_\infty \tag{C21.2.3}$$

and the x-stress under constant crack-face pressure (for the difference see section A1.4)

$$\sigma_{xx,0} = T + p = 0.4740324\,p \tag{C21.2.4}$$

Limit case a/W=1

The limit case $a/W = 1$ can be analytically evaluated for the double-edge-cracked plate. This may be done here for the two cracks loaded by a constant crack-face pressure p over the region $c \leq x \leq b$. With the geometric data of Fig. C21.4 the Westergaard stress function [C21.4] reads

$$Z = \frac{2}{\pi} p \left[\sqrt{\frac{b^2 - c^2}{c^2 - z^2}} - \arctan \sqrt{\frac{b^2 - c^2}{c^2 - z^2}} \right] \tag{C21.2.5}$$

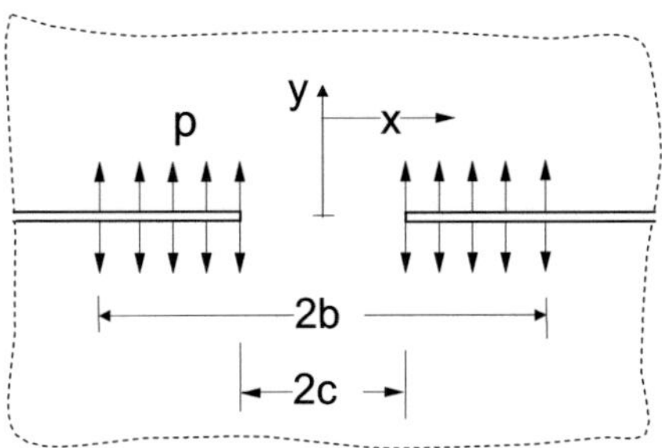

Fig. C21.4 Double-edge-cracked infinite body.

The real part of Z gives the x-stress component at $y = 0$

$$\sigma_x \big|_{y=0} = \mathrm{Re}\, Z \big|_{y=0} = \frac{2}{\pi} p \left[\sqrt{\frac{b^2 - c^2}{c^2 - x^2}} - \arctan \sqrt{\frac{b^2 - c^2}{c^2 - x^2}} \right] \tag{C21.2.6}$$

which for $(b/c)^2 \gg 1$ simplifies as

$$\sigma_x \big|_{y=0} = \frac{2}{\pi} p \left[\sqrt{\frac{b^2 - c^2}{c^2 - x^2}} - \frac{\pi}{2} \right] \tag{C21.2.7}$$

The singular stress term is

$$\sigma_{x,sing} = \frac{p\sqrt{2}}{\pi\sqrt{c}} \sqrt{\frac{b^2 - c^2}{c - x}} \tag{C21.2.8}$$

and, consequently, the regular part of (C21.2.7) is given by

$$\sigma_{x,reg} = \frac{2p}{\pi} \sqrt{b^2 - c^2} \left(\frac{1}{\sqrt{c^2 - x^2}} - \frac{1}{\sqrt{2c(c - x)}} \right) - p \tag{C21.2.9}$$

Neglecting the constant term $-p$ and rearranging yields

$$\sigma_{x,reg} = \frac{2p}{\pi} \sqrt{\frac{b^2 - c^2}{c^2}} \left(\frac{1}{\sqrt{1 - (x/c)^2}} - \frac{1}{\sqrt{2(1 - x/c)}} \right) \tag{C21.2.10}$$

and in the limit for $x \to c$

$$\sigma_{x,reg,x\to c} = \frac{2p}{\pi}\sqrt{\frac{b^2-c^2}{c^2}}\sqrt{\frac{1-x/c}{32}} \qquad\qquad \text{(C21.2.11)}$$

Finally, it results by introducing the crack length $a = b\text{-}c$ and extending $2b$ over the whole specimen width

$$\lim_{a\to b} T(1-a/b) = \lim_{\alpha\to 1} T(1-\alpha) = 0 \qquad\qquad \text{(C21.2.12)}$$

This result allows extrapolating the data compiled in Table C11.1 and C12.4. In Fig. C21.5 the product $T(1\text{-}\alpha)$ is plotted for the double-edge-cracked circular disk under constant normal stresses along the circumference (solid circles) and the double-edge-cracked plate with $H/W \geq$ 1.5 (open circles). The value given by (C21.2.12) allows extrapolating the data to $\alpha\to 1$.

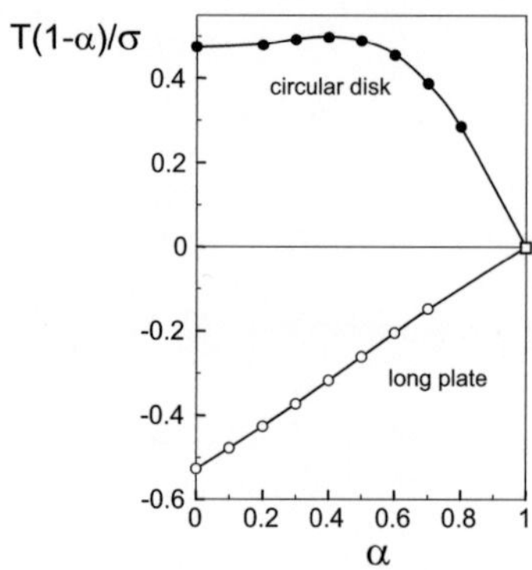

Fig. C21.5 Data compiled in Tables C11.1 and C12.4 (circles) and limit value from (C21.2.12) (square).

References C21

[C21.1] Shen, G., Glinka, G., Determination of weight functions from reference stress intensity factors, Theor. and Appl. Fract. Mech. **15**(1991), 237-245.

[C21.2] Fett, T., Munz, D., Stress intensity factors and weight functions, Computational Mechanics Publications, Southampton, 1997.

[C21.3] Fett, T., Stress intensity factors, T-stress and weight functions for double-edge-cracked plates, Forschungszentrum Karlsruhe, Report FZKA 5838, 1996, Karlsruhe.

[C21.4] Tada, H., Paris, P.C., Irwin, G.R., The stress analysis of cracks handbook, Del Research Corporation, 1986.

[C21.5] Kaya, A.C., Erdogan, F., Stress intensity factors and COD in an orthotropic strip, Int. J. Fract **16**(1980), 171-190.

[C21.6] Wigglesworth, L.A., Stress distribution in a notched plate, Mathematica **4**(1957), 76-96.

[C21.7] Fett, T., Rizzi, G., Bahr, H.A., Bahr, U., Pham, V.-B., Balke, H., Analytical solutions for stress intensity factor, T-stress and weight function for the edge-cracked half-space, Int. J. Fract., Letters in Fracture and Micromechanics, DOI 10.1007/s10704-007-9152-8.

<u>C22</u>

Zones with volume change at crack surfaces

C22.1 Green's functions for a zone in the crack wake

In many cases local strains occur near the crack surfaces. An example is shown in Fig. C22.1 where only the region behind the crack tip is affected.

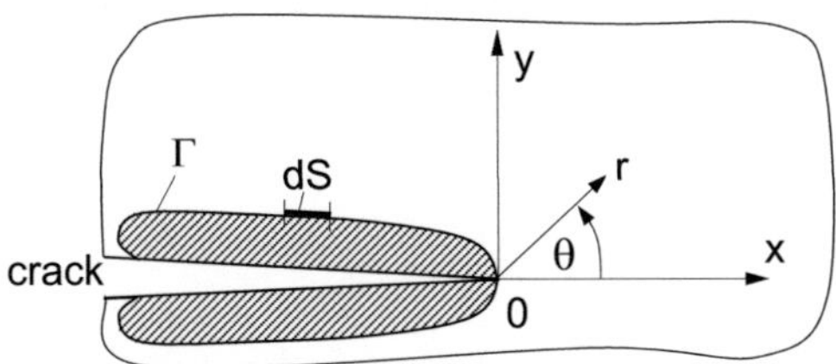

Fig. C22.1 Crack with a crack face zone undergoing volumetric strains.

Such strains in the crack surface region can be generated for instance by a volume reduction or expansion due to diffusion effects. In the special case of glass, there is experimental evidence [C22.1] for a thin hydration layer at the crack surfaces in which an ion exchange occurs.

C22.1.1 Mode-I stress intensity factor

McMeeking and Evans [C22.2] developed a procedure for the computation of mode-I stress intensity factors. The mode-I contribution results from a contour integral

$$K_{\mathrm{I}} = p \int_{\Gamma} \mathbf{n} \cdot \mathbf{h}\,dS \qquad (C22.1.1)$$

with the normal vector $\mathbf{n}$ on the zone contour and the normal pressure p defined by

$$p = \frac{\varepsilon E}{3(1-2v)}, \qquad (C22.1.2)$$

where E is Young's modulus, v Poisson's ratio, and ε the volumetric strain. Γ is the contour line of the zone and dS is a line length increment. The vector $\mathbf{h}$ represents the weight function $\mathbf{h}_{\mathrm{I}} = (h_{\mathrm{I},y}, h_{\mathrm{I},x})^{\mathrm{T}}$ with the components $h_{\mathrm{I},y}$ and $h_{\mathrm{I},x}$ [C22.2]

$$h_{\mathrm{I},x} = \frac{1}{\sqrt{8\pi r}(1-v)}[2v-1+\sin(\theta/2)\sin(3\theta/2)]\cos(\theta/2) \qquad (C22.1.3a)$$

$$h_{\mathrm{I},y} = \frac{1}{\sqrt{8\pi r}(1-v)}[2-2v-\cos(\theta/2)\cos(3\theta/2)]\sin(\theta/2). \qquad (C22.1.3b)$$

In these relations r and θ are the polar coordinates with the origin at the crack tip (see Fig. C22.1). Using the Gauss theorem, eq.(C22.1.1) can be rewritten [C22.2] as

$$K_I = p \int_{(A)} \operatorname{div} \mathbf{h}_I \, dx\, dy \qquad\qquad (\text{C22.1.4})$$

where A is the area in the x-y plane (extending above <u>and</u> below the crack plane).

For a numerical evaluation of the stress intensity factor K_I, the total zone can be divided in parts of simple geometry as shown in Fig. C22.2.

It may be of advantage to carry out the integration over y from the crack surface to the zone height b. For the case of a zone located symmetrically above and below the crack ($dA= -2b\times dx=2b\times ds$, see insert in Fig. C22.3a),

$$dK_I = \frac{\varepsilon E}{1-v}\sqrt{b}\, g_I(s/b)\, d(s/b) \qquad\qquad (\text{C22.1.5})$$

with the function g_I shown in Fig. C22.3a.

The finite element method was used to determine K. A volume strain was introduced by a thermal expansion due to a localised temperature change. The temperature inside the zone segment $b\times\Delta s$ was chosen to be $\Theta= 1°$, whereas zero temperature was prescribed in the remaining structure.

The results are given in Fig. C22.3a. The asymptotic behaviour of K_I is given by the two straight lines. An interpolation of these asymptotes, for example, is given by

$$g_I \approx -\frac{1}{4\sqrt{s/b} + 5(s/b)^{3/2} + 10(s/b)^{5/2}} \qquad\qquad (\text{C2.1.6})$$

from which

$$K_I = \frac{E}{1-v}\int_0^{s_1}\varepsilon(s')\, g_I(s'/b')\frac{1}{\sqrt{b'}}\, d(s') \qquad\qquad (\text{C22.1.7})$$

can be derived by a single integration for any arbitrarily varying zone height $b(s)$.

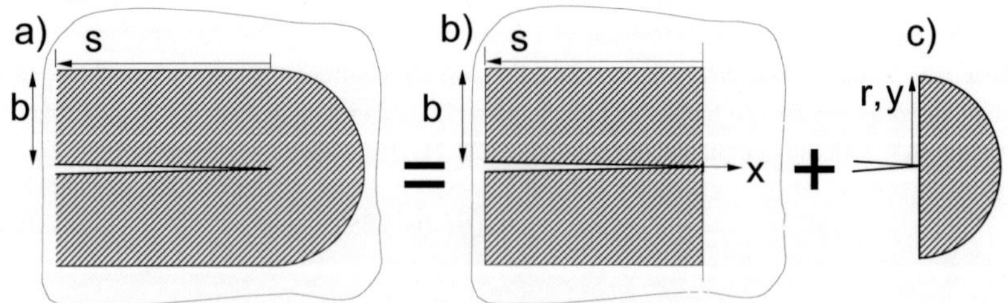

Fig. C22.2 Composition of zones by superposition of the zone part ahead the crack tip and zone parts in the crack wake.

C22.1.2 Mode-II stress intensity factor

In the case of a volume strain zone which is not developed symmetrically to the crack plane, also a mode-II stress intensity factor must result. The mode-II weight functions can be obtained from the near-tip displacement field for mode-II loading and the Rice equation [C22.3] which relates the weight function to the change of displacement for a virtual crack extension, the result being

$$h_{\mathrm{II},x} = \frac{1}{\sqrt{32\pi r}(1-v)}[4 - 4v + \cos(\theta) + \cos(2\theta)]\sin(\theta/2) \qquad \text{(C22.1.8a)}$$

$$h_{\mathrm{II},y} = \frac{1}{\sqrt{32\pi r}(1-v)}[2 - 4v + \cos(\theta) - \cos(2\theta)]\cos(\theta/2) \qquad \text{(C22.1.8b)}$$

The mode-II stress intensity factor K_{II} for a zone segment of length ds lying on one side of the crack is plotted in Fig. C22.3b, where now g_{II} is defined by

$$K_{\mathrm{II}} = \frac{E}{1-v} \int_0^{s_1} \varepsilon\, g_{\mathrm{II}}(s'/b') \frac{1}{\sqrt{b'}} d(s') \qquad \text{(C22.1.9)}$$

The asymptotes of g_{II} are introduced in Fig. C22.3b as the straight lines. An interpolation for the full range of s/b is

$$g_{\mathrm{II}} \cong \frac{1}{\frac{21}{2} + 15(s/b)^{3/2}} \qquad \text{(C22.1.10)}$$

The corresponding mode-I stress intensity factor for this non-symmetric zone is half of the stress intensity factor computed for the symmetric case (see Fig. C22.3a).

C22.1.3 T-stress

The T-stress results are shown in Fig. C22.3c. An approximate interpolation relation for the Green's function is given by

$$g_T \cong -\frac{1}{5} \frac{1}{1 + \frac{7}{8}(s/b)^2} \qquad \text{(C22.1.11)}$$

resulting in the T-stress

$$T = \frac{E}{1-v} \int_{s_0}^{s_1} \varepsilon(s') g_T(s'/b') \frac{1}{b'} ds' \qquad \text{(C22.1.12)}$$

A layer of constant height b (extending from $s_0 > 0$ to $s = s_1$ and excluding the crack tip), Fig. C22.2b, causes the T-stress

$$T \cong -\frac{1}{5}\sqrt{\frac{8}{7}}\frac{\varepsilon E}{1-\nu}\left(\arctan\left[\sqrt{\frac{7}{8}}\frac{s_1}{b}\right]-\arctan\left[\sqrt{\frac{7}{8}}\frac{s_0}{b}\right]\right)$$

$$\xrightarrow{s_0\to 0} -\frac{1}{5}\sqrt{\frac{8}{7}}\frac{\varepsilon E}{1-\nu}\arctan(\sqrt{\frac{7}{8}}\,s_1/b)$$

(C22.1.13)

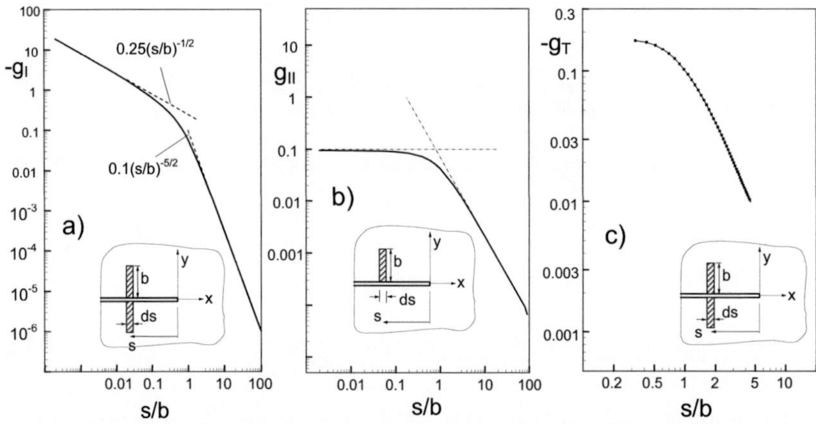

Fig. C22.3 a) Green's function g_I as the stress intensity factor for a symmetric strip-shaped zone of height b and width ds, b) mode-II stress intensity factor contribution for a single zone located on one side of the crack exclusively, c) T-stress for a symmetric zone.

C22.2 Semi-circular crack tip zone

For the semi-circular zone ahead of the crack tip (Fig. C22.2c), the stress intensity factor

$$K_I \cong 0.1214\frac{\varepsilon E}{1-\nu}\sqrt{b}$$

(C22.2.1)

and the T-stress term

$$T \approx 0.17\frac{\varepsilon E}{1-\nu} \approx \tfrac{1}{6}\frac{\varepsilon E}{1-\nu}$$

(C22.2.2)

were obtained.

C22.3 Zone of constant height

C22.3.1 Mode-I stress intensity factor

This special case of a zone of constant height can be treated simply by the direct evaluation of eq.(C22.1.1). Numerical evaluation of the integral expression yields

$$K_I = -C_I \frac{\varepsilon E}{1-v}\sqrt{b} \;\overset{s\to\infty}{\longrightarrow}\; -0.374\frac{\varepsilon E}{1-v}\sqrt{b} \qquad (C22.3.1)$$

for a constant strain ε in the crack wake (case Fig. C22.2b). The coefficient C_I can be written

$$C_I = 0.119017\left[\arctan\left(\frac{2.105\sqrt{s/b}}{1-A_1 s/b}\right)+\phi\right]+0.090393\ln\frac{1+A_2\sqrt{s/b}+A_1 s/b}{1-A_2\sqrt{s/b}+A_1 s/b} \qquad (C22.3.2)$$

with $\qquad A_1 = 1.58114,\; A_2 = 1.38285,\; \phi = \begin{cases} 0 & s/b < 1/A_1 \\ \pi & s/b > 1/A_1 \end{cases}$

In the case of a zone as shown in Fig. C22.2a, the stress intensity factor of eq.(C22.2.1) must be added.

C22.3.2 Mode-II stress intensity factor

As an example, the mode-II stress intensity factor K_{II} is computed for the case of a zone of constant height b at the upper side of the crack extending from $s=0$ to $s=s_1$ (Fig. C22.2b).

$$K_{II} = C_{II}\frac{\varepsilon E}{1-v}\sqrt{b} \qquad (C22.3.3a)$$

The coefficient C_{II} reads

$$C_{II} \cong 0.00834\left\{\sqrt{3}\pi - 6\sqrt{3}\arctan\left[\frac{1}{\sqrt{3}}\left(1-2(10/7)^{1/3}\sqrt{\frac{s_1}{b}}\right)\right]+\right.$$
$$\left. +3\ln 7 + 3\ln[7+7^{1/3}10^{2/3}\sqrt{s_1/b}] - 6\ln\left[7+7^{2/3}10^{1/3}\sqrt{\frac{s_1}{b}}\right]\right\} \qquad (C22.3.3b)$$

C22.3.3 T-stress

The T-stress for the fully developed zone (see Fig. C22.2a) with a layer length of $s=s_1$ (the crack tip excluded) results from (C22.1.13) and (C22.2.2) as

$$T \cong (0.17 - \tfrac{1}{5}\sqrt{8/7}\arctan(\sqrt{7/8}\,s_1/b))\frac{\varepsilon E}{1-v} \qquad (C22.3.4a)$$

For Fig. C22.2a with $s_1/b\to\infty$ the FE-evaluation resulted in

$$T \cong -0.1656\frac{\varepsilon E}{1-v} \approx -\tfrac{1}{6}\frac{\varepsilon E}{1-v} \qquad (C22.3.4b)$$

C22.3.4 Re-starting arrested crack

An arrested crack in glass generates an ion exchange layer as illustrated in the upper part of Fig. C22.4a. After an increase of the externally applied stress intensity factor, the crack may grow rapidly without a significant further exchange layer generation. The crack extension is shown in the lower part Fig.C22.4a. The stress intensity factor and the T-stress are represented in Figs. C22.4b and C22.4c. When the crack leaves the initial zone, the T-stress changes sign.

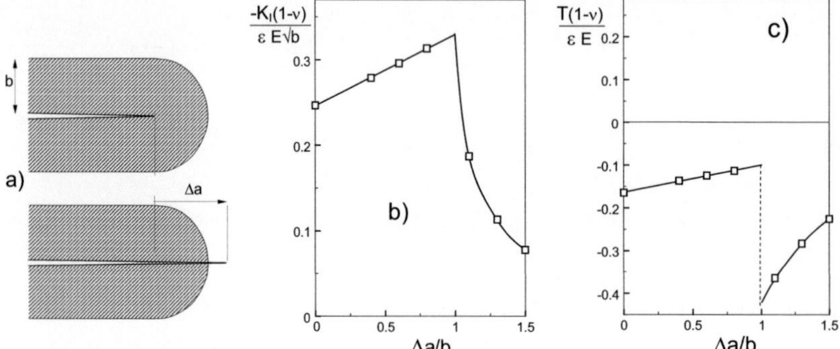

Fig. C22.4 a) Crack growth in the initial zone, b) mode-I stress intensity factor, c) T-stress.

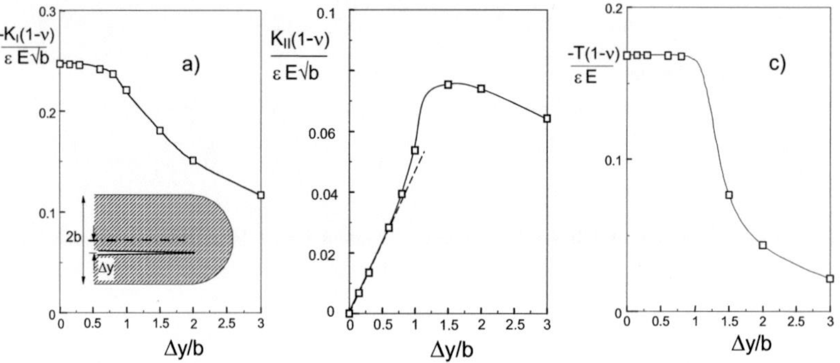

Fig. C22.5 Mixed-mode stress intensity factors caused by a misalignment of the crack within the layer, a) stress intensity factor K_{I}, b) stress intensity factor K_{II}, c) T-stress.

C22.3.5 Non-symmetrically aligned crack

Figure C22.5a shows a crack with a misalignment Δy with respect to the symmetry axis. Such cases may occur for ion exchange layers in glass due to statistical thickness fluctuations. In Fig. C22.5a, the mode-I stress intensity factor K_{I} is plotted. Figures C22.5b and C22.5c represent K_{II} and T.

For small values of $\Delta y/b$, the initial-straight line behaviour of K_{II} may be expressed as

$$K_{\mathrm{II}} \cong \tfrac{7}{150}\frac{\Delta y}{b}\frac{\varepsilon E}{1-\nu}\sqrt{b} \tag{C22.3.5}$$

C22.4 Variable layer height

The height of an ion exchange layer in glass under a constant crack growth rate (Fig. C22.6a) is

$$b = \alpha\sqrt{s} \qquad (C22.4.1)$$

giving rise to the fracture mechanics parameters

$$K_{\mathrm{I}} = -\frac{\varepsilon E}{1-\nu}\frac{4\alpha}{3\sqrt{15}}\arctan\left[\frac{\frac{4}{3}\sqrt{\frac{5}{3}}\sqrt{s_1}/\alpha}{1+\frac{5}{27}(1+4\sqrt{s_1}/\alpha)}\right] \qquad (C22.4.2)$$

$$T \cong -\frac{2}{5}\sqrt{\frac{8}{7}}\frac{\varepsilon E}{1-\nu}\arctan\left(\frac{7}{8}\sqrt{s_1}/\alpha\right) \qquad (C22.4.3)$$

(with the crack tip excluded).

In the case of a zone developed at one crack face exclusively (Fig. 22.6b),

$$K_{\mathrm{II}} = \frac{\varepsilon E}{1-\nu}\frac{4}{45}\alpha\ln\left(1+\frac{10}{7}\frac{s_1^{3/4}}{\alpha^{3/2}}\right) \qquad (C22.4.4)$$

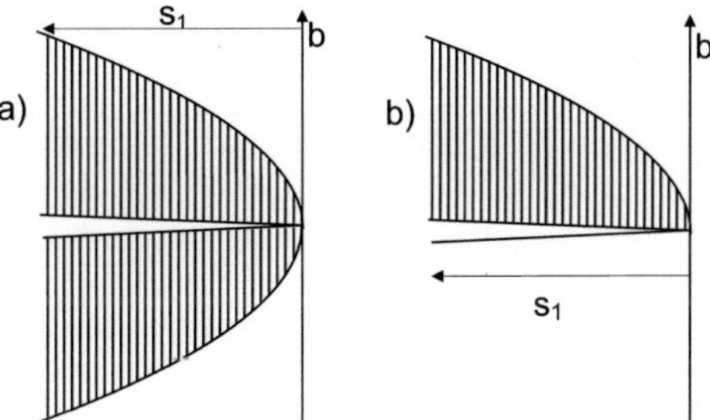

Fig. C22.6 Square-root shaped volume strain regions as occurring for cracks in glass growing at constant rate.

<u>REMARK:</u>

In Ref.[C22.4] the T-stresses for expanding zones were computed replacing the general strain by a thermal expansion. For this purpose the authors of [C22.4] used the FE-programs ABAQUS Version 6.2 and 6.4.

In the newer ABAQUS versions, starting with 6.5-6 the T-stress evaluation for identical input files then showed a change in the sign keeping the absolute values the same.

In this context it should be mentioned that ABAQUS detected an error in the T-stress program and corrected it in March 2005 (see error message v63_4854 by the *bugadmin*, March 17, 2005). This change results in a step-shaped variation of T by an amount of δT

$$\delta T = -\tfrac{1}{3}\frac{\varepsilon E}{1-\nu}$$

depending on the condition whether the crack tip is included or excluded in the volume undergoing volumetric expansion. For this fact see also Section A1.4 and eq.(A1.4.17) and Section E1 in [C22.5].

FE-results obtained with older ABAQUS versions have therefore to be corrected. This may have been done in this section for the erroneous results reported in [C22.4]. Since the sign of T-stresses is very important for path stability considerations, these inconsistencies and the necessary changes should be taken into account in the assessment of problems dealing with thermal stresses.

References C22

[C22.1] Bunker, B.C., Michalske, T.A, Effect of Surface Corrosion on Glass Fracture, pp. 391-411 in *Fracture Mechanics of Ceramics, Vol. 8, Microstructure, Methods, Design and Fatigue*, R.C. Bradt, A.G. Evans, D.P.H. Hasselman and F.F. Lange Eds., Plenum Press, New York (1986).

[C22.2] McMeeking, R.M., Evans, A.G., Mechanics of transformation-toughening in brittle materials, J. Am. Ceram. Soc. **65**(1982), 242–246.

[C22.3] Rice, J.R., Some remarks on elastic crack-tip stress fields, Int. J. Solids and Structures **8**(1972), 751-758.

[C22.4] T. Fett, G. Rizzi, Fracture mechanics parameters of crack surface zones under volumetric strains, Int. J. Fract. **127**(2004), L117-L124.

[C22.5] Fett, T., Stress intensity factors, T-stresses, weight functions, -Supplement Volume-, Universitätsverlag Karlsruhe, Voume IKM55 (2009).

C23

Tetrahedron-shaped cracks

C23.1 Complete tetrahedron

A tetrahedron-shaped crack in a plate of thickness B is shown in Fig. C23.1. By finite element computations, the Green's functions were determined for mixed-mode stress intensity factors and T-stress. Therefore, point forces P normal to the crack and Q in crack direction were applied.

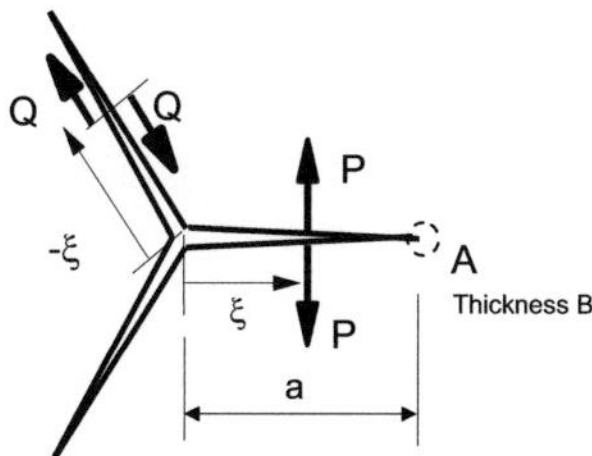

Fig. C23.1 Tetrahedron-shaped crack with normal and tangential point forces.

Under a combined crack-face loading, the stress intensity factors generally read

$$K_{\mathrm{I}} = \int_{-a}^{a}[h_{\mathrm{I}}^{(1)}(\xi,a)\sigma_n(\xi) + h_{\mathrm{I}}^{(2)}(\xi,a)\tau(\xi)]d\xi \qquad (C23.1.1)$$

$$K_{\mathrm{II}} = \int_{-a}^{a}[h_{\mathrm{II}}^{(1)}(\xi,a)\sigma_n(\xi) + h_{\mathrm{II}}^{(2)}(\xi,a)\tau(\xi)]d\xi \qquad (C23.1.2)$$

with the weight functions

$$h_{\mathrm{I}}^{(1)} = \sqrt{\frac{1+\xi/a}{\pi a}}\left[\frac{1}{\sqrt{1-\xi/a}} + A_{\mathrm{I}}^{(1)}\sqrt{1-\xi/a} + B_{\mathrm{I}}^{(1)}(1-\xi/a)^{3/2} + C_{\mathrm{I}}^{(1)}(1-\xi/a)^{5/2}\right] \qquad (C23.1.3)$$

$$h_{\mathrm{I}}^{(2)} = \sqrt{\frac{1+\xi/a}{\pi a}}\left[A_{\mathrm{I}}^{(2)}\sqrt{1-\xi/a} + B_{\mathrm{I}}^{(2)}(1-\xi/a)^{3/2} + C_{\mathrm{I}}^{(2)}(1-\xi/a)^{5/2}\right] \qquad (C23.1.4)$$

$$h_{\mathrm{II}}^{(1)} = \sqrt{\frac{1+\xi/a}{\pi a}}\left[A_{\mathrm{II}}^{(1)}\sqrt{1-\xi/a} + B_{\mathrm{II}}^{(1)}(1-\xi/a)^{3/2} + C_{\mathrm{II}}^{(1)}(1-\xi/a)^{5/2}\right] \qquad (C23.1.5)$$

$$h_{\mathrm{II}}^{(2)} = \sqrt{\frac{1+\xi/a}{\pi a}}\left[\frac{1}{\sqrt{1-\xi/a}} + A_{\mathrm{II}}^{(2)}\sqrt{1-\xi/a} + B_{\mathrm{II}}^{(2)}(1-\xi/a)^{3/2} + C_{\mathrm{II}}^{(2)}(1-\xi/a)^{5/2}\right] \qquad (C23.1.6)$$

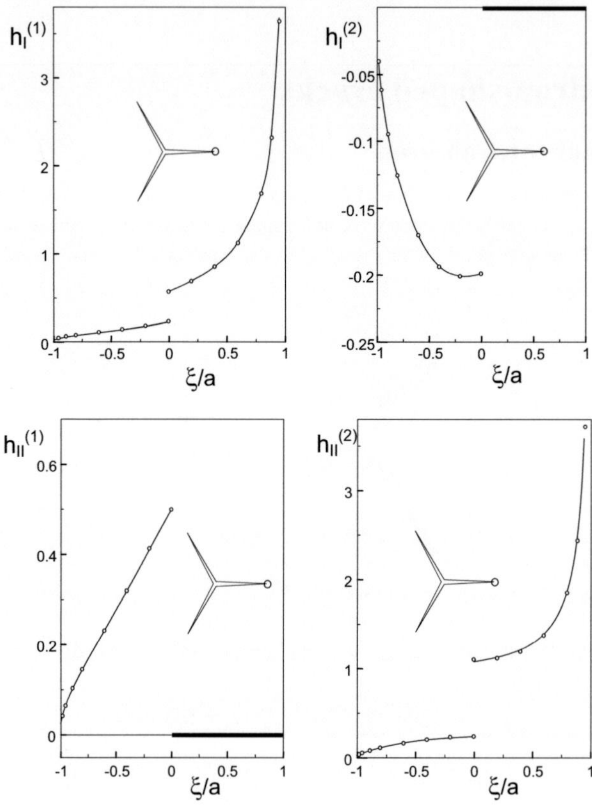

Fig. C23.2 Mixed-mode weight functions for tetrahedron-shaped cracks.

Table C23.1 Coefficients for weight functions eqs.(C23.1.3)-(C23.1.6) (computed with ν=0.25).

	A	B	C
$\xi<0$	$A_{\mathrm{I}}^{(1)}=-1.125$	$B_{\mathrm{I}}^{(1)}=0.6402$	$C_{\mathrm{I}}^{(1)}=-0.1228$
$\xi>0$	$A_{\mathrm{I}}^{(1)}=-0.041$	$B_{\mathrm{I}}^{(1)}=0.0385$	$C_{\mathrm{I}}^{(1)}=0$
$\xi<0$	$A_{\mathrm{II}}^{(1)}=1.8822$	$B_{\mathrm{II}}^{(1)}=-1.2319$	$C_{\mathrm{II}}^{(1)}=0.2347$
$\xi>0$	$A_{\mathrm{II}}^{(1)}=0$	$B_{\mathrm{II}}^{(1)}=0$	$C_{\mathrm{II}}^{(1)}=0$
$\xi<0$	$A_{\mathrm{I}}^{(2)}=-0.1879$	$B_{\mathrm{I}}^{(2)}=-0.2416$	$C_{\mathrm{I}}^{(2)}=0.0773$
$\xi>0$	$A_{\mathrm{I}}^{(2)}=0$	$B_{\mathrm{I}}^{(2)}=0$	$C_{\mathrm{I}}^{(2)}=0$
$\xi<0$	$A_{\mathrm{II}}^{(2)}=-1.900$	$B_{\mathrm{II}}^{(2)}=1.8151$	$C_{\mathrm{II}}^{(2)}=-0.4950$
$\xi>0$	$A_{\mathrm{II}}^{(2)}=0.3142$	$B_{\mathrm{II}}^{(2)}=0.5885$	$C_{\mathrm{II}}^{(2)}=0$

The mixed-mode stress intensity factors are shown in Fig. C23.2 and the T-stress in Fig. C23.3. The coefficients for eqs.(C23.1.3)-(C23.1.6) are compiled in Table C23.1. In all cases of $\xi<0$, the data correspond to load on the upper crack part (as illustrated for Q in Fig. C23.1). For the lower crack part, the symmetry of K_I and T and the anti-symmetry of K_II have to be taken into account.

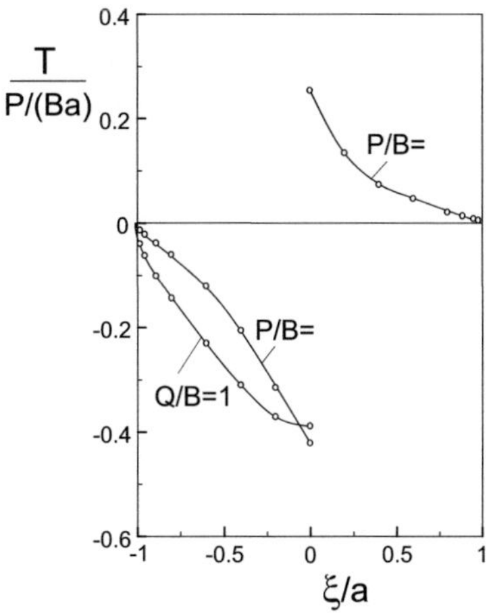

Fig. C23.3 T-stress for tetrahedron-shaped cracks.

C23.2 Incomplete tetrahedron

Figure C23.4 shows a modification of the tetrahedron-shaped crack with one crack part missing. The stress intensity factors are shown in Fig. C23.5

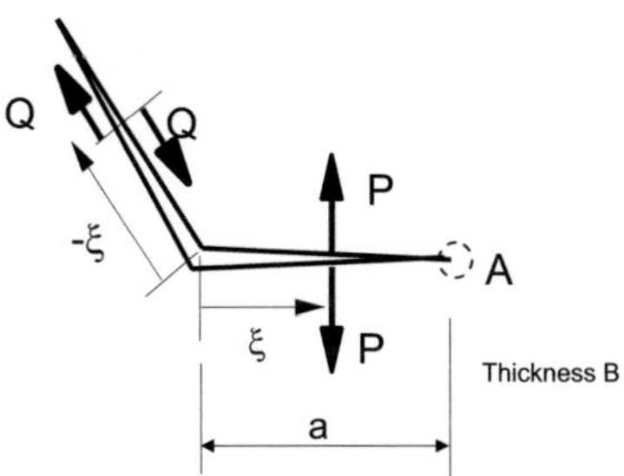

Fig. C23.4 Two cracks of equal length under an angle of 120°.

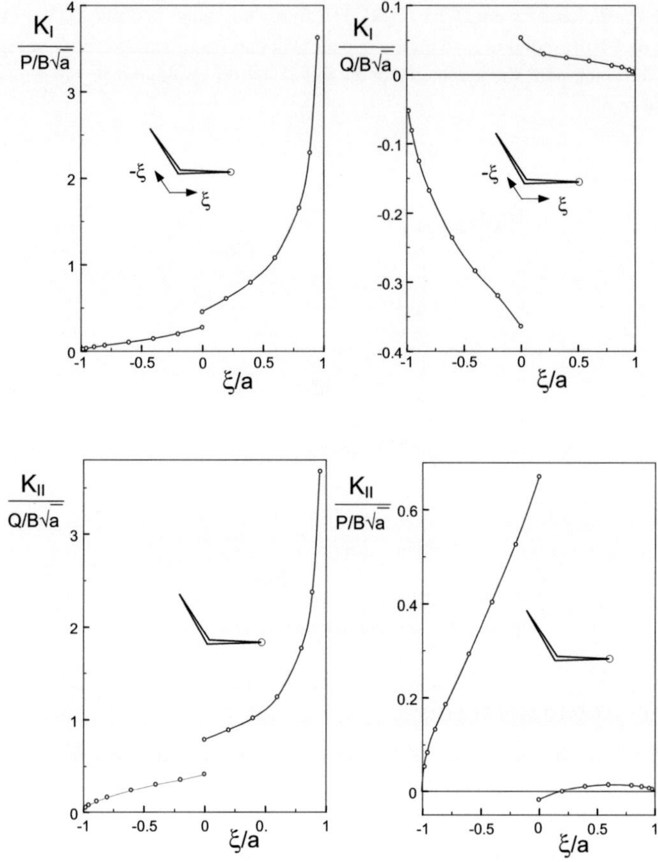

Fig. C23.5 Stress intensity factors for an incomplete tetrahedron-shaped crack.

C24

An example for kinked cracks in a finite body

C24.1 DCDC specimen with kinks at both cracks

The stress intensity factors of kinked cracks in the DCDC specimen (Fig. C24.1, for straight cracks see Section C15) were determined for $H/R=4$ and $a/R=4$ by finite element computations. The results are shown in Fig. C24.2 by the symbols as a function of the kink length ℓ and the kink angle φ.

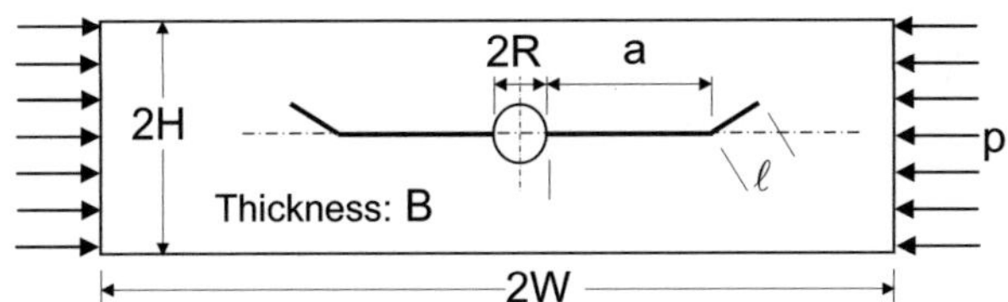

Fig. C24.1 DCDC specimen with symmetrically kinked cracks.

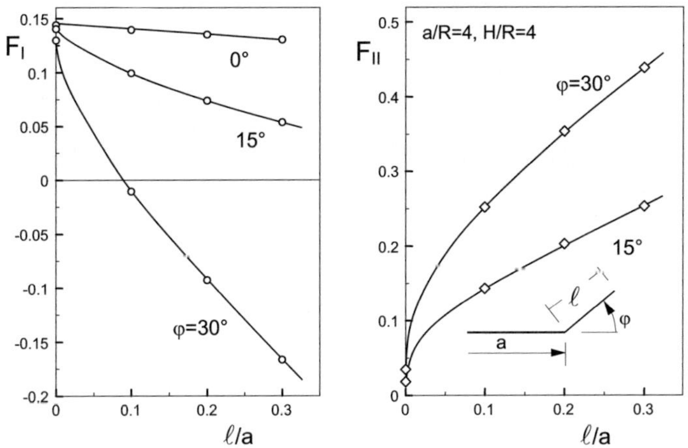

Fig. C24.2 Mixed-mode stress intensity factors for a kinked crack (Fig. C24.1). Geometry: $a/R=4$, $H/R=4$.

C24.2 DCDC specimen with a kink at one of the cracks

A kink at only one side is illustrated in Fig. C24.3. Figures C24.4a and C24.4b show the mixed-mode stress intensity factors for a kink angle of $\varphi=5°$ and the geometric parameters $a/R=4$ and $H/R=4$. The kink is located at point (A) and the stress intensity factors are evaluated for the crack tip at location (B). The mode-I stress intensity factor at (B) is hardly affected by the kink. The mode-II stress intensity factor is roughly proportional to the kink length ℓ.

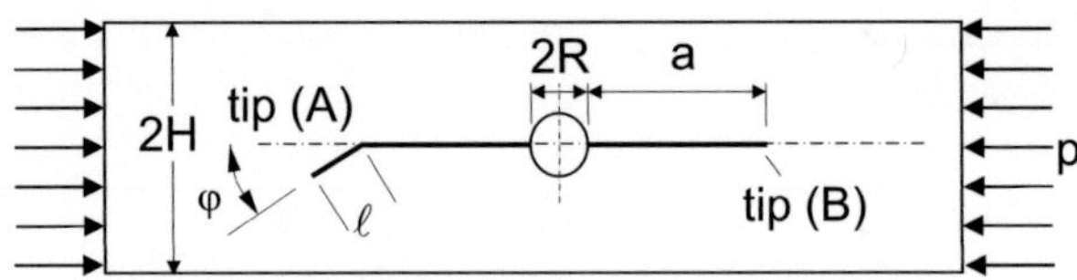

Fig. C24.3 DCDC specimen with a kink at point (A).

Figures C24.4c and C24.4d show the influence of the kink angle φ [C24.1]. The mode-II stress intensity factor K_{II} at point (B) is proportional to the kink angle. From the results of Fig. C24.5 it holds

$$K_{II,(B)} \propto (\ell/a)^2 \varphi \tag{C24.7}$$

The geometric functions for the stress intensity factors at the kink (i.e. at location (A)) are represented in Fig. C24.5. Figures C24.5a and C24.5b show the mixed-mode stress intensity factors for the crack tip located at point (A) The mode-I stress intensity factor is also hardly affected by the kink. In contrast to point (B), the mode-II stress intensity factor is roughly proportional to the square root of the kink length. Figures C24.5c and C24.5d show the influence of the kink angle φ.

The influence of the kink angle can be shown by using the solutions obtained for the semi-infinite crack in an infinite body according to Cotterell and Rice [C24.2].

Under pure mode-I loading, the stress intensity factors for the kink, K_I, and K_{II}, can be expressed roughly by the stress intensity factor for the original (straight) crack, k_I, using the simple expressions of

$$K_I = k_I\, g_{11} + T\, b_1 \sqrt{\ell} \tag{C24.1}$$

$$K_{II} = k_I\, g_{21} + T\, b_2 \sqrt{\ell} \tag{C24.2}$$

with the angular functions

$$g_{11} = \cos^3(\beta/2) \tag{C24.3}$$

$$g_{21} = \sin(\beta/2)\cos^2(\beta/2) \tag{C24.4}$$

and the coefficients b_1 and b_2 given by

$$b_1 = \sqrt{\frac{8}{\pi}}\, \sin^2 \beta \tag{C24.5}$$

$$b_2 = -\sqrt{\frac{8}{\pi}}\, \sin \beta \cos \beta \tag{C24.6}$$

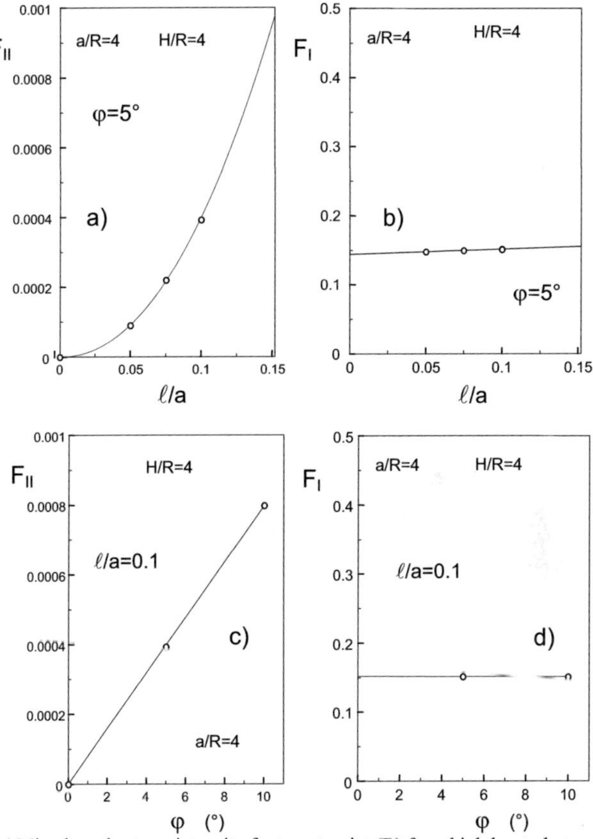

Fig. C24.4 Mixed-mode stress intensity factors at point (B) for a kink located at point (A); a), b) influence of the kink length, c), d) influence of the kink angle, (F according to eq.(2.2)).

The agreement of these asymptotic results with the finite element data of Fig. C24.5a is very good. The strong influence of the kink length in form of the $\sqrt{\ell}$ dependency is an effect of the strongly negative T-stress term (Section C15).

The near-tip solutions of the mode-II stress intensity factor K_{II} at point (A) valid for $\ell/a \to 0$ result from eqs.(C24. 2), (C24. 4) and (C24. 6) as

$$F_{II,(A)} = F_{I,0}\left(g_{21} - \frac{\sqrt{8}}{\pi}\beta\sqrt{\ell/a}\,\sin\varphi\cos\varphi\right) \tag{C24.8}$$

with the mode-I stress intensity factor $F_{I,0}$ in the absence of the kink. This asymptotic result is entered in Fig. C24.5a as the solid curve. It is in sufficient agreement with the FE results (circles).

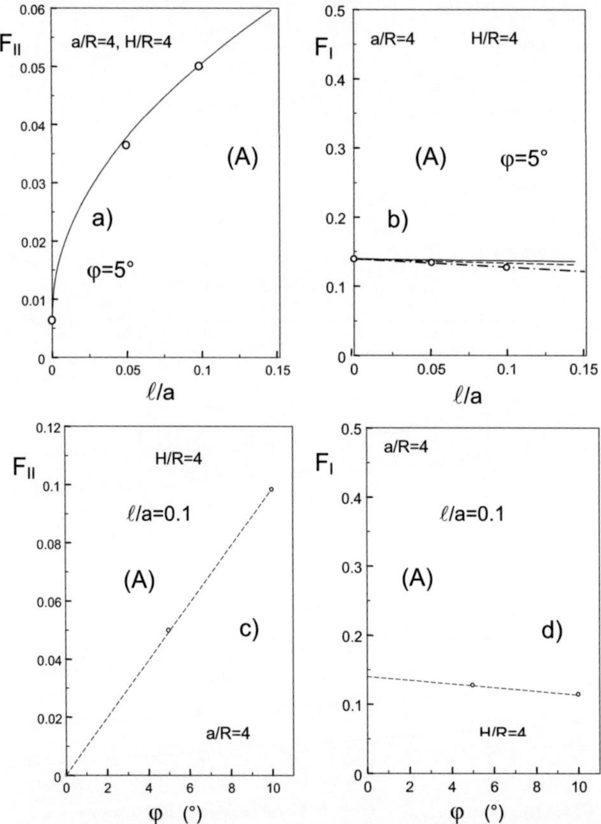

Fig. C24.5 Mixed-mode stress intensity factors at point (A) for a kink located at point (A); a), b) influence of the kink length (symbols: FE results, curves: cumputed from eqs.(C24.1)-(C24.6)), c), d) influence of the kink angle, (F according to eq.(C15.1.1)).

The mode-I stress intensity factor is obtained for $\ell/a \to 0$

$$F_{I,(A)} = F_{I,0}\left(g_{11} + \frac{\sqrt{8}}{\pi}\beta\sqrt{\ell/a}\,\sin^2\varphi\right) \tag{C24.9}$$

as shown in Fig. C24.5b by the solid curve.

The solid curve in Fig. C24.5b deviates very early from the FE results. The reason for this behaviour is the decrease of the mode-I stress intensity factor with the total crack length. This becomes obvious by replacing the crack length a by the total length $a+\ell\cos\varphi$, i.e.

$$F_{I,(A)} \cong F_{I,0}\left(\frac{a+\ell\cos\varphi}{R}\right) \tag{C24.10}$$

The result of this estimation is entered in Fig. C24.5b as the dashed line. The agreement with the FE results is slightly better. The remaining disagreement is caused by the fact, that a crack at one side of the hole of length a is present and at the opposite location a crack of length $a+\ell\cos\varphi$ has to be considered. By using eq.(C15.5.1), which now reads

$$F_{I,(A)}(a_1) \cong F_I(a_0) - 0.13\frac{\ell\cos\varphi}{a_0} \approx F_I(a_0) - 0.13\frac{\ell}{a_0} \tag{C24.11}$$

the dash-dotted line in Fig. C24.5b results. It is in best agreement with the FE data.

References C24

[C24.1] Fett, T., Rizzi, G., A fracture mechanics analysis of the DCDC specimen, Report FZKA 7094, Forschungszentrum Karlsruhe, 2005.

[C24.2] Cotterell, B. and Rice, J.R., Slightly curved or kinked cracks, International Journal of Fracture **16**(1980), 155-169.

PART D

2-DIMENSIONAL CRACKS

Stress intensity factor solutions in handbooks are mostly reported for plane stress or plane strain conditions, with the thickness dimension not entering the solution. The plane stress condition assumes an infinitely thin specimen, for instance an extremely thin sheet with a disappearing stress component in thickness direction, $\sigma_z=0$. The plane strain condition is given by an infinitely thick plate or by a disappearing strain in the thickness direction, $\varepsilon_z=0$. Although the crack always is a 2-dimensional object of dimension "length×length", these two problems are commonly referred as a 1-dimensional crack.

If the full test specimen is modelled without simplifying the boundary conditions of plane stress or plane strain, the related crack is referred to as a 2-dimensional one and the fracture mechanics problem as a 3-dimensional problem. In this case, plane strain dominates in the bulk of a specimen and plane stress occurs in the surface regions.

In contrast to the constant fracture mechanics parameters in the case of a 1-dimensional crack, these parameters must vary over the thickness.

A second, somewhat different definition of a 2-dimensional crack is a crack the shape of which is not exclusively defined by the crack length. This is true for a crack in a trapezoidal bar. Here, the angle of the trapeze is a second parameter for the description of the crack. In the example of Section D4, both definitions of the 2-dimensional crack are applicable simultaneously.

D1

Cone cracks

Under spherical contact loading, cone cracks can be initiated in brittle materials (see Fig. D1.1). Figure D1.1a shows a sphere in contact with the plane surface of a semi-infinite body. Under increasing load, a cone crack develops (Fig. D1.1b). Geometric data are given in Fig. D1.1c with the complicated parts near the contact ignored. It should be noted that in indentation fracture mechanics it is usual to denote the crack length by c instead of a.

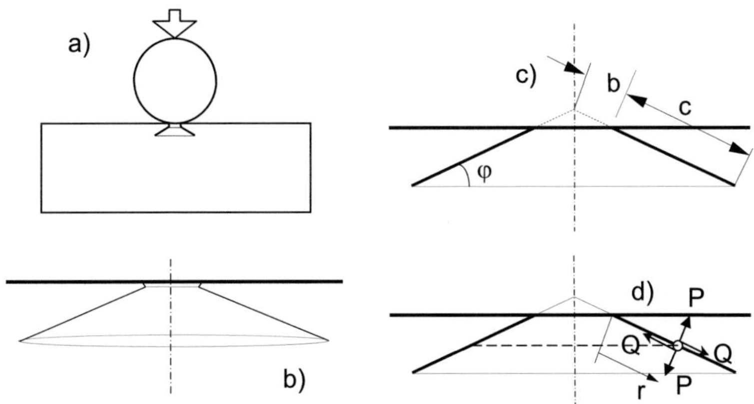

Fig. D1.1 a) Crack generation by a spherical indenter, b) cone crack, c) simplified geometry, d) couples of normal and shear line loads for the determination of the weight functions.

The stress intensity factors K of cracks can be computed by the weight function method as

$$K_\mathrm{I} = \int_0^c \sigma_n h_{11}(r,c)\,dr + \int_0^c \tau_{r\varphi} h_{12}(r,c)\,dr \qquad (\text{D1.1})$$

$$K_\mathrm{II} = \int_0^c \sigma_n h_{21}(r,c)\,dr + \int_0^c \tau_{r\varphi} h_{22}(r,c)\,dr \qquad (\text{D1.2})$$

where σ_n are the stresses in the uncracked body normal to the prospective crack plane and $\tau_{r\varphi}$ the shear stresses in this plane. The stresses below the Hertzian contact, needed in eqs.(D1.1) and (D1.2), can be taken from the analysis of Huber [D1.1].

Taking into account the cone shape of the crack with the circumference increasing with increasing r, an appropriate set up of the weight functions is

$$h_{ij} = \sqrt{\frac{2}{\pi c}}\,\frac{r+b}{c+b}\sum_{n=0}^{\infty} D_n^{(ij)}(1-r/c)^{n-1/2} \qquad (\text{D1.3})$$

with
$$D_0^{(11)} = D_0^{(22)} = 1, \quad D_0^{(12)} = D_0^{(21)} = 0 \tag{D1.4}$$

In order to determine the weight function, finite element computations were performed. The computations were performed for couples of line forces P and Q along the cone circumference for variable relative distances r/c from the crack tip, Fig. D1.1d (for details, see [D1.2]).

Figure D1.2 shows the weight functions obtained from the normal forces P (Fig. D1.2a) and from the shear forces Q (Fig. D1.2b) as the circles. It is of importance that mixed-mode stress intensity factor terms occur even under pure normal or pure shear force. From the FE results, the coefficients $D_n^{(ij)}$ were determined for ν=0.2 and 0.3 by application of a fit procedure. They are compiled in Tables D1.1 to D1.4.

There is a rather small influence of Poisson's ratio ν on the weight function components, as obvious from Fig. D1.3, where the weight functions are plotted as functions of ν.

In order to obtain the full weight function solution in the range of $15\leq\varphi\leq30°$, the data of Tables D1.1 to D1.4 may be interpolated with respect to b/c, ν, and φ. For a more simplified practical use, the coefficients were approximated as

$$D_n^{(ij)} \cong A_0 + A_1\varphi + A_2\varphi^2 + (A_3 + A_4\varphi + A_5\varphi^2)\frac{b}{c} \tag{D1.5}$$

with the coefficients A_0-A_5 compiled in Tables D1.5 and D1.6 for ν=0.2 and ν=0.3. For other ν values the weight function coefficients D may be computed from

$$D_n^{(ij)} = D_n^{(ij)}(\nu = 0.2) + 10[D_{n,\nu=0.3}^{(ij)} - D_{n,\nu=0.2}^{(ij)}](\nu - 0.2) \tag{D1.6}$$

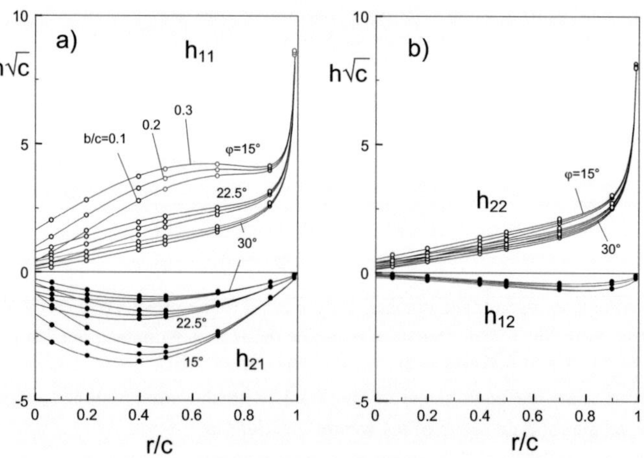

Fig. D1.2 Weight functions for b/c=0.1, 0.2, and 0.3 and several cone angles, a) results obtained under normal force P, b) under shear force Q.

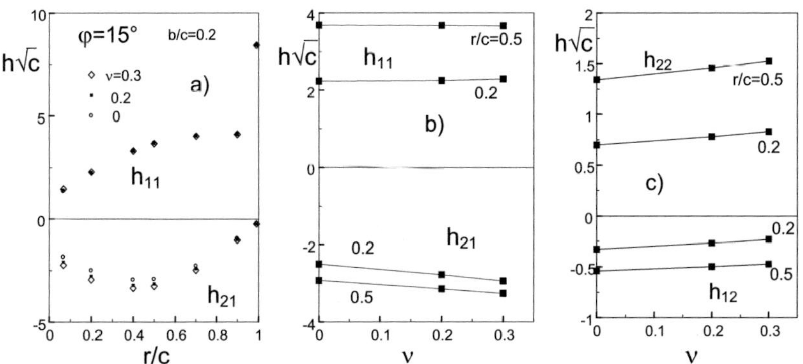

Fig. D1.3 Influence of Poisson's ratio ν on weight functions for $b/c=0.2$ and $\varphi=15°$.

Table D1.1 Coefficients of the weight function h_{11}.

φ	b/c	$D_1^{(11)}$	$D_2^{(11)}$	$D_3^{(11)}$	$D_1^{(11)}$	$D_2^{(11)}$	$D_3^{(11)}$
		$\nu=0.2$			$\nu=0.3$		
15°	0.1	6.619	9.990	-11.762	6.558	10.589	-12.776
	0.2	6.876	9.811	-10.282	6.843	10.301	-10.971
	0.3	7.115	9.642	-8.936	7.107	10.041	-9.354
22.5°	0.1	3.035	4.196	-5.438	2.971	4.463	-5.962
	0.2	3.215	4.175	-4.902	3.170	4.403	-5.294
	0.3	3.385	4.138	-4.371	3.359	4.327	-4.640
30°	0.1	1.322	2.024	-2.798	1.235	2.188	-3.127
	0.2	1.461	2.073	-2.640	1.391	2.217	-2.905
	0.3	1.595	2.098	-2.445	1.540	2.221	-2.646

Table D1.2 Coefficients of the weight function h_{22}.

φ	b/c	$D_1^{(12)}$	$D_2^{(12)}$	$D_3^{(12)}$	$D_1^{(12)}$	$D_2^{(12)}$	$D_3^{(12)}$
		$\nu=0.2$			$\nu=0.3$		
15°	0.1	2.8468	-.6314	-.7125	2.9871	-.3460	-.9470
	0.2	2.9530	-.7080	-.5530	3.0857	-.4384	-.7727
	0.3	3.0349	-.7481	-.4694	3.1606	-.4964	-.6749
22.5°	0.1	1.4316	.6574	-1.1476	1.5717	.8672	-1.3347
	0.2	1.5367	.5997	-1.0457	1.6692	.7978	-1.2197
	0.3	1.6204	.5662	-.9887	1.7460	.7505	-1.1503
30°	0.1	.7130	.9232	-.9420	.8587	1.0914	-1.1098
	0.2	.8127	.9024	-.9204	.9503	1.0601	-1.0735
	0.3	.8948	.8889	-.9092	1.0250	1.0358	-1.0503

Table D1.3 Coefficients of the weight function h_{21}.

φ	b/c	$D_1^{(21)}$	$D_2^{(21)}$	$D_3^{(21)}$	$D_1^{(21)}$	$D_2^{(21)}$	$D_3^{(21)}$
		$\nu=0.2$			$\nu=0.3$		
15°	0.1	-2.684	-18.200	9.410	-2.7593	-18.5636	8.8931
	0.2	-2.750	-17.680	8.144	-2.8254	-17.9514	7.5379
	0.3	-2.798	-17.332	7.280	-2.8698	-17.5575	6.6635
22.5°	0.1	-1.654	-8.686	4.318	-1.7606	-8.9094	4.1169
	0.2	-1.673	-8.495	3.729	-1.7746	-8.6650	3.4623
	0.3	-1.690	-8.352	3.293	-1.7851	-8.4935	3.0023
30°	0.1	-1.239	-4.841	2.299	-1.3693	-4.9709	2.2173
	0.2	-1.235	-4.743	1.953	-1.3567	-4.8435	1.8217
	0.3	-1.235	-4.664	1.680	-1.3480	-4.7482	1.5254

Table D1.4 Coefficients of the weight function h_{12}.

φ	b/c	$D_1^{(12)}$	$D_2^{(12)}$	$D_3^{(12)}$	$D_1^{(12)}$	$D_2^{(12)}$	$D_3^{(12)}$
		$\nu=0.2$			$\nu=0.3$		
15°	0.1	-2.990	3.952	-2.002	-2.9743	3.9767	-1.7737
	0.2	-2.973	3.847	-1.891	-2.9563	3.8605	-1.6732
	0.3	-2.961	3.770	-1.834	-2.9443	3.7756	-1.6281
22.5°	0.1	-2.052	1.888	-.613	-2.0142	1.9440	-.4819
	0.2	-2.042	1.893	-.628	-2.0052	1.9404	-.4994
	0.3	-2.034	1.894	-.647	-1.9990	1.9338	-.5242
30°	0.1	-1.583	1.002	-.043	-1.5091	1.0672	.0271
	0.2	-1.563	1.017	-.083	-1.4941	1.0764	-.0100
	0.3	-1.548	1.026	-.117	-1.4835	1.0796	-.0447

Table D1.5 Coefficients for eq.(D1.5), $\nu=0.2$.

n	(ij)	A_1	A_2	A_3	A_4	A_5	A_6
1	(11)	18.909	-1.081	.0163	4.970	-.212	.0031
2	(11)	33.132	-2.024	.0329	-7.017	.457	-.0070
3	(11)	-40.096	2.355	-.0373	47.411	-2.916	.0465
1	(22)	7.689	-.422	.0062	.823	.0128	-.0003
2	(22)	-6.249	.516	-.0092	-.366	-.0354	.0014
3	(22)	1.958	-.273	.0059	1.429	.0138	-.0019
1	(21)	-6.409	.332	-.0053	-1.922	.115	-.0017
2	(21)	-55.653	3.249	-.0519	15.347	-.986	.0168
3	(21)	31.798	-1.879	.0302	-32.190	1.902	-.0311
1	(12)	-6.330	.286	-.0043	.660	-.0525	.0012
2	(12)	12.113	-.706	.0112	-5.314	.4066	-.0075
3	(12)	-7.723	.496	-.0079	5.306	-.4064	.0072

Table D1.6 Coefficients for eq.(D1.5), $v=0.3$.

n	(ij)	A_1	A_2	A_3	A_4	A_5	A_6
1	(11)	18.739	-1.072	.0161	5.551	-.2404	.0035
2	(11)	35.404	-2.165	.0353	-10.505	.6799	-.0108
3	(11)	-43.896	2.575	-.0408	57.016	-3.5007	.0560
1	(22)	-6.393	.324	-.0052	-2.014	.1242	-.0018
2	(22)	-56.571	3.289	-.0524	16.889	-1.0552	.0176
3	(22)	30.073	-1.768	.0284	-32.684	1.8973	-.0308
1	(21)	7.858	-.424	.0063	.724	.0155	-.0004
2	(21)	-5.679	.494	-.0089	-.683	-.0228	.0012
3	(21)	1.520	-.257	.0056	1.678	.0036	-.0017
1	(12)	-6.322	.284	-.0041	.675	-.0517	.0011
2	(12)	12.021	-.695	.0110	-5.439	.4078	-.0075
3	(12)	-7.160	.469	-.0076	4.985	-.3894	.0070

References D1

[D1.1] Huber, M.T., Zur Theorie der Berührung fester elastischer Körper, Ann. Phys, **43**(1904), 153-63.

[D1.2] Fett, T., Rizzi, G., Diegele, E., Weight functions for cone cracks, Engng. Fract. Mech. **71**(2004), 2551-2560.

<u>D2</u>

Inclusion with an annullar crack

D2.1 The ring-shaped crack

Ceramic components can fail by unstable propagation of microscopic flaws as pores, cracks, inclusions, or other material inhomogeneities which are present due to manufacturing. Description of failure due to internal elliptical cracks or semi-elliptical surface cracks is well established. The same holds for pores.

In the case of inclusions, failure generally is modelled by failure of an assumed annular crack extending around the inhomogeneity. Reference solutions for special loading cases are given in literature [D2.1-D2.6].

D2.1.1 Weight functions for the ring-shaped crack

A ring-shaped crack of inner radius R and crack size a in a homogeneous infinite body is shown in Fig. D2.1a. The two crack tips are denoted as (A) and (B).

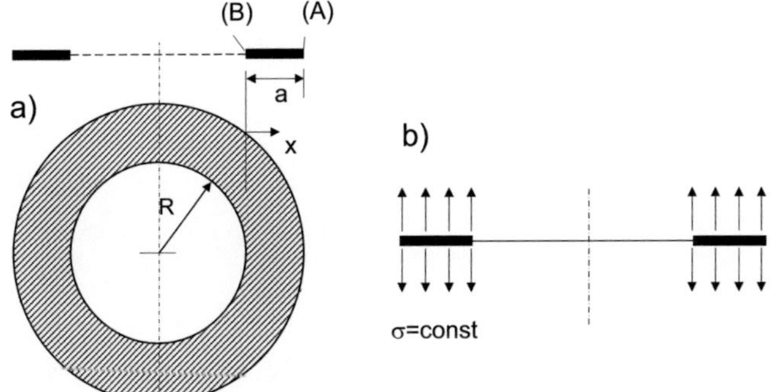

Fig. D2.1 Ring-shaped crack, a) geometry, b) reference loading case (constant stress).

Stress intensity factor solutions for the ring-shaped crack in an infinite body loaded by constant stress normal to the crack plane (Fig. D2.1b) were compiled by Rosenfelder [D2.1]. For an arbitrarily given stress distribution $\sigma(x)$ in the uncracked body normal to the crack plane, the related stress intensity factors can be computed from

$$K_{(A)} = \int_0^a h_{(A)}\sigma(x)dx \qquad (D2.1.1)$$

$$K_{(B)} = \int_0^a h_{(B)}\sigma(x)dx \qquad (D2.1.2)$$

The weight function solutions for the ring-shaped crack were determined in [D2.7] as

$$h_{(A)} = \frac{2}{\sqrt{\pi a}\,\sqrt{1-\frac{x}{a}}\sqrt{\frac{a}{R}+1}}\left(\frac{1+\frac{x}{R}}{\sqrt{2+\frac{a}{R}+\frac{x}{R}}} - \frac{1-\sqrt{\frac{x}{a}}}{\sqrt{\frac{a}{R}+2}}\right) \qquad\qquad \text{(D2.1.3)}$$

$$h_{(B)} = \sqrt{\frac{2}{\pi a}}\left(\sqrt{\frac{1-\frac{x}{a}}{\frac{x}{a}}} + 0.865\tfrac{a}{R}\exp(-C\tfrac{x}{a})\sqrt{\tfrac{x}{a}(1-\tfrac{x}{a})}\right), \quad C = 1+2.44\tanh(\tfrac{1}{40}\tfrac{a}{R}) \qquad \text{(D2.1.4)}$$

Figure D2.2 shows these results for variable values of a/R. It becomes obvious that the influence of the relative crack size is much stronger for location (B) than location (A).

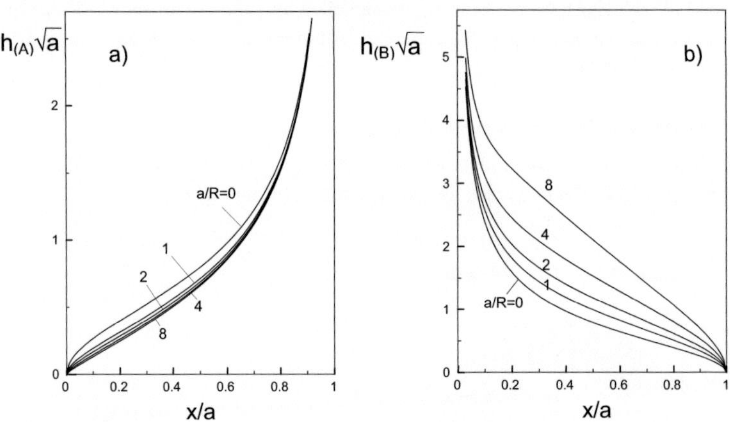

Fig. D2.2 a) Weight function for location (A), b) for location (B).

D2.1.2 Ring crack under constant load

For the reference load $\sigma=\sigma_0=$const., stress intensity factors were computed by Nied and Erdogan [D2.2] and Tada et al. [D2.3]. The data of Nied and Erdogan [D2.2] were fitted by Rosenfelder [D2.1, D2.7] as

$$K_{(A)} = \sigma\sqrt{\frac{\pi a}{2}}\left[\left(1-\frac{2\sqrt{2}}{\pi}\right)\frac{1}{1+\alpha} + \frac{2\sqrt{2}}{\pi}\right] \qquad\qquad \text{(D2.1.5)}$$

$$K_{(B)} = \sigma\sqrt{\frac{\pi a}{2}}\left[\frac{1+\frac{\sqrt{32}}{\pi^2}\alpha+\left(\frac{3}{5}-\frac{\sqrt{32}}{\pi^2}\right)\frac{\alpha}{1+\alpha}}{\sqrt{1+\alpha}}\right] \qquad\qquad \text{(D2.1.6)}$$

with $\alpha=a/R$. Figure D2.3 shows these stress intensity factor solutions as curves.

The results obtained with the weight functions (D2.1.3) and (D2.1.4) are shown by the circles. There is an excellent agreement between the weight function results and those of Nied and Erdogan.

In the 1985 edition of Tada's handbook [D2.3], closed-form expressions are given by

$$K_{(A)} = \sigma \sqrt{\frac{\pi a}{2}} \left(1 - 0.116 \tfrac{\alpha}{1+\alpha} + 0.016 \left(\tfrac{\alpha}{1+\alpha} \right)^2 \right) \tag{D2.1.7}$$

and (after correction of the sign of the α^2 term)

$$K_{(B)} = \sigma \sqrt{\frac{\pi a}{2}} \, \frac{1 - 0.36 \tfrac{\alpha}{1+\alpha} - 0.0676 \left(\tfrac{\alpha}{1+\alpha} \right)^2}{\sqrt{1 - \tfrac{\alpha}{1+\alpha}}} \tag{D2.1.8}$$

both shown in Fig. D2.3 as curves.

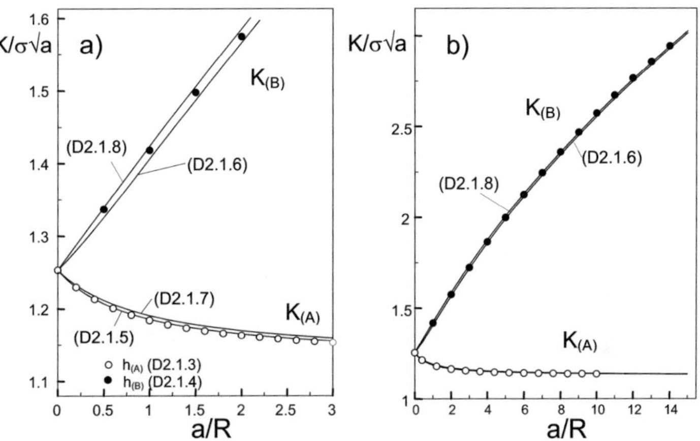

Fig. D2.3 Stress intensity factor solutions for a ring-shaped crack under constant stress.

A comparison of stress intensity factor relations for location (A) shows an excellent agreement of all solutions for $a/R>2$. The agreement of solutions (D2.1.5), (D2.1.7), and the weight function results is excellent for any a/R.

An analytical evaluation of the weight function integrals with eqs.(D2.1.3) and (D2.1.4) is possible only in special cases. Under constant stress, the integration yields

$$K_{(A)} = \sigma \sqrt{\pi a} \, \frac{\pi - 4[1 - \sqrt{\tfrac{\alpha+2}{\alpha}} \arctan \sqrt{\tfrac{\alpha}{\alpha+2}}] + \tfrac{8}{3} \alpha \, \mathrm{Hypergeometric2F1}(\tfrac{1}{2},2,\tfrac{5}{2},-\tfrac{\alpha}{\alpha+2})}{\pi \sqrt{(1+\alpha)(2+\alpha)}} \tag{D2.1.9}$$

D2.1.3 Weight function approximations from reference loading cases

The weight functions eqs.(D2.1.3) and (D2.1.4) are expected to be highly accurate. A disadvantage is the rather complicated structure of the relations. For evaluations with a reduced degree of approximation, simplified expressions may be of advantage. Such weight functions can be derived easily by adjusting h to simple reference stress intensity factor solutions [D2.4].

If the weight functions are approximated by

$$h_{(A)} = \sqrt{\frac{2}{\pi a}} \left(\sqrt{\frac{\frac{x}{a}}{1-\frac{x}{a}}} + D_{(A)}\sqrt{\tfrac{x}{a}(1-\tfrac{x}{a})} \right) \tag{D2.1.10}$$

$$h_{(B)} = \sqrt{\frac{2}{\pi a}} \left(\sqrt{\frac{1-\frac{x}{a}}{\frac{x}{a}}} + D_{(B)}\sqrt{\tfrac{x}{a}(1-\tfrac{x}{a})} \right) \tag{D2.1.11}$$

the stress intensity factors for constant stress σ result as

$$K_{(A)} = \sigma\sqrt{\frac{a\pi}{32}}(D_{(A)}+4) \tag{D2.1.12}$$

$$K_{(B)} = \sigma\sqrt{\frac{a\pi}{32}}(D_{(B)}+4) \tag{D2.1.13}$$

Coefficients $D_{(A)}$ and $D_{(B)}$ can be obtained from the solution of Tada [D2.3], eqs.(D2.1.7) and (D2.1.8), used as the reference solutions

$$D_{(A)} = -4\left(0.116\tfrac{\alpha}{1+\alpha} - 0.016(\tfrac{\alpha}{1+\alpha})^2\right) \tag{D2.1.14}$$

$$D_{(B)} = 4[(1-0.36\tfrac{\alpha}{1+\alpha} - 0.0676(\tfrac{\alpha}{1+\alpha})^2)\sqrt{1+\alpha} - 1] \tag{D2.1.15}$$

Application of eqs.(D2.1.5) and (D2.1.6) results in

$$D_{(A)} = 4\left(\frac{\sqrt{8}}{\pi}-1\right)\frac{\alpha}{1+\alpha} \cong -\frac{2}{5}\frac{\alpha}{1+\alpha} \tag{D2.1.16}$$

$$\begin{aligned}
D_{(B)} &= 4\left(\frac{1+\frac{8}{5}\alpha + \frac{\sqrt{32}}{\pi^2}\alpha^2}{(1+\alpha)^{3/2}} - 1\right) \\
&\cong 4\left(\frac{1-1.6\alpha + 0.76\alpha^2}{(1+\alpha)^{3/2}} - 1\right)
\end{aligned} \tag{D2.1.17}$$

Figure D2.4 represents the approximate weight functions computed with (D2.1.14) and (D2.1.15).

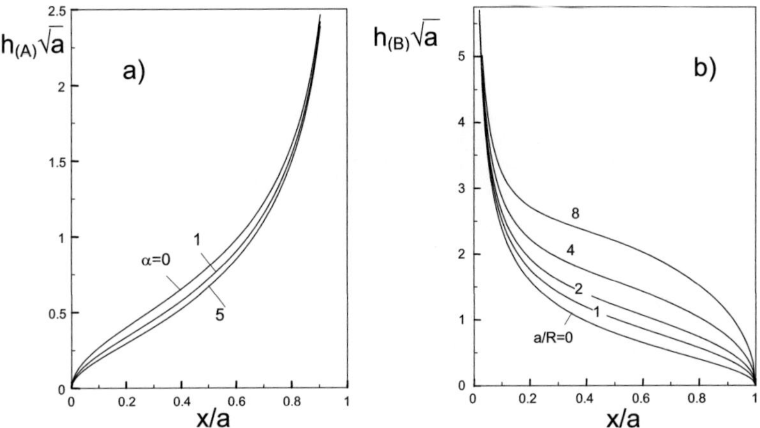

Fig. D2.4 Weight function approximations using (D2.1.14) and (D2.1.15) for the outer (a) and the inner crack tip (b).

D2.2 Spherical inclusion with an annular crack

Figure D2.5 shows a spherical inclusion of radius R with an annular crack of size a. The elastic parameters for the inclusion are E_i and v_i and for the matrix E_m and v_m.

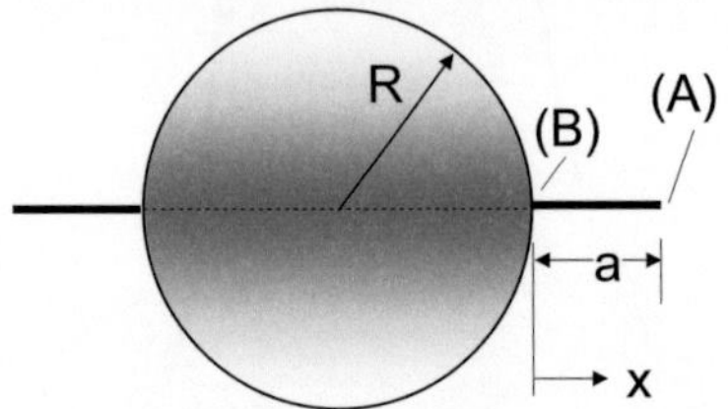

Fig. D2.5 Spherical inclusion with an annular crack.

D2.2.1 Stress intensity factor due to thermal stresses
D2.2.1.1 Stress intensity factor at the outer crack tip
The weight function derived for a ring-shaped crack in a homogeneous body was used to estimate the stress intensity factor at the outer crack tip (A) for thermal stresses caused by different thermal expansion coefficients of inclusion and matrix, α_i and α_m. At the inner crack tip (B), where the crack terminates the inclusion, a square-root shaped stress distribution does no longer appear in general. Hence, no stress intensity factor can be determined there. An exception is the special case of identical elastic parameters for the matrix and the inclusion. With the thermal stresses σ_n normal to the crack plane

$$\sigma_n = \sigma_0 \frac{R^3}{(x+R)^3} \tag{D2.2.1}$$

and

$$\sigma_0 = \frac{(\alpha_i - \alpha_m)\Delta T}{2\bar{k}} \; , \; \bar{k} = \frac{1+v_m}{2E_m} + \frac{1-2v_i}{E_i} \tag{D2.2.2}$$

the thermal stress intensity factor K^{th} at (A) can be obtained using the weight function (D2.1.3). The results are shown in Fig. D2.6 as the thick solid curves.

The approximate weight functions according to (D2.1.10) allow for simple analytical solutions. For the weight function coefficients given by (D2.1.14) and (D2.1.16), it results

$$K_{(A)}^{th} = \int_0^a h_{(A)}\sigma_n(x)\,dx = \sigma_0\sqrt{\frac{\pi a}{2}}\,\frac{1+2.134\alpha+1.284\alpha^2+0.15\alpha^3}{(1+\alpha)^{9/2}} \tag{D2.2.3}$$

and

$$K_{(A)}^{th} = \int_0^a h_{(A)}\sigma_n(x)\,dx = \sigma_0\sqrt{\frac{\pi a}{2}}\,\frac{1+\frac{3}{20}\alpha}{(1+\alpha)^{5/2}} \tag{D2.2.4}$$

In Fig. D2.6, these solutions are entered as the dashed and dash-dotted curves. Within the thickness of the curves, all these solutions coincide.

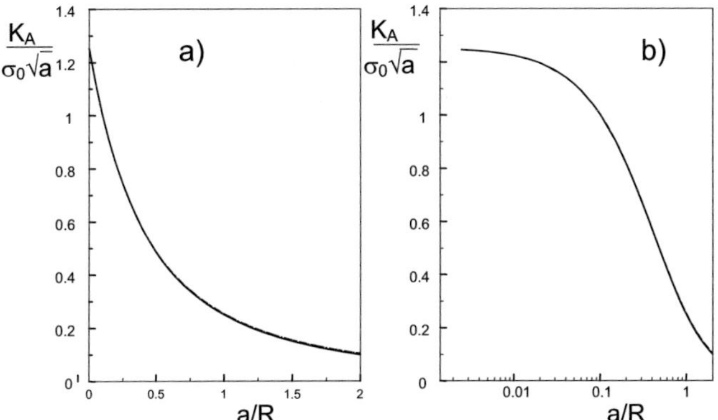

Fig. D2.6 Stress intensity factor at the outer crack tip caused by thermal stresses: (solid curve: eq.(D2.1.3), dashed: eq.(D2.2.3), dash-dotted: eq.(D2.2.4)).

D2.2.1.2 Stress intensity factor at the inner crack tip

In the special case of identical elastic constants E and ν for the inclusion and the matrix, a stress intensity factor also occurs at the inner crack tip. The solution computed with the weight function (D2.1.4) is plotted in Fig. D2.7 as the solid curve.

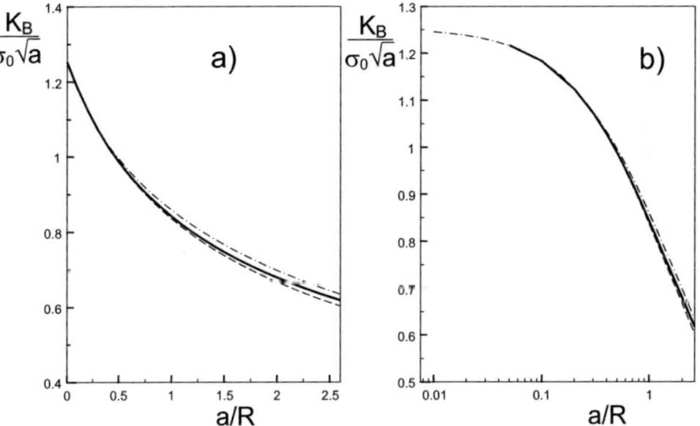

Fig. D2.7 Stress intensity factor at the inner crack tip for the case of the same elastic constants in the inclusion and matrix (solid curve: eq.(D2.1.3), dashed: eq.(D2.2.5), dash-dotted: eq.(D2.2.6)).

The approximate weight function with (D2.1.15) gives the stress intensity factor relation for this location

$$K_{(B)}^{th} = \sigma_0 \sqrt{\frac{\pi a}{2}} \frac{1 + 2.64\alpha + 2.2124\alpha^2 + 0.5723\alpha^3 + 0.75\alpha(1+\alpha)^{5/2}}{(1+\alpha)^4} \qquad (D2.2.5)$$

and is shown in Fig. D2.7 as the dashed curve. The weight function coefficient (D2.1.17) results in

$$K_{(B)}^{th} = \sigma_0 \sqrt{\frac{\pi a}{2}} \, \frac{1+1.6\alpha+0.76\alpha^2+0.75\alpha(1+\alpha)^{3/2}}{(1+\alpha)^3} \qquad (D2.2.6)$$

as shown in Fig. D2.7 by the dash-dotted curve.

D2.2.1.3 Comparison with stress intensity factor solutions from literature

Stress intensity factor results from literature were compiled by Rosenfelder [D2.1]. A simple estimation procedure was proposed by Baratta [D2.5]. This estimation is identical with the solution for constant crack-face loading with the stress taken at location $x=a$, i.e.

$$\sigma_n = \frac{\sigma_0}{(1+\alpha)^3}$$

resulting in

$$K_{(A)}^{th} = \sigma_0 \sqrt{\frac{\pi a}{2}} \, \frac{1}{(1+\alpha)^3}\left((1-\tfrac{\sqrt{8}}{\pi})\frac{1}{1+\alpha} + \tfrac{\sqrt{8}}{\pi} \right) \qquad (D2.2.7)$$

Since the thermal stress value at $x=a$ is the lowest one that appears on the crack face, this solution underestimates the true stress intensity factor (Fig. D2.8).

A solution from Ito [D2.6] resulting from finite element computations reads for identical elastic properties of matrix and inclusion

$$K_{(A)}^{th} = \sigma_0 \sqrt{a}\, 0.89 \sqrt{\frac{2+\alpha}{(1+\alpha)^5(1+\nu)}} \qquad (D2.2.8)$$

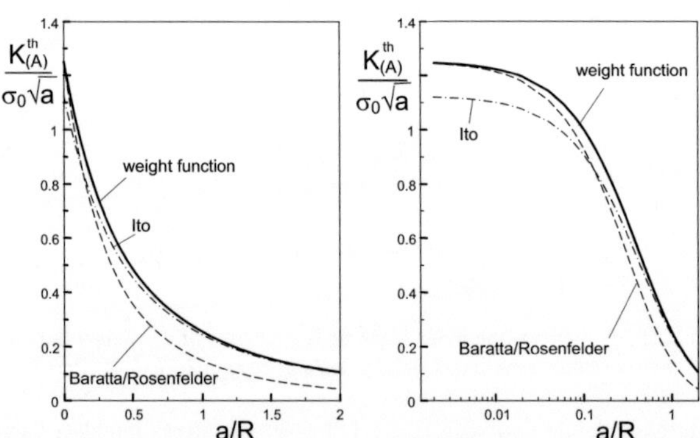

Fig. D2.8 Stress intensity factor for the inclusion with annular crack for thermal stresses.

Whereas stress intensity factors for remote stresses increase continuously with increasing crack size, stress intensity factors under thermal stresses exhibit a maximum at $a/R \cong 0.28$, as can be seen from the representation in Fig. D2.9.

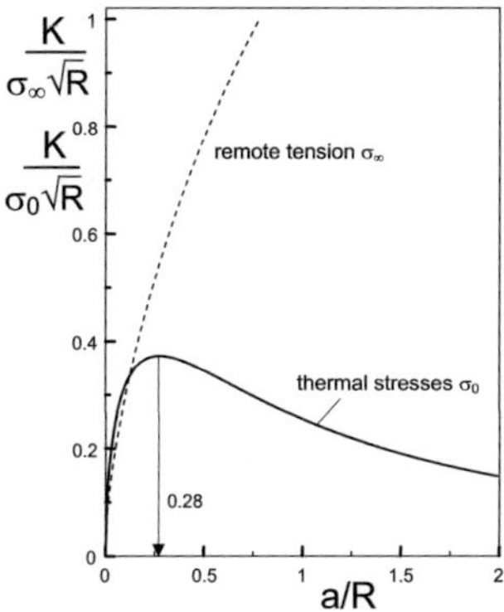

Fig. D2.9 Comparison of stress intensity factors for loading by remote tension (dashed curve) and thermal stresses (solid curve).

D2.2.2 Stress intensity factor for remote tension

Under a remote tensile stress σ_∞, the tangential stresses at the equator of the sphere (normal to the prospective crack plane) are in the matrix

$$\sigma_n = \sigma_\infty \left[1 - \frac{A}{(1+x/R)^3} + \frac{9B}{(1+x/R)^5} \right] \qquad \text{(D2.2.9)}$$

with the material parameters [D2.8]

$$A = -\frac{1}{2} + \frac{E_i + E_m - 2E_m v_i}{E_i + 2E_m - 4E_m v_i + E_i v_m}$$
$$- \frac{1}{2} \frac{E_m(1+v_i)(4-5v_m) + E_i(1+v_m)}{E_m(1+v_i)(7-5v_m) + 2E_i(4-v_m-5v_m^2)} \qquad \text{(D2.2.10)}$$

$$B = \frac{E_m(1+v_i) - E_i(1+v_m)}{2E_m(1+v_i)(7-5v_m) + 4E_i(4-v_m-5v_m^2)} \qquad \text{(D2.2.11)}$$

Also in this case the subscript "i" stands for inclusion and "m" for matrix. By using the approximate weight function with coefficient (D2.1.16), integration gives the solution

$$K_\infty = \sigma_\infty \sqrt{\frac{\pi a}{2}} \left(\frac{1+\frac{9}{10}\alpha}{1+\alpha} - \frac{1+\frac{3}{20}\alpha}{(1+\alpha)^{5/2}} A + 9 \frac{1+\frac{13}{20}\alpha + \frac{11}{40}\alpha^2 + \frac{3}{64}\alpha^3}{(1+\alpha)^{9/2}} B \right) \qquad \text{(D2.2.12)}$$

In Fig. D2.10, the numerical solution obtained with eq.(D2.1.3) (solid curve) is compared with the approximation (D2.2.12) (dashed curve) and the rough approximation derived by Baratta [D2.5]

$$K_\infty = \sigma_\infty \sqrt{\frac{\pi a}{2}} \left(1 - \frac{A}{(1+\alpha)^3} + \frac{9B}{(1+\alpha)^5}\right)\left[\left(1 - \frac{\sqrt{8}}{\pi}\right)\frac{1}{1+\alpha} + \frac{\sqrt{8}}{\pi}\right] \tag{D2.2.13}$$

and displayed by the dash-dotted curve. Whereas deviations between (D2.2.12) and the numerical solution are hardly visible, the estimation of Baratta deviates slightly.

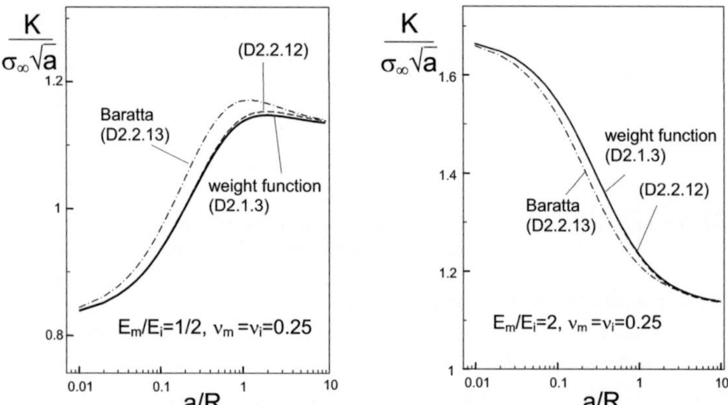

Fig. D2.10 Stress intensity factors for a spherical inclusion with an annular crack. Comparison of the numerical solution obtained with the weight function (D2.1.3) and the closed-form expressions eqs.(D2.2.12) and (D2.2.13).

D2.3 Spherical inhomogeneities

D2.3.1 Continous variation of material parameters

In many cases an inclusion or a pore may not be described sufficiently by a sphere with a sharp radius having sectional constant material parameters. For instance local agglomerations or density fluctuations in a material will result in more "diffuse inhomogeneities". This is for instance the case for glass with regions of higher density in a matrix of mean density.

Computations on stresses and stress intensity factors were carried out in [D2.7] for the radial distribution of the differences of thermal expansion coefficient α

$$\Delta\alpha(r) = \Delta\alpha(0)\exp[-(r/R)^{2n}] \tag{D2.3.1}$$

with integer n, and the Young's modulus

$$E(r) = E_\infty + \Delta E(0)\exp[-(r/R)^{2n}] \;, \quad \Delta E(0) = E_0 - E_\infty \tag{D2.3.2}$$

where E_0 is the value in the centre and E_∞ far away from the inhomogeneity. The variation of v was neglected and a fixed value of $v=0.25$ used.

The dependencies (D2.3.1) and (D2.3.2) are plotted in Fig. D2.11 for several values of $2n$. The higher the number n, the steeper is the transition from the inhomogeneity to the matrix. The sphere with sharp transition is given by the limt $n\rightarrow\infty$.

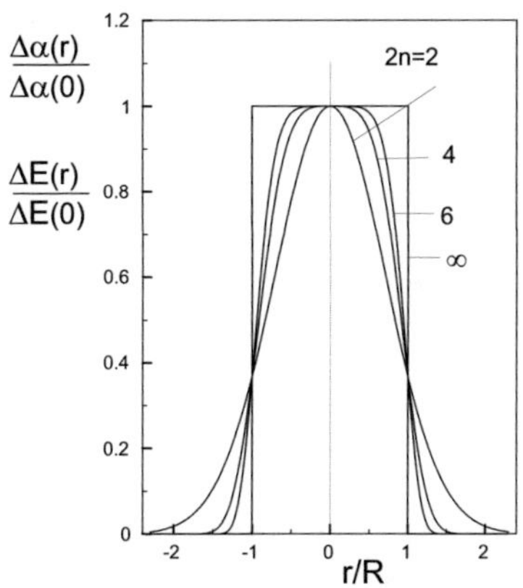

Fig. D2.11 Distribution of mismatch of thermal expansion coefficient and Young's modulus according to eqs.(D2.3.1) and (D2.3.2).

Following the analysis by Timoshenko and Goodier (Section 132 in [D2.9]) the problem of simultaneously changing thermal expansion mismatch $\Delta\alpha$ and Young's modulus E can be solved. Having in mind that for a constant temperature change $\Delta T \neq f(r)$ the elastic strains ε_{el} are given by the local differences in the expansion coefficient, it results from Hooke's law

$$\varepsilon_{el,r} = \varepsilon_r - \Delta\alpha(r)\Delta T = \frac{1}{E(r)}(\sigma_r - 2v(r)\sigma_t) \tag{D2.3.3}$$

$$\varepsilon_{el,t} = \varepsilon_t - \Delta\alpha(r)\Delta T = \frac{1}{E(r)}[\sigma_t - v(r)(\sigma_r - \sigma_t)] \tag{D2.3.4}$$

With the total strains ε_r and ε_t replaced by the radial displacements u according to

$$\varepsilon_r = \frac{du}{dr} \;, \quad \varepsilon_t = \frac{u}{r} \tag{D2.3.5}$$

the stress components read

$$\sigma_r = \frac{E(r)}{[1+v(r)][1-2v(r)]}[(1-v(r))\varepsilon_r + 2v(r)\varepsilon_t - (1+v(r))\Delta\alpha(r)\Delta T] \tag{D2.3.6}$$

$$\sigma_t = \frac{E(r)}{[1+v(r)][1-2v(r)]}[\varepsilon_t + v(r)\varepsilon_r - (1+v(r))\Delta\alpha(r)\Delta T] \qquad (D2.3.7)$$

Introducing these stress components into the radial equilibrium condition

$$\frac{d\sigma_r}{dr} + \frac{2}{r}(\sigma_r - \sigma_t) = 0 \qquad (D2.3.8)$$

yields a somewhat lengthly ordinary differential equation for $u(r)$ with the highest derivative d^2u/r^2.

D2.3.2.2 Determination of stresses for some examples

In the following considerations, the special value of $2n=2$ is used for the thermal expansion mismatch.

From the solution of eq.(D2.3.8), the radial displacements u were determined. They are shown in Fig. D2.12a. From u the strains are obtained via eqs.(D2.3.5). Inserting ε_t and ε_r in (D2.3.6) and (D2.3.7) gives the stress components. Here, the tangential stresses are of special interest. They are plotted in Fig. D2.12b in the normalisation according to

$$\sigma'_t = \frac{1-v}{E\Delta T(\alpha_i - \alpha_m)} \sigma_t \qquad (D2.3.9)$$

The maximum tangential stresses are shown in Fig. D2.13a as a function of the ratio of the Young's moduli in the centre of the inhomogeneity, E_0, and the bulk material, E_∞.

The maximum tensile stresses for $2n=2$ are by a factor of about 10 smaller than those obtained from the model of a sharp transition of material properties from the inclusion to the bulk material (represented by the dash-dotted curve in Fig D2.13a).

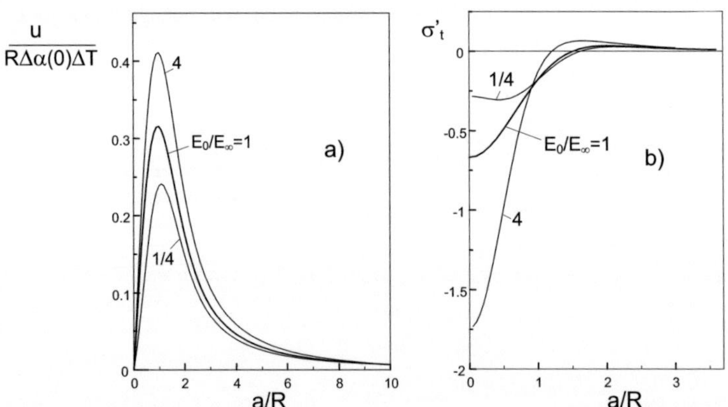

Fig. D2.12 a) Displacements $u(r)$ in normalised representaton for different ratio of Young's moduli, b) tangential stresses in normalised representaton (D2.3.9) for different ratios E_0/E_∞.

The maximum tangential tensile stresses for $2n=2$ can be expressed for $0.2 < E_0/E_\infty \le 5$ by

$$\sigma'_{t,max} \cong 0.024 + 0.01157\frac{E_0}{E_\infty} - 0.001563\left(\frac{E_0}{E_\infty}\right)^2 + 0.00029\left(\frac{E_0}{E_\infty}\right)^3 \qquad (D2.3.10)$$

The square in Fig. D3.13a represents the maximum tangential stress for $2n=4$ and $E_0/E_\infty=1$. Also this comparison shows clearly the stress reducing effect of a continuous change of material properties. Figure D2.13b shows the influence of the Young's moduli on the stress intensity factor at the outer crack tip of an annular crack, $K_{(A)}$. The location of the inner crack tip is chosen to be at $r=R$. In Fig. D2.13b, K is represented by the geometric function F according to

$$F = \frac{1-\nu}{E\Delta T\Delta\alpha(0)\sqrt{a}}\,K \qquad (D2.3.11)$$

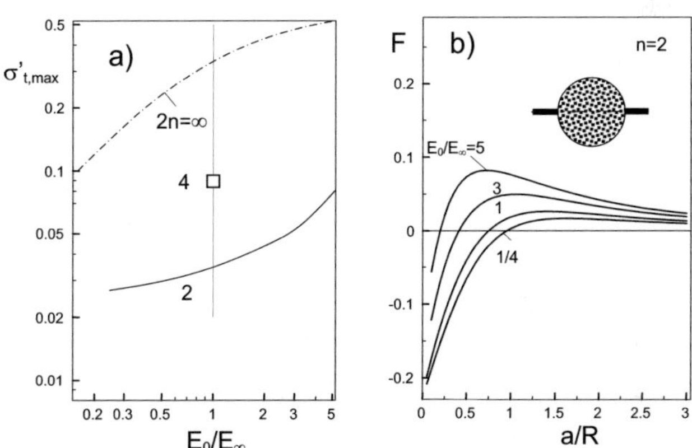

Fig. D2.13 a) Comparison of maximum tangential stresses for different values of 2n, b) influence of Young's moduli on the geometric function according to eq.(D2.3.11) for the outer tip of an annular crack.

References D2

[D2.1] Rosenfelder, O., Fraktografische und bruchmechanische Untersuchungen zur Beschreibung des Versagensverhaltens von Si_3N_4 und SiC bei Raumtemperatur, Dissertation, Universität Karlsruhe, 1986, Kalsruhe, Germany.

[D2.2] Nied, H.F., Erdogan, F., The elasticity problem for a thick-walled cylinder containing a circumferential crack, Int. J. Fract. **22**(1983), 277-301.

[D2.3] Tada, H., Paris, P.C., Irwin, G.R., The stress analysis of cracks handbook, Del Research Corporation, 1986.

[D2.4] Fett, T., Munz, D., Stress Intensity Factors and Weight Functions, Computational Mechanics Publications, Southampton 1997.

[D2.5] Baratta, F.I., Mode-I stress intensity factor estimates for various configurations involving single and multiple cracked spherical voids, Fracture Mechanics of Ceramics, Vol. 5, Plenum Press, 1983, 543-567.

[D2.6] Ito, M., Numerical modelling of microcracking in two-phase ceramics, Vol. 5, Plenum Press, 1983, 479-493.

[D2.7] Fett, T., Rizzi, G., Weight functions and stress intensity factors for ring-shaped cracks, Report FZKA 7265, Forschungszentrum Karlsruhe, 2007.

[D2.8] Goodier, J.N., Concentration of stress around spherical and cylindrical inclusions and flaws, J. Appl. Mech. 1933, 39-44.

[D2.9] Timoshenko, S.P., Goodier, J.N., Theory of Elasticity, McGraw-Hill Kogagusha, Ltd., Tokyo.

<u>D3</u>

Bending bar with trapezoidal cross section

Fracture toughness measurements on ceramic materials are mostly carried out with edge-cracked rectangular bars under bending loading. For special purposes the rectangular standard geometry has to be replaced by a trapezoidal cross section.

There are different reasons for the need of non-rectangular specimens:

- In the case of injection moulded specimens slant side faces simplify demoulding. Fracture tests without an additional specimen finishing are possible if the stress intensity factor solution would be known.

- A geometry with a strongly reduced thickness in the tensile zone of a bending bar is necessary to allow extremely sharp notches to be introduced by use of the focused ion beam method with tenable expenditure [D3.1].

The outer fibre tensile stress for a bending bar with trapezoidal cross section (Fig. D3.1) is

$$\sigma_{bend} = \frac{12M_b}{W^2}\frac{2B+b}{B^2+4Bb+b^2} \tag{D3.1}$$

where M_b is the bending moment applied and W, B, and b describing the geometry of the cross section (Fig. D3.1).
The stress intensity factors can be expressed via the geometric function F by

$$K = \sigma_{bend}F\sqrt{\pi a} \tag{D3.2}$$

with the crack depth a. Two cross-sections were studied in [D3.1] as illustrated in Fig. D3.1:

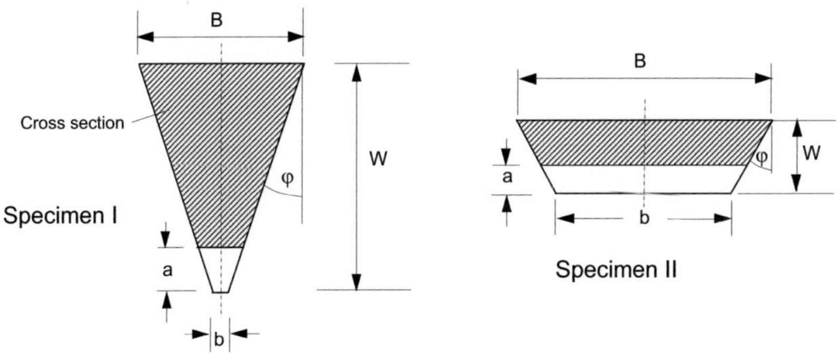

Fig. D3.1 4-point bending test with two different trapezoidal bars.

Specimen I: The geometry for this specimen is $B/W=3/4$, $b/B\leq0.2$, $\varphi\approx17\text{-}20°$. The ratio B/W was chosen to be ¾ since standard test bars have commonly cross sections of 3mm×4mm. In Fig. D3.2a, the geometric function is shown as a function of the location along the crack tip. The maximum stress intensity factor is found in the crack centre where plane strain conditions are fulfilled. Near the side surfaces the stress intensity factor decreases significantly. The maximum stress intensity factor is plotted in Fig. D3.2b versus the geometric parameters b/B and a/W.

The geometric function for the maximum stress intensity factor occurring at the symmetry axis tends to $F_{max}\cong1.135$ for $a/W\rightarrow0$ (dashed curve parts in Fig. D3.2b). This value is only 1% higher than the 2-d solution of an edge-cracked half-space.

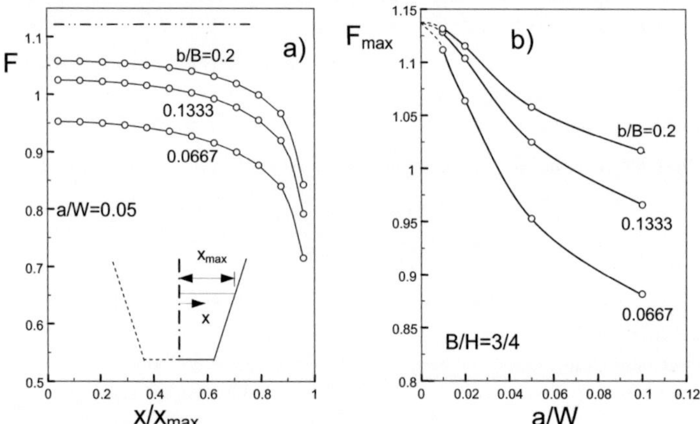

Fig. D3.2 Stress intensity factor along the crack front for specimen I (dash-dotted horizontal line: geometric function for the edge-cracked half-space, $F=1.12155$), b) geometric function for maximum stress intensity factors occurring at the crack centre.

Specimen II: The geometry is defined as $B/W=3.8$, $b/B=0.7$, $\varphi\approx30°$. The geometric functions for stress intensity factors, eq.(D3.2), occurring in the specimen centre ($x=0$) are shown in Fig. D3.3a as circles together with 2-d results for an edge-cracked rectangular bar.

For $\alpha=a/W\rightarrow0$, the well-known limit case of an edge-cracked half-space is fulfilled in the centre region of the crack. It must hold for loading by σ_{bend}, eq.(D3.1), $F=F_0$, where F_0 is the geometric function for the edge-cracked rectangular bending bar as available in handbooks (see e.g. [D3.2]). This limit case is introduced in Fig. D3.3a as the upper dashed curve.

For $a/W\rightarrow1$, the limit case of the "deep crack" is approached. The stress intensity factor must coincide with that of an edge-cracked rectangular beam of thickness B for the same bending moment M_b and the bending stress $\sigma=6\,M_b/(BW^2)$. This limit solution, F_1, is introduced in Fig. D3.3a as the lower dashed curve. For F_1 it holds

$$F_1 = F_0 \frac{B^2 + 4Bb + b^2}{2B(2B+b)} \qquad (D3.3)$$

The FE-results for specimen II can be fitted by

$$F \cong F_0 (1 - \alpha^2) + F_1 \alpha^2 \ , \quad \alpha = a/W \qquad (D3.4)$$

Having in mind the accuracy of handbook solutions in the order of about 1% and the expected accuracy of 1-2% of the FE results, one can conclude, that for $\alpha < 0.4$ the maximum stress intensity factors are sufficiently represented by the 2-d results. An extension of this approximate relation to specimen I is not recommended since in this case the length b and B and, consequently, the limit curves are too different.

Figure D3.3b shows the variation of local stress intensity factor with the distance from the symmetry line. A decrease of the stress intensity factor is obvious at the side surfaces. In the outer surface layer the validity of stress intensity factor data is doubtful since the singularity behaviour must change there.

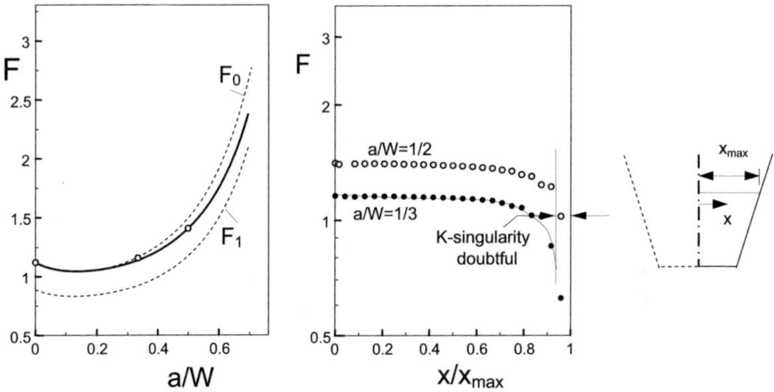

Fig. D3.3 Specimen II: a) Geometric function for the stress intensity factor at the centre, b) Geometric function for the stress intensity factors along the crack front (b/B=1.44, W/b=3/8).

References D3

[D3.1] Fett, T., Rizzi, G., Esfehanian, M., Volkert, C., Riva, M., Wagner, S., Oberacker, R., (2008), Progress in strength, toughness, and lifetime methods for ceramics, Report FZKA 7338, Forschungszentrum Karlsruhe, 2008, Karlsruhe.

[D3.2] Munz, D., Fett, T., *CERAMICS, Failure, Material Selection, Design*, Springer-Verlag, 1999.

D4

Three-dimensional analysis of the DCDC specimen

In Section C15, the DCDC specimen was described as a 2-dimensional specimen with a pair of 1-dimensional cracks. Here, some results of a 3-dimensional analysis [D4.1] are addressed. Special attention is drawn to the side-surface displacements. Figure D4.1 shows the geometry of the DCDC specimen. In contrast to Section C15, the origin of the x-axis is now at the crack tip.

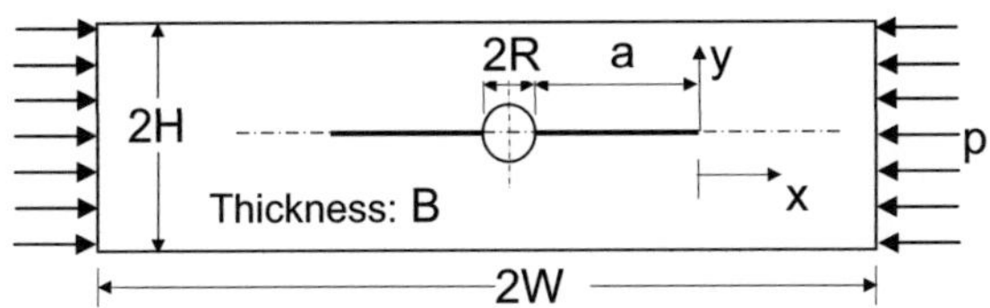

Fig. D4.1 Standard DCDC specimen (side view).

D4.1 Straight crack

D4.1.1 Stress intensity factor and T-stress

In Fig. D4.2, the distributions of the stress intensity factor K_I and the T-stress are shown as functions of the thickness coordinate z with origin in the specimen centre.

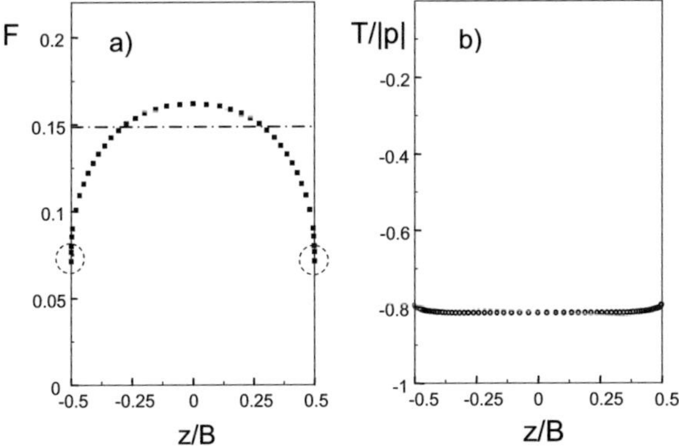

Fig. D4.2 a) Variation of the stress intensity factor along the front of a straight crack (symbols: FE results, dash-dotted line: 1-d stress intensity factor solution, dashed circles: region of disappearing square-root singularity) b) variation of T-stress.

In Fig. D4.2a, the change of the stress intensity factor along the straight crack front is shown. A very strong variation of the geometric function F defined by

$$K = pF\sqrt{\pi R} \qquad\qquad (D4.1.1)$$

is visible. F decreases to a value of less than 50% of that in the centre. The data near the surfaces cannot be expressed as stress intensity factors because the singularity behaviour changes. Whereas in the bulk a singularity exponent of 0.5 occurs, at the surface this value changes to 0.54 for a Poisson's ratio of v=0.25. Consequently, singular stresses can no longer be expressed by K.

The result for the 1-d crack (dash-dotted line in Fig. D4.2a) is roughly identical with the average of the local values.

Figure D4.2b shows the T-stress over the cross-section. The variation of this fracture mechanics parameter is small compared to that of the stress intensity factor.

D4.1.2 Strains in thickness direction

As a further parameter representing the stress state at the crack tip, the strain ε_{zz} parallel to the crack tip line was computed.

Figure D4.3a shows ε_{zz} over the thickness B at several distances x/W from the crack tip. It can be concluded that nearly over the whole thickness of the specimen the z-strains are identical. Only the near-surface strains show a dependency on x.

Figure D4.3b shows details near the side surface, z/B=-0.5. The arrows indicate the distance from the surface (Δz_d) where the deviations from the common curve for ε_{zz} are $\Delta\varepsilon_{zz}$=0.002. In Fig. D4.3c, these depths from the surface are plotted versus the distance from the crack tip.

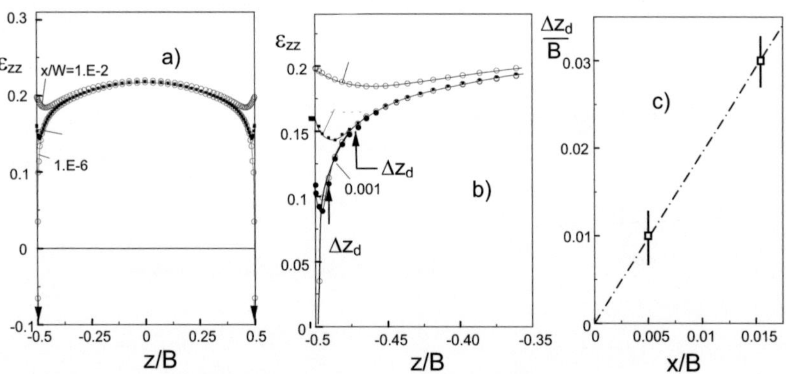

Fig. D4.3 a) z-strains over the cross-section at several distances from the crack tip, b) detail near z/B=-0.5 (arrows: first deviation from the common curve), c) depth of first deviation versus distance from the crack tip.

D4.2 Influence of a curved crack front

In experiments on glass, a curved crack front can be found (Fig. D4.4). This is a consequence of the reduced stress intensity factors in the surface region (see Fig. D4.2a), resulting in a reduced crack growth.

Figure D4.5a shows the crack approximated by straight segments. The outer crack part intersects the free surface under an angle of φ ($\varphi=90°$ corresponds to the straight crack). The next deeper part was modelled as a straight line with an intermediate angle of $(\varphi+90°)/2$. The geometric function of the local stress intensity factor is plotted in Fig. D4.5b for $\varphi=90°$, $60°$, and $45°$.

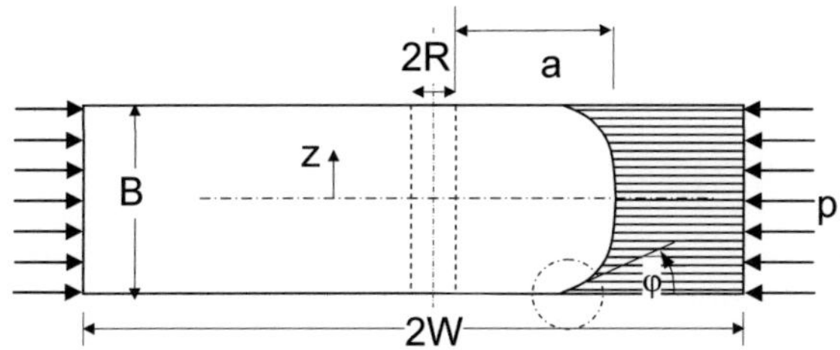

Fig. D4.4 DCDC specimen (top view) with a curved crack front ($\varphi=$ crack terminating angle).

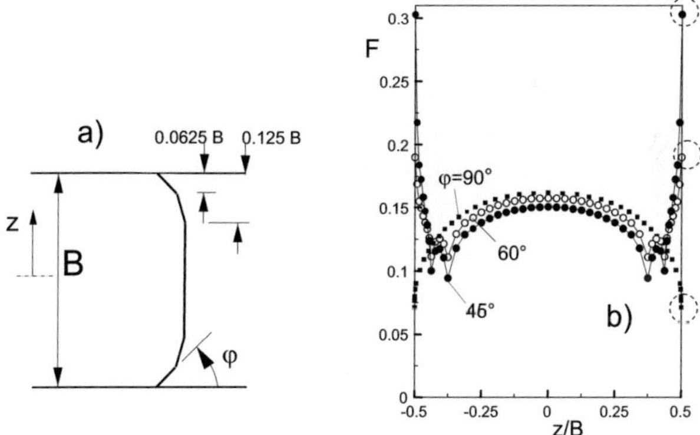

Fig. D4.5 a) Curved crack front approximated by straight segments, b) stress intensity factor along the crack front (dashed circles: doubtful data since singularity exponent deviates from ½).

D4.3 Side-surface displacements near the crack tip

For the finite element computation, a crack in the DCDC specimen, Fig. D4.4, was modeled with $a/R=4$, $H/W=0.1$, and $R/H=0.25$. The specimen thickness, B, was chosen as $B/W=0.2$, resulting in a square cross section. Poisson ratio was assumed as $\nu=0.25$.

The u_z displacements at the free surface are shown in Fig. D4.6a. For angles of 45°, 60°, and 90°, the same displacement behavior was found in greater distances from the tip (Fig. D4.6a) Figure D4.6b gives a log-log plot of the differences with respect to the value at $x = 0$, $\Delta u_z = u_z(x)-u_z(0)$, for $x > 0$.

In Fig. D4.6c, the differences in displacements, Δu_z, are plotted for a smaller distance from the crack tip. In this representation, significant differences in Δu_z are obvious. The variation of u_z becomes more pronounced for a decreasing angle, φ.

Figure D4.7 illustrates the differences in displacements along the y - axis, i.e. for $x = 0$. Fig D4.7a shows a systematic shift in the curves with the angle, φ. The curves also show an increase of the slopes with increasing terminating angle φ.

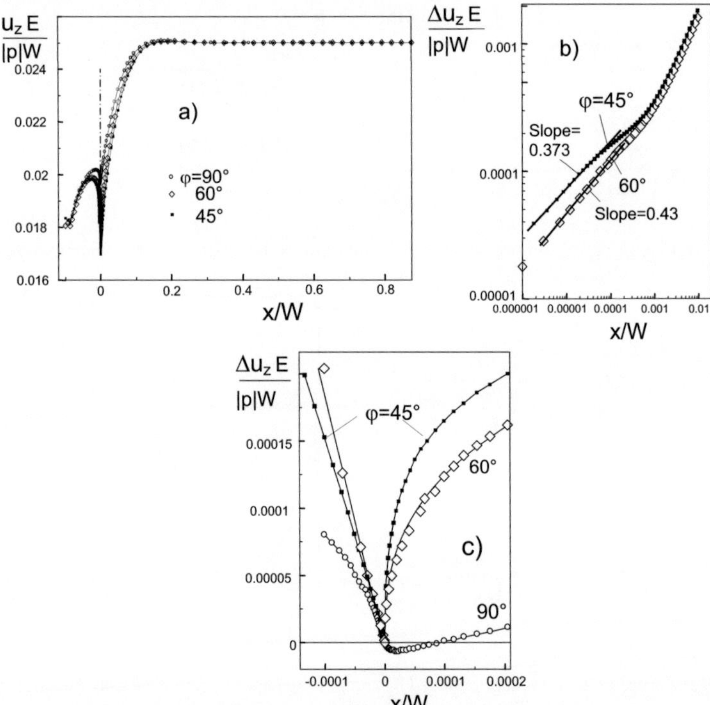

Fig. D4.6 a) u_z displacements along the x-axis obtained for φ=90°, 60°, and 45°; b) log-log representtation of Δu_z for $x > 0$, c) details for shorter crack tip distances.

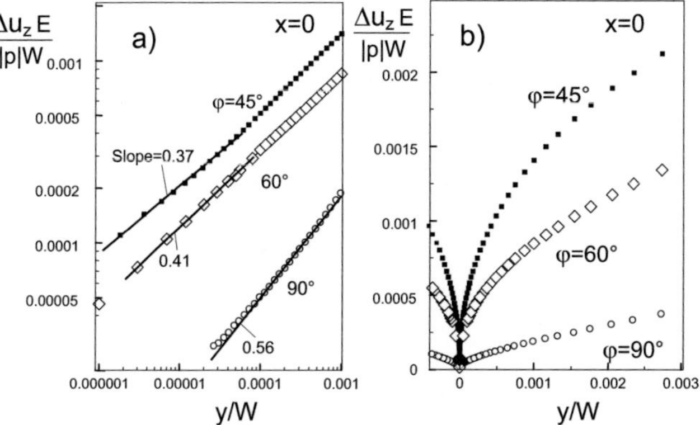

Fig. D4.7 a) Log-log representation of the differences in displacements, $\Delta u_z = u_z(y) - u_z(0)$, along $x = 0$ obtained for $\varphi = 90°$, $60°$, and $45°$; b) details for small distances y.

AFM-scans across cracks are often used for the determination of the COD of cracks under mechanical load. Such measurements are affected not only by the crack opening, but also by pure elastic surface displacements u_z outside of the crack.

The displacements for $\varphi = 45$, $60°$, and $x < 0$ found along the line $y = 0$, $\Delta u_z(x,0)$, are shown in Fig. D4.8. Along this line the near-tip displacements show a nearly linear slope, i.e., $\Delta u_z(x,0) \sim -x$.

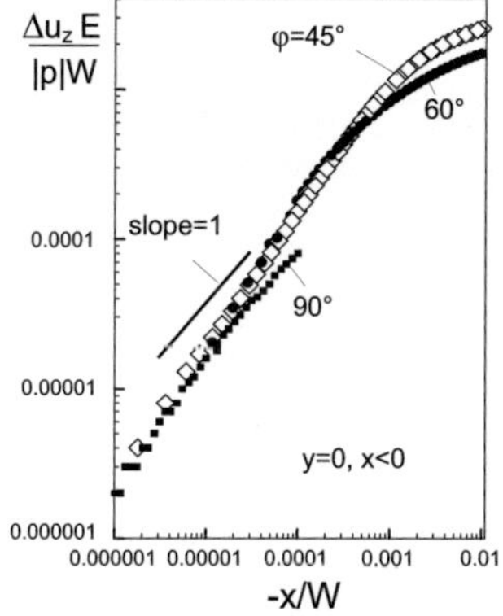

Fig. D4.8 Displacements u_z in the crack wake ($x<0$) along the x-axis.

Cross sections in several distances from the tip are represented in Fig. D4.9a for a crack terminating angle of $\varphi = 45°$ and in D4.9b for an angle of $60°$.

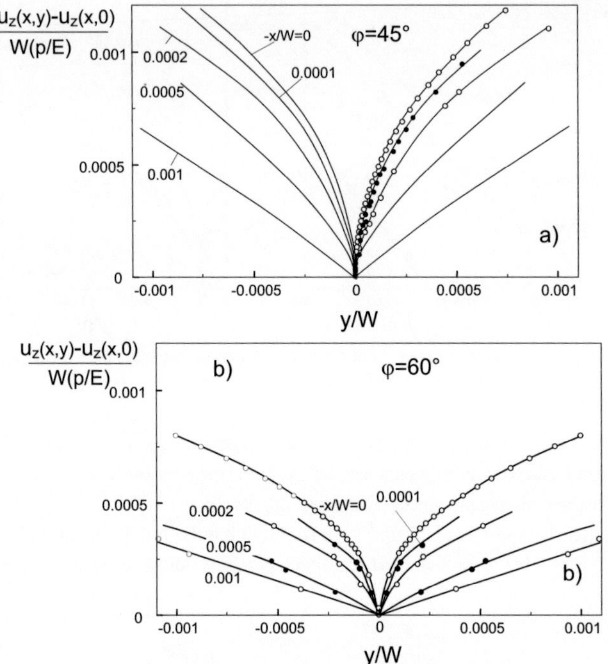

Fig. D4.9 Displacement profiles in the crack wake along section lines x = constant; a) results for φ = 45°, b) for φ = 60°.

Reference D4

[D4.1] Fett, T., Rizzi, G., A fracture mechanics analysis of the DCDC-specimen, Report FZKA 7094, 2005, Forschungszentrum Karlsruhe, Karlsruhe, Germany.